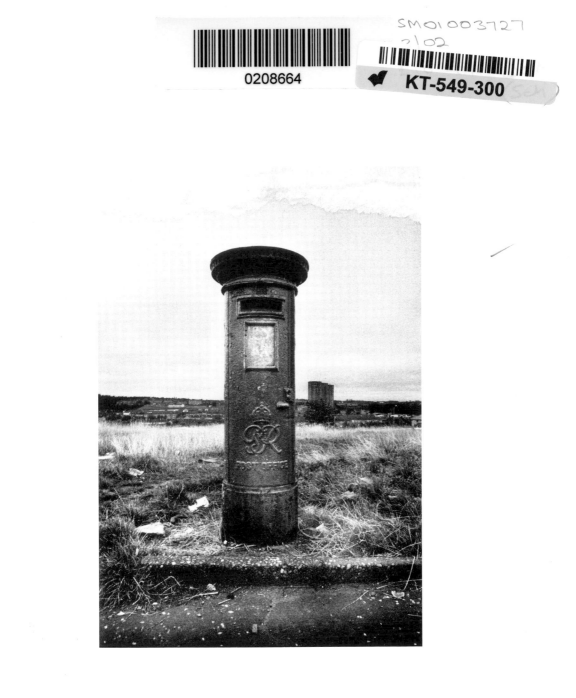

To Julie, with love and thanks

and

To the memory of
Mary Delaney, 1961–1999

NICHOLAS SCHOON

THE CHOSEN CITY

SPON PRESS
Taylor & Francis Group

LONDON AND NEW YORK

First published 2001 by Spon Press
11 New Fetter Lane, London EC4P 4EE

Simultaneously published in the USA and Canada
by Spon Press
29 West 35th Street, New York, NY 10001

Spon Press is an imprint of the Taylor & Francis Group

Typeset in Bembo by M Rules
Printed and bound in Great Britain by
St Edmundsbury Press, Bury St Edmunds, Suffolk

British Library Cataloguing in Publication Data
A catalogue record for this book is available from the British Library

Library of Congress Cataloging in Publication Data
Schoon, Nicholas.
The chosen city / Nicholas Schoon.
p. cm
Includes bibliographical references and index.
1. Cities and towns – Great Britain. 2. Metropolitan areas – Great Britain.
3. Inner cities – Great Britain. 4. Urban renewal – Great Britain.
I. Title.

HT133.S385 2001
307.76'0941 – dc21   200102804

ISBN 0-415-25801-4 (hbk)
ISBN 0-415-25802-2 (pbk)

CONTENTS

FOREWORD

WHY WON'T WE choose the city? For years we have been leaving the major conurbations, heading for the suburbs and beyond. Even London is no exception – although that city is constantly restocked by the inward migration of rich and poor from other countries. But we all know that urban flight means loss of countryside, traffic pollution and congestion. It also means loss of morale and self-respect for those who are left behind on stigmatised council estates or in poorly served neighbourhoods.

In exploring the possibilities for the 'urban renaissance' advocated by Lord Rogers' Urban Task Force, and the renewal of marginalised neighbourhoods promoted by the Cabinet Office's Social Exclusion Unit, the Joseph Rowntree Foundation has funded over sixty academic studies. We feel well rewarded by the government's National Strategy for Neighbourhood Renewal and its new housing policies that emphasise the importance of engaging with local residents and creating a mix of incomes. The Foundation has also built 'CASPAR' developments (City-centre Apartments for Single People at Affordable Rents) in downtown Birmingham and Leeds: these are succeeding in attracting and retaining the people with money and talent who are so necessary to the revival of inner cities. But neither our academic programmes nor our demonstration projects can do what Nick Schoon's book sets out to do. This is not an academic treatise or a guide to good practice. It appeals to the emotions as well as the intellect. It seeks to provoke, stimulate and enthuse, as much as to spell out the facts.

Deep-seated divisions in society, the very unequal distribution of wealth and opportunities, the hazards for the environment and problems for the wider economy, can all be predicted to get worse if there is no change in national attitudes toward our cities. Revival of city living will depend on jobs and the resources to change the urban environment must be found. Investment in clearing up derelict sites, removing eyesores, enhancing local services – education and policing in particular – are all costly but necessary. Public perceptions are at least as important as public money. This book's combination of analysis and perceptive journalism should make a very real contribution to focusing the national debate, shifting attitudes and changing minds.

Richard Best, Director
Joseph Rowntree Foundation, York

PREFACE

THIS IS A BOOK about what has gone wrong with our cities and how they might be improved. The proposals are, I guess, just about within the realms of political possibility during the next couple of decades. We can all think of excellent, sensible proposals which are politically impossible, but what's the point? (How about making Liverpool the UK's capital and seat of government? No large city is closer to England, Wales, Scotland and Northern Ireland, and Merseyside could do with the uplift).

In what follows, I've attempted to divide the complex clockwork of urban decay into manageable chunks such as crime, education, housing and transport. One danger of doing this is that you neglect the need for joined up solutions; something we are going to keep on hearing about now that New Labour has been shoo-ed in for a second term of government. And because Britain has an overwhelmingly urban population, there is another danger — that of seeing every kind of problem as a town and city problem and therefore writing a bloated, formless book which tries to solve everything.

I hope I've succeeded in skirting both of these elephant traps as well as an even bigger one — writing a book about a problem which doesn't even exist. For there is, one could argue, no overall urban problem. There are plenty of bad and poor urban neighbourhoods but there always have been and always will be. There are also a few entire conurbations and several smaller towns, mostly in Northern England and the UK's smaller nations, ex-industrial places, coastal resorts and market towns, which have gone into severe decline. That's tough, you might say, but we can't all be winners. Most of urban Britain is doing fine and the places that thrive more than compensate for the sick and dying. Edinburgh's successes outweigh Glasgow's failures. Just look at London, with its huge and growing wealth and its now briskly increasing population.

But the capital has concentrated poverty and decay running through its inner boroughs and far out to the east which far surpasses that in, say, Greater Manchester or Merseyside. And even in young, prosperous, fast growing towns you find places where deep rot has set in — as in the sink estates of Milton Keynes. There really is an urban problem made up of lots of problems and I've underplayed or altogether missed some of them in an effort to keep my subject under control. By way of apology, I should at least note the ones I'm aware of. They are perhaps predictable omissions given that the author is (i) white, (ii) in reasonably good health, (iii) lives in South East England and (iv) has spent most of his working life as an employee of unprofitable or non-profit making enterprises.

First, there's too little about differences in race and culture. British citizens from ethnic minorities are concentrated in our larger towns and cities where they suffer higher than average rates of bad and overcrowded housing and other kinds of deprivation. I argue that the tendency to segregate neighbourhoods on the basis of income is damaging and should be resisted. This must also apply to segregation on the basis of race and culture.

Second, this book could have included a chapter on health in cities. People living in failing urban areas generally have poorer health than the national average and they have also tended to receive worse health services – certainly at general practitioner level. I didn't include these issues because I did not think them among the prime reasons why people with choices in life tended to shun urban living. However, epidemiologists have interesting ideas on and arguments about these issues. There are those that believe there is some factor, or several factors, in poorer neighbourhoods, beyond the more immediate afflictions of poverty, that puts the health of all of their residents - including those on average and above average incomes - at some extra risk. And perhaps this connects to the observation that the more unequal, income-polarised conurbations, regions and entire nations (such as the USA and the UK) tend to have worse overall health and higher mortality than more egalitarian places which have an equivalent average income but a narrower gap between the rich and the poor.

Third, there's not much in here about regional issues and the great, enduring north south divide, a divide which is complex, murky and certainly more than a myth. The failings of many towns and cities are partly a cause of the wide economic disparities between the UK's regions, partly a consequence. There is an interplay of supply and demand-side factors which tend to widen these differences, to make regions which get ahead get even further ahead, and of countervailing factors which tend to narrow the gulf. The latter rarely seem to win the day, so government – at European and national level – feels compelled to intervene in order to help the straggler regions. These interventions have often failed the great provincial conurbations; that must change. What are the shapes on the map, the assemblages of towns, cities and hinterland, that function as regions and sense a strong regional identity? How can that feeling be harnessed? How should the governance and planning of cities relate to that of their hinterlands? Questions about which many books and papers have been written…but not this one.

Fourth, the fate of towns and cities is tied up with their ability to develop new skills and ideas, to attract investment in buildings, equipment and people, to launch or improve products and services, found new enterprises, grow established ones, raise productivity. Economic performance and entrepreneurship provide a crucial perspective on cities but, for the most part, it is not this book's. I'm interested in something which is closely related and just as important; how well cities perform in making people want to live in them. My perspective is mainly that of the resident, the consumer of cities. It is a vantage point from which all of us cannot help but keep a look out.

This book may be wasted on those who remain well on the political left or right in these muddled, middling days (although the limited municipalisation of land I advocate might appeal to some old lefties). As I researched, I often found the brac- ing wind of Marxist thought blowing from the pages of books written in the seventies and eighties. Town planning and urban regeneration were just ways of dealing with another crisis of late capitalism, they argued. It was the rich and powerful, working through corporations and various arms of the state, who were doing for our cities and their poorer residents. There is, of course, some truth in that. But capitalism flourishes. It has spread, it assumes more and more identities, it has found ways of enlisting more and more of us. Whatever the injustices and inequalities it entrains it seems to work tolerably well for the majority in a wealthy western nation – or, at least, better than anything else on offer. But ask yourself this: is our society so uneven and unfair because of the way we segregate our towns and cities by income, or are our towns and cities so segregated because our society is so uneven and unfair? It's both – and in what follows I focus on the former.

I'm sorry, too, if I seem too pessimistic. I found plenty of encouraging things in Britain's cities and I mention many of them. Some of the ideas in here are already well established. The general election manifestos of the two main parties in 2001 both said cities must be made better places to live in – my main theme. The word 'urban' has become trendy. But I don't think we have turned the corner between decline and renaissance.

I provide no proper treatment of contaminated land (important, but difficult to render interesting), the heritage of older buildings and urban landscapes, the 24 hour city (I'm usually tucked up in bed by midnight) and the role of gay people in urban regeneration. These omissions prevented a long book being longer and I mention them merely to show expert readers that I have at least noticed them. You experts make me nervous. I was a journalist attempting the difficult task of writing a book that could be useful to the learned and the well-informed as well as engaging the interest and enthusiasm of any general reader with some passion for cities. The specialist and the professional will, inevitably, find some passages covering familiar ground, some simplifications, garish colours and vulgar noises. The amateur may find that she or he is occasionally getting rather denser detail and argument than they want. I hope I've provided some surprises and insights for both types of reader.

I wrote this book because the Joseph Rowntree Foundation picked me and my idea for a Journalism Fellowship, giving me the time and money I needed for the task. I have visited all of Britain's largest conurbations and many smaller towns and cities, either in the course of researching this book or during my work as a journalist. I travelled around several of them on foot, by bicycle, public transport and car, in an effort to see and feel their neighbourhoods. I also read numerous books, papers and reports, many of them in those most essentially urban of things, libraries – in particular Bromley's fine central lending and reference facility and the magnificent British Library at St Pancras. Most of my sources are listed in the end-notes but there is no reference to three wonderful books so I recommend them here; Richard Hoggart's *Uses of Literacy* with its astonishing portrait of bygone working class neighbourhoods (London, Chatto & Windus 1957), Robert Fishman's *Bourgeois Utopias: the rise and fall of suburbia* (New York, Basic Books, 1987) and *Civilia*, Ivor de Wofle's (a.k.a. Hubert de Cronijn Hastings) extraordinary vision of a compact hilltop city for a million people on derelict land halfway between Leicester and Birmingham (London, Architectural Press, 1971). You will need a good library to find them.

Hundreds of people in cities around Britain - regenerators, developers, plan-ners, architects, people from central and local government, academics and residents (including my former neighbours in Hayes) – helped me, mostly with information, opinions and ideas. I want to name a few who were particularly kind in taking the time and trouble to show me around places and in helping and encouraging me at the beginning and near the end; Patrick Clarke, Susan Dunsmore, Cathy Garretty, Andy Gibb, Michael Gwilliam, Peter Hennessey, Roger Levett, Paul Keenan, Duncan Maclennan, John Oldham, Laura Padoan, Lee Shostak, Polly Toynbee and Raymond Young. Especial thanks to the Royal Commission on Environmental Pollution whose secretariat I worked for during the closing stages of this project (the views herein are, however, mine and most definitely not theirs), Richard Best of the Joseph Rowntree Foundation for his interest and support, Caroline Mallinder of E & F.N. Spon, Paul Drew and my friend David Rose. And most thanks of all to my wife Julie who coped with my long absences from family life and with the inex-cusable obsession this book became and who helped with ideas and much encouragement.

ACKNOWLEDGEMENTS

ALL OF THE PHOTOGRAPHS were taken by David Rose. Those on pages 50, 69, 72, 230 and 258 have already appeared in *The Independent* newspaper, which retains copyright and kindly permitted them to be used here. The maps on page 288 appear with the permission of the Council for the Protection of Rural England. The drawings and plan in Chapter 13 are all by Paul Drew, apart from the sketch maps on page 253 which were drawn by Arthur Schoon. The masterplan for a Northampton extension on page 317 appears with the permission of the Prince's Foundation, English Partnerships, the Department of the Environment, Transport and the Regions and CPRE. The three maps of inner London on page 332 appear with the permission of the British Medical Journal Publishing Group.

THE JOSEPH ROWNTREE FOUNDATION has supported this project as part of its programme of research and development projects, which it hopes will be of value to policy makers, practitioners and service users. The facts presented and views expressed, however, are those of the author and not necessarily those of the Foundation.

Chapter 1

THE ABANDONED CITY

Hayes still remains clean and tidy with good schools, plenty of parks, but gradually the area has become less friendly and the spread from London seems to be drifting our way. Within five years we will have moved further into Kent to give our children a better start in life.

I feel safe, secure and contented here but I don't like to dwell on the fact that we are only a few miles away from troublesome areas. If this was to encroach upon my immediate area, I would move further out.

Residents of Hayes, a suburb on the rim of south-east London

We talked pleasantly enough, until I told him that Moses' road was going to blow every trace of both of our childhoods away. Fine, he said, the sooner the better; didn't I understand that the destruction of the Bronx would fulfil the Bronx's own basic moral imperative? 'What moral imperative?' I asked. He laughed as he bellowed in my face: 'You want to know the moral-ity of the Bronx? Get out, schmuck, get out!' For once in my life I was stunned into silence. It was the brutal truth: I had left the Bronx, just as he had, and just as we were all brought up to, and now the Bronx was collapsing not just because of Robert Moses but also because of all of us. It was true, but did he have to laugh?

Marshall Berman, *All That is Solid Melts into Air*[1]

2   MOST OF BRITAIN'S BIG CITIES now have their emptying quarters, places where despair and decline have gone so deep that large-scale abandonment of homes and business premises is underway. The most spectacular and unsettling example I came across was in Possilpark, one mile out from Glasgow's thriving centre with its crowds of shoppers and office workers.

To reach it, you follow the familiar spoor of urban decay, passing low-rise, post-war blocks of council flats with big, raucous scrawls of graffiti and the easily-burgled ground floor units boarded up, then a battered-looking primary school fortified with CCTV and a high, spiked steel fence. Finally you come to a place of solitude and silence where there is no longer any housing. Here is a grid of tarmac streets, road lamps, street signs and even bus stops covering hundreds of acres, but only grass and weeds slowly engulfing levelled rubble fill the space in between.

Hundreds of council flats once stood in this part of Possilpark and were lived in for a decade or two. Then, as with tens of thousands of other municipal homes in the UK's City of Architecture 1999, they became harder and harder to let, even to poor people with few choices in life who seemed fairly desperate for council hous-ing. This neck of the woods became known as a place of isolation, disrespect and danger. Eventually the council decided, as it had done many times before, that the only way forward was mass demolition.

That happened in 1994; the site has lain abandoned ever since. Glasgow and the rest of Britain have more and more places like this. The government admits there are parts of our towns and cities that are so dejected and rejected that it makes no sense to rebuild there. Most Britons know next to nothing about them, for they have no reason ever to visit them. They offer only smashed up blocks of flats and row after row of mean houses with doors and windows covered by steel or plywood shutters. The local councils no longer bother to keep the streets clean and the few cars you see are often burnt out. You don't see many people about, and those you do often look poor. If mass demolition of the housing in these areas has not begun, then it appears as if it soon will.

And, in most of these areas, a fresh start is unlikely. No new and better homes to last longer this time. There is no prospect of any real redevelopment because there is no confidence in these neighbourhoods being a place where anyone – rich, poor or middling – would choose to live.

Yet the heart of Glasgow, with its blatant prosperity and endless entertain-ments, its hundreds of thousands of jobs, splendid streets and celebrated architecture, is a 20-minute walk away from Possilpark. The drains and utility mains which lie below those vanished homes, the network of roads which once connected them to the rest of the city, an urban infrastructure worth millions of pounds, lie idle.

Meanwhile, the national debate about where to build the more than four million new homes Britain is forecast to need over the next quarter century becomes louder and angrier. What chiefly excites the pressure groups, the press, the public and the politicians – in roughly that order of causation – is the threat to the UK's remaining countryside from this frenzy of construction. Two questions being asked are whether government has got its projections right – can the demand for new homes *really* be that large when population growth is so slow – and how much of that development can be squeezed into existing towns and cities instead of obliterating greenfield land?

Apart from covering an area of fields and woods as large as Greater London in bricks, mortar and tarmac, what further environmental harm will be done by these millions of new houses built in the countryside? Other types of development – shops, workplaces, leisure facilities – will spread with them. More roads will have to be laid. There will be many more car journeys to work and to play, to schools and shops, accompanied by more noise, air pollution and congestion. The area in which the glare of street lights puts out the stars at night will grow.

These fears are quite justified, but they miss the worst thing about this unending spread of housing. It is not only the damage it does to the countryside that matters. A much more precious environment is being damaged in the process – the urban one. This, after all, is the home of millions of people, whom only the most fanatical environmentalist would say matter less than wildlife and scenery.

Since the industrial revolution began in these islands, the history of most of its cities has essentially been one of desertion by people with choice and money. Most Britons appear to hate them, getting out if they can, escaping to the countryside or small towns or suburbia. If they can't, then they dream of a day when they will be able to. The cities were where almost all our wealth was created, where the middle classes were formed and grew. But even as their populations exploded, as they sucked people from the countryside into their overcrowded, stinking and lethal centres, anyone who succeeded in life was getting out. With them went their standards and money, children and talents, voices and votes. Over a hundred years ago Britain's large towns and cities had settled into smaller versions of the structures you can still find today – a commercial core with offices, shops and a few grand public buildings to demonstrate the city fathers' civic pride, encircled by residential suburbs segregated by class. The dense, low-income housing tended to be nearest the centre with more spacious suburbs for the better off further out.

The rapid suburbanisation of the past 200 years had to happen. Without it people's overall standard of living and life expectancy could not have been raised. Midway through the nineteenth century the population was rising fast and more

than a million urban poor were crammed together in squalid courts, cellars, tenements and back-to-back houses. Reformers fretted about epidemics, sanitation, drunkenness and depravity; improvements were slowly pushed through. But for most people, a move into a less densely populated area further from the city centre was the way to better housing and health, and by the end of the century the innermost areas of the big cities had begun to depopulate.

We continue to demand new suburbs because we are still fleeing the city. The population of Britain's eight largest conurbations* fell by two million between 1961 and 1991.[2] Each day they suffer a net loss of around 300 people to smaller towns and the countryside.[3] People perceive that as well as being a place of crime, deviancy and people who don't know how to behave, the big city has failing schools, 'immigrants' (who have often lived here for two or more generations) and unbearable road traffic. The stock of existing suburban and rural and small town homes is nowhere near large enough to satisfy the demands that arise from this shunning of the urban. So new suburbs spread into the countryside and abandonment continues. The cities are left with poorer people, less money, diminishing social capital. The blight spreads out from the inner city, seeping into tree-lined streets where large houses become dilapidated and big gardens turn to scrub. What was once the best of suburbia is degraded and devalued, people with choices in life no longer want to live there, and so still more new suburbs are needed.

Britain's housebuilders and planners, the chief defenders of the status quo, see things differently. They point out that for the past 50 years we have had a complex, democratic system that negotiates and plans the growth of our towns and cities. It is not just the prosperous who have escaped from the cities; the system has decanted hundreds of thousands of working-class families into suburban-style council housing on the urban fringe or into verdant new towns deep in the countryside. Furthermore, they say, it is unfair to accuse a spanking new, three-bedroom house on a greenfield estate on the edge of town of causing urban decline. The real culprit is the changing structure of the economy, or advances in technology, or plain poverty, or something else. An ordinary, hard-working family will buy that new house, cherish it and make it a home.

But we do not really have a planning system, we have organised, formal attempts to control and slow the spread of towns and cities. And while low-income families were moved out from crowded homes near the centres during waves of slum clearance, this only worsened the prospects of many of them – and deprivation remains an abidingly inner city phenomenon. And yes, that new house on the

---

*These eight are Greater London, the West Midlands, Greater Manchester, West Yorkshire, South Yorkshire, Merseyside, Tyne and Wear and Greater Glasgow. A third of Britain's population lives in them. London and Glasgow are essentially large, built-up masses; the others are more diffuse structures, each made up of several cities which retain distinct identities and are to some extent still separated by thin wedges of countryside, river and estuary. I am often going to consider them as units, so let me acknowledge now that the residents of, say, Coventry and Wolverhampton don't think of themselves as West Midlanders. Each conurbation does, however, form its own distinct urban region – big, mainly built-up areas surrounded by countryside. The West Midlands is

edge is contributing to urban decline. If it could have been built in an inner city and sold to a family with a middling income, it would be improving rather than worsening urban prospects.

Isn't suburbanisation rather like slash and burn farming? And don't the quotations at the beginning of this chapter, from residents of the suburb where I lived, sum up the process? The countryside is consumed by development and for a few decades the land is valuable. Then, all too often, it starts to go wrong and the money moves off to cover new countryside, leaving barren ground behind.

Surely we should not to treat our urban environments this way. We can no longer justify the urban exodus by saying that people and homes are overcrowded, that they need more room to breathe and that, if they get it, the population as a whole will be better off. The link between density and wretchedness was broken decades ago. Some of Britain's most deprived people live in spacious, grassy council estates on the edge of cities while some of its richest live in high density apartments at their heart. (The most deprived electoral ward in England turns out to be in Wythenshawe, a leafy, medium-density suburb of council semis built on Manchester's southern edge between the wars. It was designed to be England's third garden city.[4])

People are sometimes pulled to the suburbs or to smaller towns outside the big conurbations because they can purchase more house and garden space; it is a consumer option. But what is pushing them away from the cities is as important and that, primarily, is the presence of poorer people. The greatest shortages in the city are not of space and greenery, clean air and quiet, but of trust and security, of earned incomes and self-esteem. Each decision to leave further concentrates the population who lack money and choices, which makes the inner city an even less desirable place to live.

Yet the core of most of Britain's big cities enjoyed a revival through the 1990s, thanks to sustained growth in the service economy. Waves of National Lottery money and public spending have helped things along.[**] The city centres have survived the booms in out-of-town shopping, leisure and working which have devastated some smaller towns. People with money still want to work, meet, shop and play in city centres, and by and large they have become better places for all of these activities. Their great squares, public buildings and railway stations have been refurbished. That most wonderful of things, the tram, is making a comeback and whisking people in and out of them. They offer more shops and restaurants, bars and cafés, and better ones too. There are spectacular new concert halls, exhibition centres and galleries.

centred on Birmingham, West Yorkshire on Leeds, South Yorkshire on Sheffield, Merseyside on Liverpool and Tyne and Wear on Newcastle-upon-Tyne.

[**]And, in the case of Manchester, a huge IRA bomb explosion in 1996 which brought about a much praised redevelopment of its main shopping centre.

National Lottery-driven renewal; the British Museum's Great Court, overleaf

On the back of this has come a surge in city centre living led – as always – by London. Each year thousands of new homes are completed in the core of Britain's cities, by converting redundant offices, warehouses and shops or building new flats and town houses on canal banks, docksides and factory sites. This is highly fashionable and shows no signs of slowing but it is tiny compared to the flow to the suburbs and beyond. When young, prosperous loft dwellers plan babies they almost always get out of town. The city centre housing boom does not spread out into the tatty, sad badlands of the inner city which lie just beyond, into the council estates and the remaining rows of small Victorian and Edwardian terrace houses.

In a sane world the inner belt lying one to two miles out from the centre of Glasgow, Birmingham or Manchester should be where their more prosperous and successful citizens lived in grand houses. These would be the most desirable addresses, with the highest house prices. Work and play in the city centre would be just five or ten minutes away. The rest of us, with less spending power, would have to live in cheaper property further out, wasting more of our less valuable time in commuting to the centre. (Which is what happens, to some extent, in London, a very unusual British city.) But instead the inner belt is a doughnut of deprivation where the poorest citizens predominate. They often have little reason to travel to the city centre. They have no jobs to commute to and their lack of spending power makes them unwanted.

Some people will argue that all urban areas, from small towns to the largest conurbations, are bound to become divided into residential areas defined by income. Nothing can be done about this; it is as inevitable as the differences in incomes and wealth. They are missing the point. The bigger a British city, the more – in general – it has of more than its fair share of poverty, not just in its inner city but across the conurbation. Take Greater Manchester. Its population of 2.6 million is spread among 214 roughly equally populated electoral wards. Of these, 109 – more than half of its wards – belong among the most deprived fifth of all wards in England. But only eight in the whole of Greater Manchester belong to the least deprived fifth of all wards in England. The conurbation has a great excess of the deprived and a great deficit of the prosperous.[5]

The long exodus of people with choices from the larger cities is sabotaging Britain's chances of becoming a fairer, more meritocratic country. The mainstream left may have abandoned the idea of equality but it still stands for equality of opportunity, provided its pursuit does not offend focus groups. If UK plc is to remain a rich, first world country, its supply of bright, talented, hard-working people has to be maximised; we have to be open to talent. You are allowed to be poor, but only if you are lazy and unenterprising as well as stupid. When the electorate finally ejected the Conservatives in 1997 it had, somewhere near the front of its collective mind, the idea that New Labour stood for a fairer Britain. This would be achieved not so much by the old-fashioned method of redistributing income but by altering incentives and by social investment. At the heart of it all was Tony Blair's slogan,

'Education, education, education'. Improved state schools would enable the entirety of each new generation to fulfil its potential.

But the polarisation of our cities, the gradual abandonment of them by people with choices, make this quite impossible. It polarises the schools themselves. They are in the front line of urban decline, both its victim and its cause. This is such a big problem that it deserves, and gets, a chapter of its own later in this book. If a school is perceived to be a good one by local parents, its pupil numbers rise, it gets more money from the government and can expand – the money follows the pupils. Mothers and fathers who care about their children's education want to move into the catchment area of a good local school. Wealthier, owner-occupier parents have a much better chance of making this move than poorer ones living in council and housing association homes. This can push up house prices around the school, meaning you have to be wealthier still to make the move needed to get your children into it. Head teachers call it 'selection by mortgage'. Middle-class children whose parents are willing and able to fund helpful extras such as computer hardware dominate the intake. They have good reason to value education and they expect their kids to come away with clutches of qualifications. Most teachers would prefer to teach in this kind of school so head teachers find it easier to pick the best of them, which makes the school still more attractive to parents.

The process tends to move in the opposite direction for schools serving declining inner city neighbourhoods or big, post-war council estates further out. The intake is dominated by children from poor households often lacking a home environment that encourages learning. The output has few or no qualifications. The school, and the quality of its teaching, are dragged down by its troubles. The government struggles to improve such schools and there are 'superheads' who have turned around some of those facing the worst of circumstances. But they are swimming against the tide. If our schools were less polarised between private and state education, and between very good and very bad state schools, their pupils would do better overall.

It is not just the unevenness of our education system that is sabotaging any chance of moving towards a meritocracy. Growing up in a deprived, declining urban area means shrunken horizons, stifled hopes and stunted opportunities. Of course, in a liberal democracy that craves new talent, some people escape from slums to enjoy brilliant careers, make their names and fortunes. Some young people who triumph over adversity do so because their innate quality wins through. Others may succeed because hardship nurtures some talent that would never have blossomed under an easier upbringing.[6] But both types are exceptional. For every child from a bad neighbourhood who succeeds, there are dozens who could have been contenders but never had a chance. Raised in a wealthier home by confident parents they would have found happiness in work, in learning and travelling, bringing up children and making a home. Instead they will have shorter, emptier, lives. They are more likely to suffer mental and physical illness, to be victims of crime or to become criminals, and to be bad parents. They will have no idea about how good life in a rich Western country can be for the majority of its citizens. But they will know enough, through television and visits to town centres and dealings with authority, to realise that they are missing out on a great deal and that the rest of society veers between blanking them out and holding them in contempt.

Swathes of our conurbations seem to be hopeless after decades of desertion and population loss. The extra costs to society of trying to cope with this failure and turn things around are colossal.* The bills for crime and insurance are much higher than they would otherwise be. So are those for policing (by officers who usually commute from the suburbs), for convicting offenders and then locking them up. People in deprived urban areas receive higher health and education** spending per capita than the population as a whole, and their need for state benefits – to help with rents, unemployment, low incomes, disability, council tax – are much greater. Their children are more likely to be placed in care, their families are more likely to need the help of social workers. The local councils that cover such areas are given extra government grants because they are so needy. All this extra public sector money is spent trying to compensate for the private sector's disinvestment, but it does not succeed. As the needs of the local people and the scale of the dereliction carry on rising, the area may succeed in bidding for some of the money the government earmarks for urban regeneration. By then the rot has usually gone too deep.

This process of pumping in more and more taxpayers' money as conditions deteriorate fails to turn things around because people who have money and choices do not want to live there. Employers do not want to have their workplaces there. And crucial state services such as schooling, policing and health services tend to deteriorate too because these are difficult places in which to provide them, and because residents' complaints go unspoken or ignored. Urban regeneration schemes that only attempt to make the local people better off – by giving them training, or better housing and schools, or subsidised jobs – will often disappoint. If, for instance, such a scheme gives a family steady employment, it will use the extra income to leave the area.

So it's as important to attract outsiders with jobs and money who have a choice about where they live to move into the area. It means spending some of the regeneration money not on poor local people but on outsiders, for example, in subsidising developers to build homes for sale. You may find this unjust, but a growing number of councils and other organisations involved in regeneration now accept it and have changed their priorities.

---

*A government-commissioned study found that public spending in the most deprived wards of three cities (in Liverpool, Nottingham and the London borough of Brent) was nearly 50 per cent higher per head than in their least deprived wards. Public spending across these cities was also substantially higher, per capita, than the national average – 30 per cent in the case of Liverpool. G. Bramley, M. Evans and J. Atkins, *Where Does Public*

Local and national government need new policies to prevent the concentration of poor people in inner city and social housing ghettos. (Social housing is jargon for council and housing association homes; both have now acquired a tragic reputation for being anti-social.) If they succeeded there would be big savings in public expenditure and less crime and unemployment. Britain would become a less ashamed, more relaxed nation. If, however, we do not shift towards pro-urban, pro-social mixing policies, then we will lose a great deal more countryside under suburban sprawl and gain plenty more of roads and traffic. Worse than these, though, will be the widening gap between rich and poor. Our cities will become more like the USA's grotesquely divided conurbations than they already are and the bills will go on rising.

Arguably the greatest disappointment of Labour's first term in office for 18 years was the slow progress in reducing relative poverty and absolute inequality. The number of people in households with very low incomes in 1998/99 (defined as less than 40 per cent of average incomes after housing costs) was nearly nine million or one in seven citizens – half a million higher than in 1996/97 and *four times* the level of the early 1980s.[7] And this despite several years of solid economic growth and shrinking unemployment.

The government could claim that its introduction of a statutory minimum wage and other tax and benefit-based income supplements would sharply reduce poverty and inequality in the coming years. Its problem is that it has to make expensive interventions in people's incomes in order to compensate for what appears to be an underlying increase in earnings inequality. What is causing the gap between prosperous and poor to widen?

It is partly the premium salaries commanded by skilled, educated people in a global economy that depends increasingly on knowledge, information and creativity. It is partly the legacy of years of high unemployment which crushed individuals and families and left many unable to benefit from the boom. But high unemployment also crushed entire neighbourhoods while planning and housing policies have concentrated poor people and made their prospects worse. The gap is widening because poverty clings to places as well as to families.

We need to prevent it from doing so, but this involves going against the grain of individual freedom and choice. The majority of Britons who can afford their own housing want to live in suburbs, smaller towns or the countryside and they usually choose to get as far away from the poor as their budgets allow. This brings us back to the widely shared ideal of a meritocratic Britain and the huge, never-to-be-resolved conflict surrounding it. We want equal opportunity for everyone but extra for our children and ourselves; one crucial way of striving for that extra is by

Spending Go? A Pilot Study to Analyse the Flows of Public Expenditure to Local Areas, London, Department of the Environment, Transport and the Regions, 1998.

**Up to the age of 16. Beyond that, the much greater middle-class take-up of sixth form and higher education opportunity comes into its own.

choosing to live in a more affluent area. Any policies which forced the prosperous to live closer to the poor, which made families settle in places they do not wish to, would be impossible to implement. This would be another kind of social injustice and democratic politics would, quite rightly, rule it out. People have to be attracted into cities by an urban renaissance, by improvements to the physical and social environment that benefit everyone.

But, at the same time, there has to be much more restraint on building over greenfield sites. The penultimate chapter sketches a radical reform of Britain's planning system that could do that. The form of what is built also has to change – higher densities (but not overcrowding, nor smaller rooms, nor an end to private gardens) and streetscapes which look as if they are designed for people, not their cars. More terraced houses and apartments (do not call them flats), more mixing of homes, workplaces and shops, less space wasted on over-broad roads, roundabouts and functionless grass verges.

The agenda is the re-urbanisation of Britain, and it has at last moved into the political mainstream. For years it was the preserve of environmentalists and a few radical architects and planners who have concentrated on buildings, transport and public space rather than on people's feelings about cities and each other. Sometimes the urban revivalists have displayed a woeful lack of realism. *Going to Town*,[8] a cheerful pamphlet published in 1998 by the Council for the Protection of Rural England and the Civic Trust, has plenty of good ideas. But it seriously underplays some of the most important forces driving those with choice out of cities. Its opening paragraph puts 'ill-equipped and overcrowded' schools in a list of urban ills, side-stepping something far more important and intractable – that they are seen to be bad schools where you do not get many or any GCSEs. Crime and poverty in cities receive little more than a mention.

Designing buildings and masterplanning entire urban quarters that look wonderful is a much more controllable, predictable process than turning urban fortunes around. Artists' illustrations and models usually show little people contentedly shopping, strolling and sipping coffee on terraces. But will real people with real money want to live in the city?

In 1999, a government-appointed Urban Task Force, chaired by the superstar architect Lord Rogers of Riverside, published a superb report advising ministers on how to achieve an urban renaissance.[9] It was packed with 105 firm recommendations and numerous further proposals, most of them excellent, which government ought to get on with implementing. Yet, to my mind, the task force erred at the outset in stating that urban regeneration must be 'design led'. It will be led by people, by public and private money, and by addressing the inequalities between inner city and outer suburb, between declining conurbations and booming hinterlands, between north and south.

Above all, we have to achieve a good social environment in our towns and cities. The failure to give this proper weight in the debate to date explains why urban regeneration has remained a minority taste despite being a political priority for decades. It is not surprising that the most important players in this game, the house-buying public and the housebuilding industry, have not been won over. That will only happen when there are clear signs that the problems of poor schools, crime and polarisation can be overcome.

We need to bring jobs and wealth into deprived areas by encouraging people with money to live in them, or next to them. And we need to stop areas from becoming deprived because residents who make money choose to get out of them. By people with money I mean home-owners and potential home-owners, a category that includes most British adults.

Much of what this book argues for can only be done by central and local governments and their various arms, with the support of – or at least with minimum resistance from – business people, investors, developers and housing associations. It is, after all, for governments to lead and organise when the pursuit of self-interest goes against the collective interest. But there are other compelling reasons why it falls to government to rock the boat, to fashion a new, more compact and socially mixed kind of suburbia and to interfere with the deep-rooted, long-running tendency to quit urban areas. Whitehall and Westminster and the town halls are, effectively, landlords to Britain's poor. It is the social housing they financed which dominates the most deprived parts of Britain. And it is they who allow suburbia to carry on spreading out from towns and cities by granting planning permission for developers to build on greenfield sites. This facilitates the flight from cities, thereby contributing to the urban decline which the government then spends huge sums trying to reverse. The state helped make this mess and it is up to its neck in it.

Even so, the state cannot do it alone. People's attitudes to cities will have to change. I ought to say something about my own. Until 1993 our family's home was a small, terraced, late Victorian house in the inner London borough of Greenwich. It was in a street and an area with a mixture of social housing and owner occupation, densities were fairly high, and the journey to work by public transport was quick and easy. Then we moved eight miles further out to an archetypal inter-war semi on the very edge of London, in Bromley, to an area of almost uniform owner-occupied housing a quarter of a mile from Green Belt countryside. Densities were low, commuting to the centre was long and tedious. If that made me a hypocrite, then at least I was a contrite and dissatisfied one. The main reason for moving out of inner London was the same as many other people's; that we believed that our children would get a better state education. We missed Greenwich, which was a livelier, more interesting and convenient place than Hayes, Bromley.

Attitudes and culture can change. Many of Britain's big cities still have beautiful, desirable residential areas near their hearts such as Glasgow's West End and Bristol's Clifton. More and more children have been leaving their suburban homes for big city universities and finding that life there can be, well, interesting. And each year an increasing number of Britons visit continental cities like Amsterdam, Paris

and Barcelona where urban living is more the norm for all classes. They come away wondering why our own conurbations are so segregated and drab. There are, furthermore, more and more people who will not join in the national fantasy of owning a rural cottage. Among them are Britain's millions of citizens from ethnic minorities. They do not see the countryside as a long lost home; it is a place where white people still stare at them.

So there is hope and the possibility of change. An urban renaissance could turn out to be more than a government slogan. We can revive our towns and cities by making them places where people with choices would want to live. We need to think about sharing the city, of achieving a social balance, for there would be no overall gain if an urban renaissance merely forced the poor to shift, concentrating them somewhere else. Before looking at how these things can be done, we have to try to understand how our cities became so divided and unloved. That requires some history.

West End, Glasgow, facing

Chapter 2

DARKSHIRE AND COKETOWN
THE APPROACH TO 1900

I am always haunted by the awfulness of London: by the great, appalling fact of these millions cast down, as it would appear by hazard, on the banks of this noble stream, working each in their own groove and their own cell, without regard of knowledge of each other, without heeding each other, without having the slightest idea how the other lives – the heedless casualty of unnumbered thousands of men. Sixty years ago, a great Englishman, Cobbett called it a wen. If it was a wen then, what is it now? A tumour, an elephantiasis sucking into its gorged system half the life and blood and the bone of the rural districts.

Lord Rosebery, Liberal Chairman of the London County Council,
quoted by Ebenezer Howard in *Tomorrow*, 1898[1]

The whole of the island – set as thick with chimneys as the masts stand in the docks of Liverpool; that there shall be no meadows in it; no trees; no gardens; only a little corn grown upon the house tops, reaped and thrashed by steam; that you do not even have room for roads, but travel either over the roofs of your mills, on viaducts; or under their floors, in tunnels; that, the smoke having rendered the light of the sun unserviceable, you work always by the light of your town gas: that no acre of English ground shall be without its shaft and its engine.

John Ruskin, *The Two Paths*, 1859[2]

A Lancashire village has expanded into a mighty region of factories and warehouses. Yet, rightly understood, Manchester is as great a human exploit as Athens.

Benjamin Disraeli, *Coningsby, or The New Generation*, 1983 [1844][3]

20   I HAD A HISTORY TEACHER who would wearily tell the class, several times each term, that nothing was ever really new. Every thing had already happened, somewhere, sometime, before. His favourite example concerned the hippies who, by the early 1970s, were waning. Hippies were old hat, he would say. They had been around in the French Revolution; *les incroyables* were famed for their outrageous hairstyles and clothing. The daftness of this argument made us break out of our adolescent torpor. What about, say, moon landings (also waning in the early 1970s)? Surely they were genuinely original? After a moment's thought he pointed out that the Chinese had invented rocketry nearly a thousand years earlier.

The debate about cities is also nothing new. Arguments about the evils of the urban environment, the sprawl of suburbia and the implicit abandonment of the poor by the rich have been going on for centuries. Even so, things change; there are fresh opportunities to get cities right. A lightning tour through the past two centuries, stopping at four stations along the way, allows us to recognise how our cities became so segregated and why they keep on growing outwards while large parts of their innards rot. It also helps us to see what these new possibilities are. Our first halt is at 1900, to fling a backward glance over the nineteenth century. From then on we roughly halve the length of time crossed to each successive station, focusing more and more closely on the approach to the year 2000.

Over the years, one history has been about the city within. On the inside, the city is a stressful, unhealthy, overcrowded environment where both physical and social conditions are bad, especially for the poor. But people have had to put up with this because of the sheer necessity of the city to industry, trade and economic growth, and perhaps to culture and government as well. The need for businesses and workers to be in the city forces up land values, making landlords and developers squeeze too much onto every available square foot. Growth engenders more growth; hence the repulsive nineteenth-century imagery of the wen (cyst) or tumour. The great issues have been how to better house the poor and how to improve the urban environment.

Another history has concerned the city without, mourning the loss of beloved countryside resulting from the disgorgement of suburbs that transport improvements allowed. And fearing that huge areas of Britain would eventually be entirely built up. And launching attack after attack on these growing suburbs. They have faced repeated accusations of being badly planned, wasteful, and of acting like parasites on the cities they grew from. They have been criticised, again and again, as boring, escapist dormitories for small-minded, inward-looking people. Often this has been motivated by snobbishness, a case of people sneering at their perceived inferiors who have tried to climb one rung up the social ladder by moving to the suburbs. The salvoes have come from aesthetes raging against vulgar tastes, against feeble, backward-looking architecture and from left-wing thinkers dismayed by the social uniformity of the suburbs and the shunning of the poor that their development implied. You can see their point, but there is something futile and misguided in much of this criticism. It does not really engage with the fact that suburbanisation is a long-running mass movement that has bettered millions and millions of lives.

At the dawn of a new century we can be confident that no previous hundred years has witnessed such awesome change. In 1900 the British establishment felt similarly privileged. It was their exploding cities which most impressed, appalled and divided them. The world had never seen such things. The nation's population nearly quadrupled in the nineteenth century and by 1851 over half of the people were living in towns and cities, an urbanisation landmark which the United States would not reach until 1920. The first national Census, in 1801, found only one British town with a population above 100,000 – London with 959,000 inhabitants. By 1901 there were twenty-five, nine of them with over a quarter of a million people, while the number of Londoners had swollen to 6.5 million. The proportion of the population living in towns and cities had risen from a third to more than three-quarters.[4]

Good Victorians struggled to make sense of these cities, to bring order, civili-sation and justice amid their darkness. And it was darkness. The colours of the Victorian city are mostly dark greys, browns and blacks – smuts and smoke, soot-encrusted brick, grime, excrement, stagnant streams and rivers, fog. The idea of a permanent urban night was a popular literary metaphor. Some railed at the ugliness of cities, some wished or forecast that they would fade away. The greatest worry was the urban poor, who were regarded with as much fear as pity.

Rapid population growth, combined with the physical separation of the classes that was the hallmark of nineteenth-century cities, turned the poor into a great unknown. Who were these people who no longer went to church, who drank, who lived in such squalor and sometimes died like flies? Was a new, degraded race being bred in the cities? If there was a war, would they be any use as fighting men? Would Chartists and socialists inflame them into revolt? The poor – and especially those at the very bottom of the heap, the Irish immigrants – were something less than human in the minds, speeches and pamphlets of both reformers and reac-tionaries. The favourite term to describe poor housing was a rookery, accommodating a croaking, breeding mass of scavenging beings perched one on top of each other.

The shadow of Victorian urban poverty was cast down our century and can still make us shudder today. There is something akin to a national guilt about it, passed down through the nineteenth-century's great novels, through today's media and political debates, school history lessons and textbooks. We still have the habit of thinking of the most deprived city dwellers as subhuman. How else could a word like 'underclass' gain such wide currency, or tabloid journalists refer (among themselves) to the roughest and poorest of their readers as 'pond life'?

But industrialisation and the explosion of cities that was part of it did bring about an overall rise in living standards. People 'got on' and the middle class expanded. Death rates came down, from around 26 per 1,000 people each year in 1800 to 16 in 1900.[5] As increasingly prosperous consumers the Victorians demanded single family homes in new suburbs and got millions of them. They installed the public transport systems that made longer journeys to work – well beyond walking distance – possible, first to the middle classes and then to the upper working classes. The semi-detached home, the halfway house between country villa and terraced town dwelling that became the twentieth-century's chief suburban symbol, was invented early in theirs.[6] They belatedly engineered large reductions in the risks of disease that resulted from concentrating huge numbers of people. By the end of their century they were close to accepting that the free market would never provide a huge section of the poor with acceptable homes and that large-scale subsidy – by charities or the state – was needed.

In 1800 houses were creeping out along the roads that radiated from the larger towns and cities, sometimes joined together in terraces, while more and more detached villas were appearing in the surrounding market gardens and meadows to add to this ribbon development. Places that we would recognise as suburbs already existed but they were few and small. The term was already in common use; Chaucer had made a disparaging reference in his *Canterbury Tales* written more than 400 years earlier.* Both rich and poor were to be found living outside the city, the former at some distance, the latter right next to it. Travelling in each day by private carriage or on horseback to attend to one's business while sleeping in a fine country house miles outside town was nothing unusual. But the land fringing the built-up area had, for centuries, been a place for undesirables who needed to be close to the city but whom the townspeople, or the authorities, wanted to keep out. The most foul-smelling industries could be found there along with frowned on entertainments and people breaking the closed shop of the urban craft guilds.

For most of the nineteenth century, the growing numbers of industrial, increasingly urban, labourers needed to live very close to the new factories, workshops, docks and mines. They walked to work and it was quite common to go home during meal breaks or for wives or daughters to bring food to the factory.

---

* *Where dwelle ye? if it to telle be.*
'*In the suburbes of a toun,*' *quod he,*
Lurkynge in hernes and in lanes blynde,
Where-as thise robbours and thise theves, by kynde,
Holden hir pryvee, fereful residence.
　From the Prologue of the Canon's Yeoman's Tale

**When Sir Titus Salt had his new alpaca mill built in the countryside near Bradford in 1853 he also commissioned 850 homes (mainly two-bedroom terrace cottages), 45 almshouses, 2 churches, a school, bathhouses, shops, a 6-hectare park, steam laundry and a hospital. Saltaire had twenty-two streets and took almost 20 years to build.

New housing often had to be put up near the new workplaces and a few factory owners took pride in providing clean, well-ventilated homes built to last, sometimes on a grand scale.** But most built as cheaply and as densely as possible, while the bulk of labour had to rely on what the rental market rather than employers would provide. Even if single cottages were originally constructed to house single families, relatives and other lodgers would crowd in. It was common to have half a dozen adults and children sleeping in each small room, several to a bed. The most desperate slept in damp, badly ventilated cellar rooms.

Some of the large merchant houses along the main streets in the old town centre would be subdivided for numerous poor families. The yards and gardens behind the big old houses would become jam-packed with tiny terrace homes. These would enclose a small court connected to the street by a narrow entrance, often covered and tunnel-like. The dozens of residents in each court would share one or two latrines unconnected to a sewer. Large parts of the old centres turned into stinking mazes of these chaotic courts connected by alleys one or two yards wide. Further out from the town centre, on greenfield sites, two-storey terraces of back-to-back housing were thrown up along new streets. The front side of each house would open directly onto the street, the back would share a windowless wall with the row behind.

There might be some fairly high density building for the wealthy near the town centre, in imposing, speculatively-built terraces which sought to make a collective rather than an individual statement about the high rank of their inhabitants. In London, aristocratic developers had already begun to lay out exclusive estates between the old, walled city and the Palace of Westminster in the seventeenth century. This trend continued in and around the West End through the nineteenth century, creating tracts of grand homes and neighbourhoods that have stayed extremely wealthy to this day. In other burgeoning cities the old elite and the *nouveau riche* were more likely to decamp into the countryside proper, building detached villas and mansions set amid large gardens. Smaller and medium-sized industrial towns came to be dominated by the new workers' housing scattered around the mills and mines that brought them there. But in the very largest cities and the great ports, where commerce, finance and administration grew alongside industry, the suburban growth was more variegated. There was a larger market for middle- and upper-class suburbs.

Karl Marx's collaborator Friedrich Engels sums up the social geography strikingly in his description of mid-century Manchester. At the time it was Britain's second

Other rare examples of large nineteenth-century industrialists providing carefully planned housing with amenities for their workers on greenfield sites were soap tycoon William Lever's Port Sunlight, Birkenhead, constructed from 1888, and chocolate magnate George Cadbury's Bournville outside Birmingham from 1894.

largest city with a population of around 400,000 and the greatest manufacturing centre in the world:

> all Manchester proper, all Salford and Hulme, a great part of Pendleton and Chorlton, two thirds of Ardwick and single stretches of Cheetham Hill and Broughton are all unrivalled working-people's quarters, stretching like a girdle, averaging a mile and a half in breadth, around the commercial district. Outside, beyond this girdle, lives the upper and middle bourgeoisie, the middle bourgeoisie in regularly laid out streets in the vicinity of the working quarters, especially in Chorlton and Ardwick, or on the breezy heights of Cheetham Hill, Broughton and Pendleton, in free wholesome country air, in fine comfortable houses passed once every half or quarter hour by omnibuses going into the city.[7]

He goes on to condemn the 'hypocritical plan' that hid slums behind rows of shops lining the thoroughfares leading into town. Today, Manchester's deprived inner city girdle has widened itself by a couple of miles, and his fine comfortable houses have either vanished under later council housing or been subdivided for tenants who are not of the bourgeoisie. Shops and other commercial premises along the main roads still do a pretty good job of concealing sad neighbourhoods with littered streets and abandoned houses from passing traffic.

Engels' long description of Manchester in *The Condition of the Working Class in England from Personal Observation and Authentic Sources* bristles with anger and disgust. As his title suggests, he was challenging his contemporaries to doubt his findings. The young German was convinced that England was at the cutting edge of the most stupendously important period in world history, and that Manchester was on the very edge of that edge. But what is most startling about this urban portrait painted more than 150 years ago is that its main features are so recognisably modern; the separation of rich and poor, the wealthy, outermost suburbs, then a 'doughnut of deprivation', then a large commercial district at the core with heavy traffic, brightly lit shops, big offices and hardly any homes remaining because high property values had squeezed them all out.

This is the essence of the twentieth-century British and North American city. It anticipates the influential 'ecological model' of the modern conurbation developed by the University of Chicago sociologist E.W. Burgess in the 1920s and taught to generations of geography students. The further out you travel, the more affluent the suburbs become. In Burgess' scheme of things the most recently arrived immigrants, the poorest of the poor, live just outside the commercial core, in the oldest, most dilapidated and overcrowded properties – as the Irish did in mid-nineteenth-century Manchester, followed by the Pakistanis and Afro-Caribbeans in big British cities after the Second World War.

In pre-industrial cities rich, poor and middling lived close to each other, with the finest dwellings for the wealthiest inhabitants nearest the centre. Victorian cities separated social classes as never before and Manchester seems to have done this earlier than most. Historians have argued that there were other British towns where the middle and upper classes stayed near the centre for longer (in London, they never left), and where rich and poor remained intermingled well into the nineteenth century. Maybe so, but by 1900 segregation was the urban norm.

The image of suburbs falling into neat, concentric rings based on their class ranking is, of course, an oversimplification. A more realistic diagram for the big British city at the turn of the century is of these rings broken by wedges which had been shaped by public transport routes and barriers like rivers, railway lines and parkland. Some working-class neighbourhoods extended to the city edge, some middle-class enclaves stretched to the heart of the city. Often the wealthier neighbourhoods tended to be on better drained, higher ground and out to the west, where the prevailing winds kept the smoke from factories and houses away. For our purposes, what matters is that when families with incomes and wealth chose to move in the city, it tended to be outwards. Today they still do.

By 1900 the capital and the nation's other big cities had disgorged numerous suburbs, each larger than every British town excepting London had been 100 years before. The four parishes that made up the old London borough of Camberwell had, for instance, seen their combined population climb from 40,000 at the 1841 Census to 260,000 in 1901.[8] What had been quite separate villages in 1800 were entirely engulfed.

The process of suburbanisation looked something like this. The ribbon development along the roads leading out of cities would thicken from individual houses to terraces. Entire streets lined with homes would go up on the fields behind these roads and within two or three decades the meadows and market gardens would vanish completely. Next the big gardens surrounding mansions that had been built decades or centuries earlier with usually urban-made fortunes would be sold off as building plots, and often the mansions themselves would be demolished to make way for housing. This obliteration was the work of tens of thousands of small, speculative building firms which, considering their numbers, managed to turn out remarkably uniform housing.[9] They used pattern books, they had a pretty similar idea of what their market wanted and as the century passed they relied increasingly on mass-produced materials. Their products would usually be for rent and the builders would commonly be leasing their plots for between 30 and 100 years from landowners – often aristocrats, sometimes fairly humble smallholders – on condition that they build houses on them. When these building leases expired, the land reverted back to the original landowners (or whoever had since bought the freehold) who could then profit by raising the rents. Sometimes the building land was sold freehold, and sometimes on the 'fee farm' or 'chief rent' basis, in which the land was granted in perpetuity but was still subject to fixed annual payments to the vendor.

A complicated suburbanisation market developed with various players. There was the landowner, the housebuilder, the developer acting as middleman and laying down streets and sewers before selling off building plots, the investor (often a moderately wealthy individual rather than a bank or some other financial institution) providing capital for the builder or eventually buying the completed homes as an investment. The same person or organisation might take several of these roles in creating a suburb. As the century passed, a few made colossal fortunes, larger companies became more common and the expanding building societies financed the construction of huge numbers of new houses, mostly for rent rather than owner occupation.

Suburbanisation was a major industry that pushed towns and cities out into the countryside in fits and starts as recessions came and went. Old paths and roads, field and woodland boundaries determined the shape of the buildings plots, so while order and harmony might prevail among the new streets on any single plot, the overall layout of the growing town looked chaotic. Land was supplied relatively cheaply and with few restraints, allowing a particularly English taste for single-family houses to be satisfied. One home per family was what the Victorian reformers advocated on the grounds of moral and physical health, and in the second half of the century it was supplied in bulk to the middle and much of the working classes but never to the poorest families.

In Scotland, however, four- to five-storey tenements with one or more homes on each floor were the norm. These handsome stone buildings, some built for high-income tenants, are what makes cities north of the border so obviously different from English ones. This was probably due to the widespread use of fee farm – or feu farm as it is known in Scotland – tenure. The vendor of land for housing development would negotiate as high a permanent annual rent as possible, which put pressure on the purchaser to maximise the number of tenants and so build higher. Once tenements began to appear instead of cottages and two-storey terraces, this cemented the high price for development land around the growing Scottish towns, reinforcing the high-rise habit.[10]

The new suburbians were hoping to escape epidemics and high infant mortality. Drainage, clean water and fresh air were key selling points on the edge of cities in which cesspits overflowed, cellars became flooded with ordure and 'varying attenuations of sewage' from shallow boreholes masqueraded, as the pioneering suburban historian H.J. Dyos so nicely put it, as drinking water. The 1845 auction particulars for a plot of building land in Camberwell read:

> Bath Road already runs through the estate and has a Famous Sewer already constructed from one end to the other, and there is a deep gravely soil throughout this locality forming also a natural drainage which, together with the salubrious air for which the neighbourhood is proverbial, have gained for Peckham its present celebrity for promoting health and longevity.[11]

The better sort of Victorian suburbia: tenements in Glasgow's West End, facing

By 1900 it was clear that middle-class suburbs could be victims of their own growth. New housing in a leafy setting almost surrounded by countryside could, a few decades later, find itself deep inside the built-up area. Factories and workshops crept in. The roads became lined with shops and public houses and increasingly congested. More and more of the 'lower orders' could be seen on the streets. Public transport, which helped make the suburbs, could also be their undoing. At first the new horse omnibuses, which arrived on the scene in the 1830s, were patronised almost entirely by the expanding middle classes. So, too, were the suburban railways, which appeared from the 1860s (a couple of decades after the intercity network was laid down) and made little early impact outside of London. The new transport systems allowed people to live miles from their workplaces without the need to own horses and have the grooms and stables that went with them. But once horse-drawn trams were introduced from the 1860s onwards, to be followed by electric and steam trams some 20 years later, public transport made the streets it ran through increasingly noisy, busy and less desirable places to live in. It gave the better-off working-class and lower middle-class people the chance to live further out from the city centre, but their arrival *en masse* could begin to lower a suburb's status.

For all of these reasons a suburb built for the middle classes might easily and quickly become a place where better-off families no longer wanted to settle. As the trees and fields dwindled away, as the neighbourhood became more crowded and noisy, developers would find themselves having to go downmarket, building smaller, cheaper properties on the remaining open land. Instead of providing room for servants in the attic the new homes would offer just two or three bedrooms per family, aimed at the lower middle class or artisans. In the most congested areas, next to polluting factories and hemmed in by railway lines, canals and busy thoroughfares, such housing could almost immediately become a slum, with an entire family with four or more children occupying one or two rooms. There was no shortage of impoverished people prepared – or compelled – to take such cramped accommodation. They continued to pour into the big cities from smaller towns and the countryside, and those already there were also being forced out of the centre by high rents and the slum clearances which made way for railway stations, grand boulevards, great office buildings and hotels.[12]

Just as an earlier generation of poor, rural immigrants might have crowded into a mediaeval merchant's house in the town centre, a large suburban house built speculatively in the 1840s for one of the city's professionals or merchants could – by the 1880s – have become a shabby, overcrowded dwelling for half a dozen or more families emerging from the inner city.

The upper middle classes were moving further and further out as public transport networks expanded and families a couple of rungs lower down the social ladder moved in, or had their own more modest streets and homes built for them. A chorus of sneers broke out in periodicals and books as suburbia spread. Who did these new classes think they were, these clerks and tradesmen and their immediate superiors who had entire neighbourhoods thrown up for them, formless sprawls which lacked a history or a cultural life? In a popular, middle-class domestic manual of 1898, *The New Home*, Mrs C.S. Peel wrote: 'I must confess honestly that the suburbs of any large town appear to me detestable.'[13] They were, she judged, consolation prizes for 'those people who yearn for the pleasures of the country . . . and whom cruel fate prevents from living in the real country'.

The poor cooped up in the city centre had, of course, suffered a much crueller fate. Through the second half of the nineteenth century many were benefiting from improvements in their dangerous urban environment. Progress generally came in the wake of hundreds of separate Acts of Parliament that would apply only to individual towns and cities, followed by legislation that covered the entire nation. One consequence was the gradual construction of powerful, coherent local councils in place of a chaos of smaller, more or less ineffectual bodies. The new laws and regulations followed local petitioning and decades of high-pressure national campaigning by three generations of physicians and statisticians, civil servants and politicians, churchmen, journalists, enlightened captains of industry and the occasional Royal.[14]

The crucial sanitary, housing and administrative reforms which began in the 1840s flowed from a mass of hearings, lengthy reports and copious disease and death statistics processed by Parliamentary select committees and Royal Commissions. The starting point for these great reforms is usually taken as the 1842 *Report on the Sanitary Condition of the Labouring Population of Great Britain* by Edwin Chadwick, Secretary to the Poor Law Commissioners. Helping the process along were epidemics, the hard numbers on astonishingly high urban death rates clearly linked to poverty and overcrowding, and written descriptions which conveyed the utter degradation of a large part of the nation to a comfortable minority which never had any occasion to step into the slums.

Today even the most dry and numerical of these reports still has the power to stun, such as this one from 1847, when the Statistical Society of London's investigators surveyed Church Lane off Oxford Street.

House No. 4 – Two Parlours on Ground Floor. Size of front room, 14 ft long, 13 ft broad, 6 ft high; size of windows, 3 ft 4 in by 2 ft 2 in. Size of back-room, 11 ft 2 in long, 9 ft 4 in broad, less than 6 ft in height . . . number of families, 5 comprising 4 males above 20, 9 females above 20, three of them single, 2 males under 20, 4 females under 20, total 19. Number of persons ill, 2, deaths in 1847, 1, measles. Country, Irish; trade, dealers and mendicants. State of rooms and furniture, bad, dirty; state of windows, 6 whole panes and 10 broken. Number of beds, 6. The door of this room opens into the yard, 6 feet square, which is covered over with night soil [excrement]; no privy, but there is a tub for the accommodation of the inmates; the tub was full of night soil. These are nightly lodging-rooms. In the front room one girl, 7 years old, lay dead and another was in bed with its mother, ill of the measles.[15]

Forty years later, pamphlets and newspapers were still telling similar horror stories of poverty and overcrowding. The reformers made progress but it was slow. They had to run to stand still because the cities were always growing as people poured in and new slums formed as old ones were cleared. Legislation was often ineffectual because it tended to be permissive rather than mandatory, allowing – not compelling – local councils to do things. For instance, while it was possible for local councils to ban back-to-back housing in their own areas from 1858, many did not. In 1886 these formed 71 per cent of Leeds' housing stock, and that city did not ban their construction until 1909.

There were those who argued that the poor were to blame for the conditions they lived in; their squalor was innate. Giving them clean, spacious, well drained and aired accommodation was a waste of time because they would neglect it, stop up the ventilation, fill any extra space with refuse and have too many children, rapidly converting it to an overcrowded slum. The reformers replied that it was hardly surprising that people forced to live in such appalling conditions were immoral, promiscuous, drunken and enfeebled. What the urban poor needed more than anything were fresh air, drains, clean water and sunshine. But an even more persuasive argument for reform was that the cholera, typhus, dysentery and tuberculosis that decimated the worst-off slum dwellers moved on to take middle- and upper-class lives.

The long, regimented rows of new homes being built for the working classes on the urban fringe at the end of the century were a vast improvement on the courtyard and back-to-back hovels that had preceded them. Even the cheapest new houses would have four or five rooms, piped water and an outside lavatory of its own draining into a sewer. The front doors opened directly on to the pavement and there would be a tiny, enclosed yard out the back, behind which ran a narrow alley. This is the by-law housing of *Coronation Street*, still found in most English inner cities amid the post-war council estates. It stemmed from the Public Health Act of 1875 which enabled councils to set minimum standards covering such things as the amount of space behind the house, room dimensions and sewer connections. Because councils across England tended to set the same bottom limits, as laid out in central government's model by-laws, we still have thousands of streets of very similar late Victorian housing.

The unpaved, undrained urban roads that were still common in the first half of the century had also vanished by 1900. The mounds of human excrement which had piled up, waiting to be carted out to the fields, were long gone and so too were the pigs which were raised in the poorest and filthiest neighbourhoods. Councils had won more and more powers to either compel owners to improve the worst, older housing nearer the city centre or to demolish slums. The largest redevelopment of all took place in Glasgow where the city council, under its 1866 Improvements Act, was allowed to rebuild 88 acres of the mediaeval city centre laced with narrow, winding alleys. The urban regeneration powers it was given would be the envy of a modern council. Glasgow was allowed to increase taxes on its citizens and borrow huge sums (the equivalent of hundreds of millions of pounds at today's prices) to buy sites, demolish buildings and then sell the land to developers. When private developers failed to take up cleared and vacant land, the council started to build working-class homes in tenements with ground floor shops.

As the century closed, Britain was nearing a high point of local government autonomy. The path to municipal glory was marked by vast town halls, libraries, museums, bathhouses and grand parks. All-powerful big city councils provided tap water and disposed of sewage, supplied their citizens with gas and electricity, administered poor relief and took over the running of the new electric trams. And now, thanks to those trams and suburban railways, more and more labourers could live further than walking distance from their workplaces. The government and local councils believed special discount fares could help to decongest the city centres and allow working-class people to move out to less crowded housing. From the 1860s onwards a growing number of suburban rail and electric tram operators were compelled to offer these workmen's fares early and late in the day. Even without these subsidies, public transport was becoming increasingly affordable.

But the inner slums persisted, even grew. A large proportion of the poor were trapped in the city, competing for jobs that were highly insecure and extremely low paid. There was little to prevent them being overcrowded; it was the obvious way for landlords to maximise their income from dilapidated buildings in bad neighbourhoods. Charity, in the shape of the pioneering housing associations, built flats and cottages for the working classes, often on sites cleared of the most overcrowded and dilapidated slums. Much of this charity was of the kind we would now call 'ethical investment'.

Societies were formed to give do-gooders a safe investment with a fairly low rate of return (around 5 per cent) in the shape of new homes for rent for people on low incomes. By 1890 these philanthropic trusts and model dwelling companies had built about 25,000 homes.[16] Councils were also allowed to build housing for the working classes under a succession of Acts of Parliament passed in the second half of the century, notably the Housing of the Working Classes Act 1890. These acts gave them powers to raise construction funds from the rates (local property taxes) or to borrow money at the lowest possible interest rates. But few chose to use these powers, and by 1900 only a few thousand council homes had been built by a handful of councils.

The new homes built on sites cleared of slums could accommodate fewer tenants than the overcrowded rookeries they replaced. Sometimes the slum landlords had to be compensated for their demolished property. For these municipal and philanthropic housing schemes to be financially viable, the rents charged to tenants had to be pitched at a level that only better-off working-class people in steady jobs could afford, even though they were being subsidised by ethical investors or council ratepayers. The hope was that when they moved into their new homes, the inferior housing they left behind would become available to the needier, poorer working classes who were living in even worse accommodation.

State-subsidised housing was, then, on the agenda at the turn of the century as a means of combating the great urban ills of overcrowding and poverty – even if only some 12,000 council tenements and cottages had been built by then.[17] Meanwhile the suburbs, a cure for those who could afford it, continued to spread ever outwards.

Some observers were far-sighted enough to see that the cities would soon be pushing into the countryside even more rapidly as public transport expanded. The little market town of Bromley, which had recently become a London suburb, was home to two great visionaries of urban dispersion – H.G. Wells and the exiled Russian revolutionary, Peter Kropotkin. The former foresaw the distinction between town and country breaking down almost entirely in his prophetic *Anticipations*, published in 1902.[18] The latter advocated the dispersal of most manufacturing into rural villages, with advances in technology and mass education allowing people to divide their working hours between intensive agriculture and machine-aided, electrically powered industry.[19] But this story's most important turn-of-the-century urban visionary wanted a completely fresh urban start. Ebenezer Howard was advocating the building of garden cities deep in the countryside.

Howard's ideas, and subsequent interpretations of them, have had an enormous influence on Britain. Like many before him he rejected the great conurbations – and because he won such a following that rejection ended up hurting the cities. But the key principles underlying his garden cities – closeness to the countryside, walkability, high density, an exalted public realm, the community controlling and in some sense owning its town – remain at the centre of any hopes for an urban renaissance today. I warm to Ebenezer Howard because, like most journalists, he was a complete amateur. Very few of the underlying ideas behind garden cities were his; instead he brought other people's proposals for co-operative land ownership, social reform and town planning together into one striking proposal.

The London-born son of a confectioner, he made his early living as a shorthand writer, a pioneer farmer in the American West and a law court reporter in Chicago. A slim book outlining his ideal community was published in 1898, when Howard was 48, with the title *To-morrow: A Peaceful Path to Real Reform*.[20] It was reissued in 1902 under the title, *Garden Cities of Tomorrow*. Between the two dates he helped found a Garden City Association that aimed to turn his vision into reality.

He wrote that the big cities were overcrowded, polluted and costly to live in. They were places where people could feel lost among the vast crowds. But their job opportunities, bright lights and entertainments were still pulling the masses in from a depopulating countryside that offered very low wages and little hope for a better life. The solution was, as ever, a third way – garden cities where 'all the advantages of the most energetic and active town life with all the beauty and delight of the country may be secured in perfect combination'.

The garden city was also, he argued, a third way between socialism and individualism. It was an experiment which he hoped would lead to a gentle revolution. A group of partly philanthropic, partly hard-headed investors would buy a 2,400-hectare rural estate and there build a 400-hectare circular town with a diameter of 1.5 miles. There would be 5,500 homes for a population of 36,000, plus 2,000 people living and working on the surrounding farmland which would be dotted with woods, allotments and the odd convalescent home and asylum. This town would have six broad boulevards (to be named after greats like Columbus and Newton) radiating, spoke-like, from a spacious park at its core. The key municipal buildings – town hall, hospital, concert hall, library, and so on – were to be situated at the very centre. An enormous, glass-roofed parade of shops and a winter garden would encircle the park. Outside this would be housing arrayed along circular avenues with the broadest of these ring roads, Grand Avenue, lying halfway out to the edge of town. It was to be lined on both sides with crescents of imposing houses while schools and churches would be distributed along the middle of the road.

On the town's outer rim, served by an outer circular railway, was the industrial area of factories and workplaces; the dreaded urban 'smoke nuisance' was to be overcome by dispersing the polluters. One of Howard's intricate, spidery diagrams of a segment of his garden city shows a boot manufacturer, cycle works and jam factory located here. The furthest citizens would ever need to walk from home to anywhere in the garden city was less than one and a half miles and if that proved too much there were to be some electric tramlines. A short stroll from anywhere in town would take you either out to the countryside or in to the central park.

All the rents from the entire rural and urban area would be paid to a single elected council, the sole freeholder. This rent income would, over a thirty-year period, pay off both the interest and the capital needed for the original purchase of farmland as well as covering the building and maintenance costs of the garden city's roads, drains, schools, libraries, and so on. There would also be enough money from these rents to provide old age pensions for its citizens and give them sickness and accident insurance cover. Howard's garden city would be a municipal welfare state with a mixed economy, for the land for the factories, homes and shops would be leased to private businesses, co-operatives and building societies. He costed it all out and estimated that the original investors would get a 4.5 per cent return on their capital. It would work financially because farmland could, at that time of agricultural depression, be bought very cheaply.

Once the beautiful new town was established and growing (six pie-slice segments to be built, one after the other), the value of the land would soar, lifting rents high enough to pay for social services and pensions while maintaining the garden city's infrastructure and communal buildings. This method of capturing the 'betterment' of the land for the community was one of Howard's most important concepts. Urbanisation is the collective act of hundreds of thousands; more and more people and businesses crowd into a city, wanting and needing to be there to

THE COUNTRYSIDE DID NOT, however, come into London. Instead the capital poured out across the countryside faster than ever, as did all of Britain's large towns. In *Garden Cities of Tomorrow* Ebenezer Howard had pointed out that London, the world's greatest city at the start of the twentieth century, covered an astonishing 120 square miles. But already the most crowded inner areas were seeing their populations fall rapidly as people moved to the suburbs. By the mid-century the number of Greater London citizens had grown by only one-fifth since the 1901 census but the land area covered by the capital had more than quadrupled.[3]

While millions of poor people continued to live in old, decaying housing near the city centres, a large and growing proportion of the population was able to afford spacious, single family homes with gardens front and back. Builders put up millions of speculative homes on greenfield sites on the urban fringes during the first half of the century because farmland was cheap and planning controls lax. They were mostly detached or semi-detached and their outward appearance was vaguely traditional, making various nods and gestures to the past. That is what the house-buying public felt most comfortable with. By and large, they still do today.

And so the growth of cities carried on in much the same way as it had in the previous century – getting on meant getting out. But what was new and revolutionary in the twentieth-century's cities, what gave the outwards spread such spectacular acceleration, was petrol and diesel and electrical power. Cars and buses, along with a shorter working day, enabled people to make the longer journeys necessary for deep suburban life. In London the suburban railways, which gradually electrified, and the expanding Underground were just as important as road transport to the city's continual disgorgement. Sometimes the tracks and termini leapfrogged miles ahead of the urban fringe and had to wait years before becoming engulfed by new housing.[4] Outside of London the electric tram initially led the suburbanisation charge, but it soon began losing out to the bus. By 1951 it seemed headed towards extinction.

Why have we stopped at this particular year, just beyond the century's halfway mark, to look back? One reason is that a census was taken, the first for 20 years because war had caused the cancellation of the count due in 1941. The 1951 Census found Greater London's population to be just under 8.2 million, a peak from which it was then to decline partly, but not entirely, as a result of government policy. Up to then the numbers living in the nation's great cities had kept on growing, even though they had drastically lowered their population densities by pushing out suburbs and increasing the built-up area. Now they were about to begin a population decline with emigrants moving clear outside.

The other reason for choosing 1951 is that it witnessed the final defeat of the great post-war Labour government after it had struggled back into office with a tiny majority at the 1950 election. During its six exhausting years in power it gave the state a much more powerful role in determining the future of Britain's big cities. Through the first half of the century central government and local councils had intervened increasingly to improve the urban environment and the housing of the urban poor. Those efforts had to be put on hold for the ten years during which two total wars were fought. When the Second World War ended in 1945 the new government and most of the country believed the state could and should intervene much more. Clement Attlee's administration took decisive action in expanding council housing, in beginning to disperse the urban population and in setting up a nationwide planning system that was intended to contain and bring order to city sprawl. The all-party consensus on these policies was to last for a quarter century after Labour's defeat at the polls in 1951.

Let's take planning first. By the Second World War the notion that there should be planning controls on new development in town and country, and that cities should grow in an organised way, had widespread support. The concept had a long history, with examples of large-scale, well-thought-out town planning schemes having been drawn up and implemented before the nineteenth century. These were usually the result of powerful and wealthy individuals and elites deciding what kind of development they wanted on their own land. Edinburgh's celebrated New Town, a northwards extension that started in the 1770s, is an example of city fathers taking the lead in planning the orderly growth of their town. But prior to the twentieth century, town planning was the exception rather than the rule.

Parliament passed Britain's first planning legislation in 1909 and this allowed (but, as usual, did not compel) town and city councils to draw up and enforce land planning schemes for nearby countryside likely to be covered by housing, provided they first obtained approval from central government. German cities had had such powers for decades and a large part of the Edwardian establishment feared superior Teutonic organisation in industry, social services and armaments was allowing Germany to overtake Britain. The Victorians had insisted new housing should meet minimum standards of spaciousness and hygiene which covered its interior and immediate surroundings; now the new planning law built on this by giving coun cils powers to determine where new housing was built, at what sort of density, and what provision should be made for roads and open spaces. It was, essentially, a continuation of the Victorian sanitary agenda – clean water, sewerage, fresh air, sunshine and greenery for the masses. But the new planning powers were very limited. Any landowner whose holdings lost value as a result of a council's planning scheme was entitled to compensation, which discouraged councils from using their new authority. On the other hand, since a planning scheme could also increase the value of farmland by designating it for new housing, the law allowed councils to recover half of this increase or 'betterment' from the landowner.

Over the next 30 years there were three other Acts of Parliament that chopped and changed these powers, sometimes adding to them, sometimes reining them back. For a while councils with populations above 20,000 were compelled to draw up plans for land likely to be developed, but this was then made voluntary once again. The councils and their newfangled town and country planners (the profession did not exist before the twentieth century) took very much longer to draw up these plans than the government had envisaged – something they still stand accused of today. By the outbreak of the Second World War there was a patchy, half-baked system of planning control covering more than half of England's land area. Most of this had been laid out by individual councils, with a few joint efforts to plan more sensibly across the wider canvas of the city and its hinterland of countryside and smaller towns.

The new laws certainly failed to stop suburban sprawl. And unlike earlier forms of town planning, such as Edinburgh's New Town or the gridiron street pattern of central Glasgow, the urban development carried out under this early legislation is nowhere admired today. By the end of the 1930s there were, however, the rudiments of a thin, slightly moth-eaten Green Belt drawn around London to prevent the 25-mile-wide capital spreading out even further. Local councils on the fringe of the megalopolis had been buying up farmland to stop more homes being built, helped by grants from the London County Council under its own special Act of Parliament.

The Second World War changed everything and gave planning teeth. The Blitz left tracts of the big towns and cities open for redevelopment. During the conflict a coalition government and its armed forces exercised state power in economic and logistical planning on a scale never seen before. Now planning, including spatial planning, could help to win the peace and build a better, fairer Britain. The location of desperately needed new homes and industries, the dispersal of people and factories from the overcrowded city centres, were to be planned across the entire nation rather than left to the free market or the scrappy, voluntary efforts of local councils.

A Royal Commission on the Distribution of the Industrial Population had reported in 1940, just before the Luftwaffe's bombs started falling. Chaired by an industrialist, Sir Anderson Montague-Barlow, with Britain's foremost town planner Professor Patrick Abercrombie as a member, it rehashed the criticisms that had long been made of the unplanned growth of huge cities – the very high cost of land at their centre, traffic congestion and the huge, collective waste of time spent commuting in crowded public transport. It pointed out how vulnerable such gigantic concentrations of population and industry, often one particular type of industry, were to air attack and economic depression. The commission advocated state intervention to disperse industry and population out of the largest cities in order to secure a more rational, diversified distribution of industry and its employees. Abercrombie, whose influence was to grow during the war, was part of a dissenting minority calling for still tougher planning controls and a more radical dispersal of industry. The ideas were not new; what mattered was that it was a Royal Commission that was backing them, drawn from the great, good and wise, beholden to no political party or narrow interest. At the same time there was enormous public enthusiasm for planning. A wartime Pelican paperback by Thomas Sharp, *Town Planning*, sold a quarter of a million copies. The development of cities could no longer be left to muddling through, he declared. 'Plan we must – not for the sake of our physical environment only, but to save and fulfil democracy itself.'[5]

This pro-planning ferment led to the creation of a Ministry and a Minister of Town and Country Planning in 1943 and the passing, before the war's end, of two Acts of Parliament. One extended planning control to the nation's entire land area, albeit only on an interim basis. The other enabled local councils to make compulsory purchases of privately-owned urban land in areas where there was extensive bomb damage, or where there were narrow, tortuous streets and decrepit buildings. This latter, nicknamed the 'blitz and blight act', was intended to give councils the power to quickly redesign and rebuild 'obsolete' or obliterated urban areas.

Given this increased state power, what was going to be planned? Two grand designs for London and its surroundings commissioned by local and central government and drawn up under Abercrombie's supervision,[6] plus a string of ministerial handbooks and government guidance documents specified the shape of things to come. These all drew on the prolific thinking about urban forms and road design that had gone on between the two world wars in Britain, Europe and the USA. For the unplanned sprawl of the suburbs and the ineffectual planning legislation of the inter-war years had provoked masses of writing, drawing and debating about what ought be done. Within a couple of years of the Second World War ending, Britain was committed to a new urban order of ring roads and roundabouts, of town and city centres with shops and offices but no homes, of massive demolition of closely packed Victorian homes to make way for millions of new council houses and flats at lower, regulated densities.

The new thinking adopted by the wartime and immediate post-war governments was that the busiest roads leading into and through towns should no longer be lined with shops, houses and side roads because all of the comings and goings these generated clogged the steady flow of traffic. The main street, busy with both pedestrian and wheeled traffic, was a dangerous anachronism. People and vehicles must be separated to the maximum possible extent. Primary routes should run, uninterrupted, through the city set amid strips of greenery. A series of ring roads would combine with the spoke-like main roads to handle the traffic passing through the town or around its edges. The rapid growth in road traffic must not be slowed; instead, the traffic had to be better managed. In those days of low (but fast-rising) car ownership, no one who mattered considered curbing the car and favouring public transport instead.

Homes should be well separated from industry; factories and warehouses were best confined to their own industrial zones outside the city centre. Gradually, quarter by quarter, district by district, the jumbled, unplanned mess of the big city would be torn down and reconstructed to offer space, order, modernity.* If lots of people had to live in a neighbourhood, then they would occupy big, well-spaced blocks of flats rather than cramped rows of terraces. This wartime thinking reached its highest and most beautiful expression in Abercrombie's two great, state-commissioned plans for the capital, *The County of London Plan* and the *Greater London Plan 1944*. Both are well worth an afternoon's browsing for anyone wanting to understand the post-war planning disasters in Britain's cities. They are huge, hardback volumes stuffed with hopes and promises, with handsome black and white photographs showing the best and worst of the existing city, intricate colour diagrams and maps setting out the geography of the capital and its hinterland, and instructions for its reconstruction. Here are gentler, more practical versions of the modernist hallucinations of Le Corbusier and his disciples. Every conceivable aspect of urban improvement is covered, from the precise phasing of the demolition and reconstruction of a large slum quarter to the need for enormous helicopter landing pads on the roofs of London's main stations.

Many of these ideas, transformed into official government guidance, gradually worked their way into the fabric of every medium-sized and large British town and city during the post-war years, obliterating the familiar and sometimes cherished, throwing up new barriers to movement. We are still struggling to undo the unintentional harm they have done. Lately Birmingham, for instance, has been gradually unlocking the stranglehold of its 'concrete collar', the 1960s' inner ring road that cut the centre off from the rest of town. The Second City is hugely improved as a consequence.

Legislation, in the shape of the Town and Country Planning Act of 1947, was as important for the future of Britain's cities as these ideas and official guidance. This momentous law had been promised in Labour's manifesto and underlies today's system of planning and development control. It ordered councils to draw up map-based plans setting out what type of development should be allowed across their fiefdoms, zoning certain areas for industry, others for housing, and so on. Any proposal for development would have to be submitted to the local county or county

*Mixed uses, now seen as one of the greatest virtues of urban neighbourhoods, were reviled as one of their greatest defects. 'Mention the disadvantages in convenience and amenity that result from indiscriminately mixed uses', said a question in a 1948 examination paper for the University of London's Town Planning Diploma.
**Abercrombie's *Greater London Plan 1944*, commissioned by the government, envisaged reducing the

borough council, which would decide whether to grant or refuse permission. If permission was denied, the frustrated developer could appeal to the government to over-rule the council's refusal. Government ministers were also given wide powers to intervene if they saw fit, reserving planning decisions for themselves or setting up public inquiries. The new legislation also introduced the system of Industrial Development Certificates, which enabled government to control the location of significant new industrial premises.

Thousands of landowners could now claim that the new law had removed their freedom to do what they wanted with their land. A compensation fund was set up to pay them off. But once the new law was enacted, the value of any piece of land was going to depend on whether the local council or central government granted or refused planning permission for development. For that reason the 1947 Act introduced a levy under which 100 per cent of any increase in the value of the land was collected by the state. The legislation represented a huge extension of state power, as important as the post-war nationalisations of coal, steel, electricity, gas, railways and the health service.

By 1951, the government also had plans to decant millions of people from the inner areas of big cities, not into low-density suburbs on the urban fringe but many miles further away.** One way in which this was to be done was through the building of new towns; nationalised versions of the garden cities Ebenezer Howard had planned half a century earlier. His campaigning had spawned a movement led by the Town and Country Planning Association and the creation of two small garden cities in Hertfordshire, at Letchworth and then at Welwyn. Now, during the pro-planning years of the war and its immediate aftermath, it seemed obvious that the job of creating attractive, spacious and largely self-contained communities of 30,000 to 50,000 people, with their own industries and a balance of social classes, should fall to the state rather than to a combination of volunteering and ethical investment. As the final report from the Planning Minister Lewis Silkin's advisory committee on new towns put it in 1946:

> if the community is to be truly balanced, so long as social classes exist, all must be represented in it. A contribution is needed from every type and class of person; the community will be poorer if all are not there, able and willing to make it.[7]

Abercrombie's *Greater London Plan of 1944* had proposed ten sites for 'satellite towns' scattered around the outer rim of a development-free Green Belt surrounding the capital. In 1946 a New Towns Bill quickly passed through Parliament with all-party support, enabling central government to buy the farmland needed for new

population of the County of London (made up of the boroughs which today comprise inner London) by a million through the building of satellite towns and encouragement of out-migration.

Crawley, West Sussex, a Mark I post-war new town designated in 1947, overleaf

towns, and create powerful, state-owned corporations to develop each of them. That year Stevenage was formally designated as the first. By 1951, sites for eight new towns had been selected around London (one of which embraced and expanded Ebenezer Howard's Welwyn Garden City) and for six elsewhere in the country. Most were intended to help reduce population densities in London and Glasgow (which was to have its own satellite new town at East Kilbride) while the rest were linked to coalfields, industrial expansion or urban decongestion elsewhere in Britain.

The new towns had a difficult first few years. They were bound to be expensive, so other government ministers opposed them during the exchange rate crisis and public expenditure cutbacks of those early post-war years. With the exception of London, which was supportive, the local councils running big cities feared the new communities would pull away people and jobs. A group of local residents fiercely opposed the designation of Stevenage. The tyres on the ministerial motor were let down when Silkin went there to explain himself and a legal challenge was launched which went all the way up to the House of Lords before the government finally won. But despite Labour's loss of office in 1951, new towns survived as one of the great standard bearers for the new, planned and more egalitarian Britain. The first few residents of Stevenage New Town moved in early that year.

The survival of another standard-bearer for this new Britain, the council house, was never in question. It had already become an important part of national life. Some two million council homes were built in the first fifty years of the twentieth century and by 1951 nearly one in six Britons were living in them. A few tens of thousands had already been built prior to the First World War because it had become all too clear that the free market could not deliver anything close to adequate, uncrowded housing for millions of the poorest people.

Private sector building of new homes for the working class ground to a standstill in the property slump leading up to the Great War. There were new land taxes, local councils were setting higher building standards, thereby raising construction costs, and they, along with the philanthropic housing societies, were now starting to provide competition for the housebuilders by putting up their own cheap homes. The situation worsened during the First World War. Very few new houses were built (although a few excellent, government-funded estates for munitions workers were constructed) while next to nothing was spent on maintaining the existing stock.

And then something happened which was to make private landlords join developers in the retreat from investing in working-class housing. Legislation was rushed through Parliament in 1915 to freeze rents at their pre-war level. This law followed the outcry caused by war-profiteering landlords who took advantage of the flood of munitions workers pouring into the big cities by pushing up rents. In Glasgow there had been a rent strike that year which had threatened to grow into a general strike across Clydeside.

By the time peace came at the close of 1918 the widespread overcrowding and dilapidation that had existed before 1914 had grown much worse. With building costs soaring and interest rates high in the post-war economic turmoil, it seemed impossible for the private sector to construct the half a million or so new homes which were estimated to be needed immediately. As returning soldiers struggled to find work, as strikes and trade union militancy grew, the government was under huge pressure to subsidise a crash house-building programme. Prime Minister David Lloyd George promised during his 1918 election campaign to build 'habitations fit for heroes who have won the war'. Legislation was passed the year afterwards that compelled local councils to survey the housing needs of local people, then plan and implement building schemes to meet those needs. The government also gave councils a subsidy to encourage construction.

That first national scheme of council housebuilding was intended to last only a few years. Reducing the risks of a Bolshevik uprising was one of the government's motives. The programme ran into trouble from the outset because of the soaring costs of building and the economic problems in the years immediately after the war's end; by 1922 it was being severely cut back after a belated start. Its architect, the Minister of Health Dr Christopher Addison, fell out with Lloyd George and resigned. But despite the rapid changes of government during the inter-war years and the shifts from Labour to Tory to coalition administrations, council housing was here to stay. State-subsidised council house construction was revived and maintained by both Labour and Conservative governments into the 1930s, but sometimes the Tories favoured subsidising the construction of private homes as well.

The decision in favour of councils building masses of homes for rent to low-income families was a fateful one. Over the next fifty years renting from private landlords, which was how about 80 per cent of all households kept a roof over their heads on the eve of the First World War, was to dwindle steadily.[8] People became owner-occupiers or council tenants instead. The municipalities were chosen to house the working classes because they offered both nationwide coverage and local accountability. They were a power in the land, controlled largely by the same political parties that sought to dominate Parliament. And they had a track record; before the Great War they had already built many thousands of homes. Their rivals, the private landlords, were discredited and despised because of their profiteering from overcrowded, under-maintained slums and their attempted wartime rent rises. For the most part they were small business people owning a few dozen properties and they had little political clout. The wartime rent controls remained in place and these discouraged landlords from looking after their properties and investing in new ones. Renting a home from the private sector soon became the kind of tenancy most Britons would least like to have. It usually meant living in a run-down, often shared house in a poorer part of town; the only compensation was that it was usually the cheapest type of tenure.

The great majority of the new, inter-war council homes were built on greenfield sites on the edge of cities, laid out at much lower densities than the by-law housing which had been thrown up around the turn of the century. The new breed of town planners and the advocates of garden cities were determined that this workers' housing should be much superior to those uniform, soot-stained terraces. It was to be more spacious, with sunshine penetrating deep into both the front and rear of the home. The houses should not be in long, monotonous rows but broken up and set back a small distance from the street with front as well as rear gardens. Their looks should have something of the country cottage and the traditional village about them. Instead of terraces laid out on a stiffly geometric, gridiron pattern of streets, the streets should wind their way among the homes, offering a variety of vistas.

Examples of the right sort of stuff were already to be seen in the estates that enlightened industrialists had built for their workers at Port Sunlight on the Wirral and Bournville, outside Birmingham, at the close of the nineteenth century. But it is the architectural and town planning partnership of Raymond Unwin and Barry Parker which is credited with transforming the 'garden suburb' from an ideal into something approaching a standard for all new housing in the first decade of the twentieth. Their first major commission was the Rowntree family's New Earswick estate of low-income homes near their York cocoa factory, followed by the socially mixed and pioneering Hampstead Garden Suburb in North West London and then the first of Ebenezer Howard's garden cities at Letchworth. Building work on all three began before 1910. Unwin and Parker looked to the past for their inspiration, rejecting housing for industrial workers which itself looked industrial and divorced from nature. In fact they rejected the urban entirely; all their work was on greenfield sites and attempted to leave something of the countryside flowing among the new buildings.

Unwin was asked to join a wartime committee chaired by an eminent MP, Sir John Tudor Walters, which was advising the government on the design and layout of the new mass housing so desperately needed once the conflict was over. Its report, delivered in 1918, urged that local councils, with government subsidy, should build to a much higher standard than that which the private sector had achieved to date. Each new home should have a bath, an indoor toilet and three or more bedrooms and terraces should be no longer than eight houses in a row. The building would be mostly on cheap, undeveloped land on city outskirts with public transport links to the centre. Twelve single-family houses to the acre (30 to the hectare) should be the highest density, a maximum that Unwin had advocated in an influential pamphlet published before the war. The government adopted standards close to those of the Tudor Walters Report in the inter-war subsidised housing programmes, although there were later modifications to save costs.

Council estates sprang up around towns and cities all over the land. Some of them were colossal, such as the London County Council's Beacontree satellite town in Essex, 11 miles out from the capital's centre, which had a population of around 100,000 when building finished in 1934. The intention was to bring the suburban living enjoyed by the more affluent to the masses but there were problems from the outset. The rents were usually substantially higher than those for the more crowded, inferior inner city homes rented from private landlords. So the new tenants were drawn mainly from the better-paid, more skilled working class while the poorest were left behind in the city. Even so, some estates suffered high turnovers with tenants frequently quitting because unemployment or unsteady earnings left them unable to afford the rents.

Many of the estates did not live up to the high, Unwinesque ideals of the Tudor Walters Report. They often looked like monotonous, low-density tracts of cheap housing isolated from the rest of the city with boring street layouts and little in the way of shops or communal life. The report's warning against building large areas of uniform housing with tenants all from the same social class were ignored. Pioneering sociologists and opinion pollsters documented the anomie and anxieties found in some of these new communities. And while the new council tenants may mostly have been from the better off working class, they still faced hostility and snobbery from some people living in private housing nearby. In north Oxford, the builders of a private estate built tall walls topped with iron spikes across two roads that linked their development to a council estate next door. The fortifications followed complaints from the owner-occupiers about graffiti and packs of children and dogs from the council housing roaming their streets; property values were claimed to be at risk. The Cutteslowe Walls divided the two areas for 25 years.

More than a million council homes were built between the wars. But hundreds of thousands of the cheap Victorian dwellings that predated the building by laws were still in use in courts and back-to-back terraces, and it was common for working-class families to share lavatories and taps and have only two or three rooms of their own. In the early 1930s government subsidies were switched from building new council homes for general needs to slum clearance and rehousing slum residents. By the time the Second World War began a quarter of a million dwellings had been torn down and replaced by a slightly larger number of houses and flats. (It was during the 1930s that the government first offered councils extra subsidies for building flats, and they began to appear in most English cities.) The official estimate in 1939 was that slum clearance had crossed the halfway mark, with about another 230,000 homes requiring demolition and replacement. The Luftwaffe was soon to deal with many thousands of those.

Like the First World War, which had ended just 21 years earlier, the Second put the state under enormous pressure to build masses of new homes once it was over. As well as the losses caused by bombing (half a million homes were either destroyed or so severely damaged as to be not worth repairing), more than five years passed in which very few new houses were built while those standing had next to no maintenance and repair work carried out. There were also more people needing housing. The demographic doldrums of the 1930s, when birth rates had slumped to a record low, were over; marriages and procreation climbed fast during the war. The 1945 Labour government decided that local councils should meet the great bulk of this demand. Aneurin Bevan, the Minister of Health responsible for housing, said:

> If we are to plan, we have to plan with plannable instruments, and the speculative builder, by his very nature, is not a plannable instrument . . . We rest the full weight of the housing programme upon the local authorities, because their programmes can be planned, and because in fact we can check them if we desire to.[9]

So councils were given more subsidy per council house than ever before and 900,000 were built between 1945 and 1951. Meanwhile the government restrained private housebuilding so as to maximise resources for council homes. The output was less than the government had hoped and planned, for once again post-war economic problems slowed down the programme. Even so, it was the fastest rate of council house construction there had ever been. Had Bevan's expansionist policy been able to continue, smoothly, for another few decades, then council housing would have become as ubiquitous as the National Health Service and state schools – something which the majority of citizens relied on.

It was not to be. The private sector had dominated the growth of towns and cities through the first half of the twentieth century and it was to again in the second. The greatest housing legacy of the years 1900 to 1951 is not the council estate but the vast area of low-density, semi-detached suburban housing built speculatively between the wars for the free market.* The suburban semis of the 1920s and 1930s, mostly built for sale rather than for rent, put Britain far down the road to becoming today's nation of home owners; by 1939 about a third of Britain's homes were owner occupied.

Where did the money to build and buy all those houses come from during those inter-war decades of slump and mass unemployment? In fact there was overall economic growth during those 21 years and bricks and mortar might often have seemed a preferable investment to volatile stocks and shares. Furthermore, the recession did its real damage in areas dominated by traditional heavy industry; London, the South East and the Midlands fared very much better, attracting investment and gaining jobs in the new industries of motor vehicle manufacturing,

---

*According to the government's *English House Condition Survey 1996*, London, The Stationery Office, 1998, the inter-war semi remains the country's most abundant type of housing, even though the last were built more than half a century earlier.

electronics and electrical goods and other light engineering. J.B. Priestley's *English Journey*, a harrowing and beautiful portrait of inter-war England, begins with him marvelling at the new factories along London's Great West Road, 'all glass and concrete and chromium plate'. They were, he wrote:

> tangible evidence, most cunningly arranged to take the eye, to prove that the new industries have moved south. They also prove that there are new enterprises to move south. You notice them decorating all the western borders of London. At night they look as exciting as Blackpool. But while these new industries look so much prettier than the old, which I remember only too well, they also look far less substantial. Potato crisps, scent, tooth pastes, bathing costumes, fire extinguishers; those are the concerns behind these pleasing facades . . . But if we could all get a living out of them, what a pleasanter country this would be.[10]

Outside the 'distressed areas' there were plenty of people from the expanding middle classes and among the most highly paid manual workers who could afford the low-interest mortgage repayments needed to buy a brand new home of their own.

The housebuilding industry, which was gradually coagulating from myriad tiny enterprises into larger, slicker firms, went for the hard sell with show homes, promotions and masses of newspaper advertising stressing the new affordability of home ownership, its prestige, and the sheer healthiness of suburban life. Some builders even ran bus services from their new estates to railway stations, although the buses had a tendency to disappear once all the houses had been sold. Names such as Wates and Wimpy came to prominence during the 1930s' housing boom. Builders collaborated with the building societies to come up with schemes which required purchasers to put down the lowest of deposits.[11]

About three million homes were built by the speculative builders between the wars, one in six of them with the help of government subsidy. Between 1934 and 1938 they were completing more than 250,000 each year, a rate which has never been surpassed by either the public or private sector since. This new suburbia gobbled up countryside much more rapidly than its Victorian predecessor and was much more hated. The Council for the Protection of Rural England, which today remains one of the most effective campaigners against housebuilding on greenfield sites, was founded in 1926 during the early years of the boom. The assault on sprawl came from all quarters – modernist architects sneering at the half-cocked borrowings from the past, nature lovers, writers of the left and right. 'In fifty years time there will, in southern England, be neither town nor country but only a

Semi-detached houses built between the wars in West London, overleaf

single dispersed suburb, sprawling unendingly from Watford to the coast,' forecast the philosopher and social commentator C.E.M. Joad in a collection of save-the-countryside essays called *Britain and the Beast*.[12] The dwellings themselves were 'mean and perky little houses that surely none but mean and perky little souls should inhabit with satisfaction'.[13]

You could, however, look back on those inter-war years as a golden age of housing. For a decade or so everything came together – a supply of cheap, fairly unrestricted land, along with the shorter working hours and the transport revolution that enabled people to live miles from their workplaces. With these essentials in place, the free market performed magnificently, giving millions of people a splendid increase in their house and garden space and providing new homes whose value for money and production rate have never been bettered.[14] Today, despite being more than 60 years old the speculative builder's inter-war semi is still popular and cherished through much of its range, although the suburbia it defines is starting to look worn and faded in parts. It has become the national archetype and I was sitting in my own when I wrote this. But it turns its back on the city.

Chapter 4

OVERSPILL AND HIGH RISE
1951 TO 1976

Enough has been said to demonstrate that old 'town' and 'city' will be, in truth, terms as obsolete as 'mail coach' . . . We may for our present purposes call these coming town provinces 'urban regions'. Practically, by a process of confluence, the whole of Great Britain south of the Highlands seems destined to become such an urban region, laced all together not only by railway and telegraph, but by novel roads such as we forecast . . . and by a dense network of telephones, parcels delivery tubes, and the like nervous and arterial connections. It will certainly be a curious and varied region . . . perhaps rather more abundantly wooded, breaking continually into park and garden, and with everywhere a scattering of houses . . . the old antithesis will cease, the boundary lines will altogether disappear; it will become, indeed, merely a question of more of less populous.

H.G. Wells, *Anticipations of the Reaction of Mechanical and Scientific Progress upon Human Life and Thought*, 1902[1]

WE GET BACK ON BOARD our history train and stop a quarter century later, in 1976. Motorways and ring roads festoon the cities. High-rise council flats have replaced most of the industrial chimneys that used to poke out of them. There have been two decades of massive slum clearance. The air is cleaner and clearer, thanks largely to the disappearance of those chimneys and the decline of coal burning in household hearths. An improved environment, millions of new urban homes – things ought to be looking up. But 1976 is the year in which the government recognises that the inner city is in deep trouble.

By then, all of Britain's major cities were losing population.[2] People were moving clear out of the large cities and their suburbs to smaller towns in their hinterland. The shire counties of England, especially those around London, were now the most economically buoyant areas of Britain with the fastest rate of population growth. This beyond-suburbanisation trend was witnessed across much of the Western world and in 1976 geographers came up with a long name for it – counter-urbanisation.[3] In September that year Peter Shore, the Labour government's Secretary of State for the Environment, warned that inner cities had become victims of a creeping abandonment. In a speech in Manchester which marked a turning point in government attitudes he pointed to the falling populations of these areas, the disproportionate loss of skilled workers as industry closed or moved out, the fact that the local people left behind were not getting any benefit from the growth in white-collar jobs taking place in city centre offices. The result was chronic high unemployment, physical dilapidation and economic depression.

The awareness of this crisis had been dawning for a decade. By the late 1960s hundreds of thousands of New Commonwealth immigrants in low-paid jobs were concentrated in a few inner areas. Often denied council housing, they lived in run-down, ageing private housing which they rented or purchased cheap. Now there were ethnic tensions and racism to add to the problems of urban deprivation, while across the Atlantic there had been colossal and bloody race riots in several American cities.

The British government's Home Office had run modest programmes since 1968 in an attempt to alleviate some of the disadvantages inner city communities suffered. The Shore speech marked a major increase in effort. His words heralded a White Paper (an official government document setting out plans for new policies) published in 1977, *Policy for the Inner Cities*, new legislation – the Inner Urban Areas Act of the following year – and extra funding at a time of general cutbacks in public spending.[4] The extra Department of the Environment money made available was derided by lobby groups (and the shadow environment secretary, one Michael Heseltine) as being not nearly enough given the scale of the problems – a criticism repeated at each new urban regeneration programme since then. But most of the increased spending and investment the inner cities desperately needed was meant to come from other mainstream government spending programmes being bent towards the inner cities, and from the private sector.

This package of money, legislation, new powers for local councils and partner- ships between local and central government, health authorities, police and local voluntary groups was Britain's first serious attempt to turn the economy of deprived inner areas around, as opposed to merely improving or replacing housing. It failed, like most of the attempts that followed.[5]

This speech also marked the beginning of the end for policies to decant people from cities. Funding for new towns was cut back and in the Spring of 1976 the building of Stonehouse New Town, near Glasgow, was abandoned despite the first houses having been completed and the first families having moved in. (Even so, most of the new towns that the government was already committed to kept grow-ing and the most successful of them, led by Milton Keynes, are still expanding today.) It was also in the 1970s that Birmingham and the Greater London Council ended their support for state schemes that encouraged their citizens to move to new council housing in officially designated 'overspill' towns dozens of miles away.

By then, dispersion did not need to be helped along by housing policies; it was happening of its own accord, and with a vengeance. The main driving forces had been the soaring quantity of cars, vans and lorries and the government's readiness to build the new roads and motorways needed to carry them. Transport change was, as ever, the most important factor in urban change; more and more people were able to live further and further from their jobs. People preferred homes in the sub-urbs, smaller towns or the countryside. Employment was headed the same way. During the inter-war years, as road transport took over from rail, new industry had sited itself on suburban trading estates beside the main roads leading out of town. After 1945 the urban exodus accelerated. The most convenient and cheap location for a factory or warehouse would be near an interchange on one of the new motor-ways. The new, larger docks that could handle container cargo were being constructed well away from the centres of the great port cities. Airports, gigantic new power stations and other fast-growing industries created jobs miles outside London and the Victorian industrial cities. Like some vast spin-drier the post-war economy sucked employment and new housing out of town.

The state had speeded up the rate of revolutions since 1951 with new towns and other overspill measures. A further fifteen new towns had been created to add to the fourteen approved before the end of 1951, while the Conservative govern-ment's Town Development Act of 1952 allowed big cities to make arrangements with much smaller towns to build new housing for their out-migrating citizens. Families from London went as far as Swindon in deepest Wiltshire, nearly 80 miles away. Post-war governments also had the power to tell companies where to locate any large new factories and plants they wanted to build. Ministers and civil servants tried to use this authority to divert investment to regions hit by industrial decline and to the new towns. Through much of the 1960s and 1970s there was also a government scheme to repel new office developments from London, encouraging firms to choose more rural locations instead.

While jobs and homes were being pushed outwards from the big cities, the government, pressure groups and much of the public wanted to stop those cities spreading into the surrounding countryside. The main weapons against sprawl were the new nationwide planning system and the designation of Green Belts, first surrounding London and Glasgow and then most of the nation's largest cities.[6] They now cover one-eighth of the land area of England, and at first glance they appear to have worked. The belts have stayed green because very few applications to build on them have been granted planning permission.

But widen your view and you find that development has merely leapfrogged beyond the Green Belt, aided by road building and rising car ownership.[7] Factories, warehouses, offices and homes that would have gone up on the fringe of the city were instead built several miles further outside, just beyond the outer edge of the Green Belt. This more far-flung development – exurban rather than suburban – was still within fairly easy reach of the metropolis. The people who lived in the new homes commuted across the Green Belt; the firms that set up shop in the shires outside the city still had strong trading links with it. The zone just beyond the Green Belt was a place where you could benefit from being close to the city without having to be inside it.

This leapfrog development begat further counter-urbanisation. New housing and businesses in the shires lured in other residents, developers, industry and commerce from within the city or from further afield. When houses are built in bulk, jobs and new businesses follow because the new residents require their own local services and form a potential workforce for employers moving into their area. The growths in the local population and the local economy fed each other, sucking in people and enterprises.

The same kind of self-sustaining process created the compact, overcrowded Victorian industrial cities. But after the Second World War a combination of Green Belts, vastly improved roads and widespread car ownership meant this cycle of growth would now revolve across a much wider area than a city just a few miles wide. Once people had cars they were no longer tied to public transport networks that poked out of cities like spokes; they could commute 20 miles or more in any direction. Putting tracts of countryside around the cities off limits for development merely spread the growth further out. You can see this most clearly in a belt between 20 and 50 miles from London's centre, where an almost complete ring of prosperous towns – from Chelmsford to Guildford, from Maidstone to Maidenhead – have experienced rapid, sustained growth after the Green Belt was drawn around the capital. This is really London's growth, although the people living and working in these places would not recognise this. In *The 100 Mile City* the architectural writer Deyan Sudjic has vividly pictured the modern city as a gigantic force field, discharging shopping centres, office developments and housing estates into its hinterland as it earths great bolts of its own electricity.[8]

So if big cities could keep on spreading in this more dispersed way while parts of their interiors went into decline, was the new town and country planning system any use? It depends on your view of what planning is for. It certainly tidied up development and, as the leading planning academic Peter Hall has pointed out, it made the view from aeroplane windows much neater. Transatlantic air travellers reaching the United States

> would be bemused by the scale of the development, by the apparently endless sprawl of the suburbs in the east-coast megalopolis, by the vast network of freeways that linked them; travelling east [across Britain and Western Europe] they would be equally surprised by the relative puniness of the development, by its toytown like quality, by the planned precision of the almost geometrical break between town and country.[9]

New development now took the form of neat, compact shapes appearing on the edge of existing settlements, rather than springing up in the open countryside or forming ribbons along roads. Green Belts appeared to contain the big cities while planning control ensured that the smaller towns outside would sprawl in a more refined and orderly manner. Here was a system that succeeded in saving the appearance of the countryside in a densely populated country. But while the greenery remained, the real countryside, as a place with its own social and economic life separate from towns and cities, was dying. By 1976 most of rural, lowland Britain was, in essence, an extremely low-density suburb. The great majority of its inhabitants either worked in towns or cities or, being retired, did not work at all. The fusion of town and country that H.G. Wells had prophesied at the dawn of the century had happened, but Green Belts and planning control disguised it.

They also made both land and housing more expensive by restricting the supply of greenfield building sites.* But the state and the British people got no benefit from this rise in site values because in 1953 the Conservative government had scrapped the betterment levy, the development tax that Attlee's post-war administration had introduced. Rising land prices meant that new private homes and gardens were laid out a little less spaciously than they had been in the housebuilders' golden inter-war years. Even so, land was not expensive enough to prevent the developers putting up a higher and higher proportion of space-wasting, more expensive detached houses

---

*This has been a constant refrain of the housebuilders. One firm commissioned a consultants' study that claimed that the London Green Belt was adding £3,600 to the price of an average house in south-eastern England in 1989.

and fewer semis and terraces. If an Englishman's home was his castle, then it really ought to stand alone.

By 1976, it had become clear that the word 'planning' was too bold, too purposeful, to describe the system of development control put in place in 1947. It was certainly more democratic than allowing developers to site new buildings wherever they could afford to buy land, because elected councillors on local councils or government ministers were now responsible for planning decisions. Yet, in a capitalist democracy, planning also has to serve capital. It has the job of allocating sites to businesses in ways that try to reconcile their interests with those of the wider community; what is not in question is that a steady stream of new sites must be found.

Most local councils are extremely keen and sometimes desperate to gain new industries, shops, offices and jobs in their patch and they will compete with their neighbours to get them. Nowhereton District Council battles with nearby Anyham a dozen miles down the road to get the big retail development that will turn it into *the* sub-regional shopping centre. They cannot both wear that crown. The two councils' control over what is built within their own areas, their willingness to attract and oblige developers, are weapons in their competitive struggle. So when capital was flowing out of the inner areas of the big cities, along with people and jobs, the planning system did not obstruct them. There were murmurings of discontent in the shires, worries about over-development and the loss of green fields, but by and large most councils welcomed new investment and employment heading their way, and gave it the necessary planning approvals. Urban decay was someone else's problem, not theirs.

When new kinds of development that relied entirely on private car use started arriving in Britain from North America in the 1970s, some council planners were quick to greet them as the wave of the future and grant planning permission. By 1976 retail sheds surrounded by car parks and 'science parks' (low-density office and industrial estates built for high technology companies) were just beginning to appear on the edge of towns and cities. It was usually impossible to site this kind of development near the city centre because it needed lots of cheap land for car parking spaces. A new phase of counter-urbanisation had begun.

The big cities' chances of beating off the challenge from the rising, exurban Britain should have been boosted by the huge sums invested in rebuilding their inner areas in the post-war years, in clearing slums and constructing millions of new council flats and urban roads. This was the boldest, most expensive pro-urban policy that Britain had ever had. But by 1976 it was clear that much of this reconstruction was a disaster. The demolition of council tower blocks began before the 1970s had even ended. Public buildings had to be blown up after less than 20 years of use with their construction debts nowhere near paid off. Today these grand topplings are still taking place, still pulling in the crowds. How could the government and local councils have made such costly mistakes?

The answer lies in the huge post-war housing shortage and the belief that the slum problem was best handled in bulk, by mass demolition and rebuilding. The 1951 Census registered widespread overcrowding (more than two million homes were still shared by at least two families) and huge numbers of households lacking basic amenities; over a third still had no bath while 6 per cent did not even have a water tap indoors. The Tories were elected that year with a manifesto commitment to build a combined total of more than 300,000 council and private houses a year. With Harold Macmillan as Minister of Housing that figure was soon reached, then surpassed. The accomplishment rested partly on making the new council houses smaller and cheaper than Aneurin Bevan had wanted when he ran the national housing programme. Even so, the great bulk of those early 1950s' council homes were houses rather than flats.

Through the 1950s and the 1960s housing in Britain became a political numbers game, with the two main parties seeking to outbid each other over the number of homes – both council and private – that would be built if they were elected. They were under pressure to make such promises because of the post-war increases in marriages and births which produced a brisk population growth. There was also a trend for households to become smaller and an expectation that in a civilised, modern country every couple, with or without children, was entitled to have a home of its own.* At the same time, both Labour and Tory governments wanted to clear great tracts of dilapidated Victorian housing within the inner cities.

Dispersion was one way of meeting the demand for new housing, building hundreds of thousands of council and new town corporation homes outside the major cities or on their fringes. But from the mid-1950s onwards, more and more resources were devoted to building big blocks of flats, often on slum clearance sites, within the cities.

This was partly a matter of fashion. Council architects and councillors favoured flats because of the sheer size and height of the blocks, the way their hard, simple shapes appeared so very modern, so different to what had gone before. These 'streets in the sky' became the new symbols of civic dynamism. There were, however, more pragmatic reasons for the great blunder. Building upwards, with flats replacing houses, was the obvious way of reducing overcrowding in individual dwellings without having to depopulate the big cities too drastically. It allowed fairly high population densities *and* open space. Abercrombie's wartime *County of London Plan* had envisaged 40 per cent of the capital's population eventually living in flats.

Perhaps the most important reason for the switch from houses to flats was because the state was in a hurry. The industrialised, prefabricated methods used in

*This actually became government policy in 1945 under Winston Churchill's caretaker government of the immediate post-war period, just before Labour's landslide election victory.

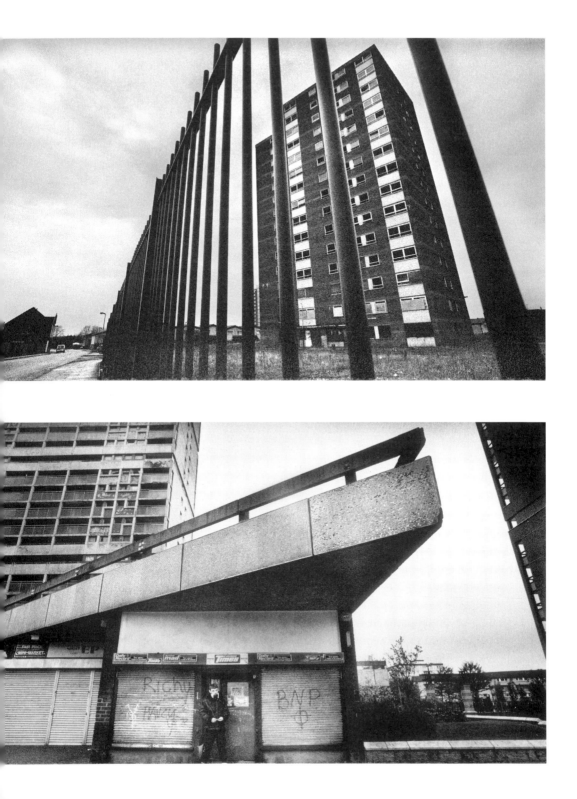

constructing estates of slab blocks and high rise towers allowed each home to be built more quickly than was the case with traditional, low-rise housing. Output could also be raised if the maximum number of homes was squeezed onto each building site that became available. Identifying and purchasing building land were two of the most time-consuming parts of the process, all the more so now that the planning system and Green Belts were restricting the supply of greenfield sites on the urban fringe. Each flat cost more to build than a conventional house and high rises were more expensive still but speed was of the essence. From 1956 to 1967 government gave councils an extra subsidy for blocks more than five storeys tall.

Brute politics were also a factor in the tragedy of the towers. High residential densities were a means of holding on to power. The Labour councillors who ran most of the big cities were alarmed at the prospect of their working-class voters being dispersed beyond their borough boundaries by slum clearance and rebuilding. Supplying plenty of new council homes within Labour-voting, inner city wards seemed a sound strategy for retaining their electorates and their dominance.

The building of slabs and towers became a mighty industry with a great deal of hard selling to councils and some corruption. The same housebuilding firms that had built the sprawling, owner-occupied suburbs between the wars now reaped fresh profits from building for the state. Keith Joseph, the leading Tory thinker who was Housing Minister in the early 1960s, was to gloomily confess: 'I was just a "more man". I used to go to bed at night counting the number of houses I'd destroyed and the number of planning approvals [for new ones] that had been given . . . Just *more*.'[10]

Glasgow ordered over 300 high rises. They sprang up from near the heart of the city to way out on its fringes among the tens of thousands of drab new three-storey tenements which composed the bulk of the city's gigantic peripheral estates. High rises were utterly unsuited for families with children and unpopular with many tenants almost from the day they were built. They had more things that could go wrong in them than ordinary housing (lifts and rubbish chutes, for instance) and because they poked hundreds of feet into the sky, their exteriors were more costly and difficult to maintain. Design defects and the failings that arose from rushed, shoddy construction soon became apparent. They were hard to keep warm, the rain penetrated and condensation trickled down their cold walls.

Then, in 1968, an entire corner of the twenty-three-storey Ronan Point block in Newham, east London collapsed after an elderly lady lit her gas early one morning. A leak caused an explosion that blew out the load-bearing walls, causing bedrooms and living rooms above and below to fall, one after the other. Five people were killed and seventeen injured. That disaster, and the findings of the inquiry which followed it, brought high-rise housing construction to an end. In 1979 the ten-storey Oak and Eldon Gardens tower blocks in Birkenhead became

Tower blocks in Manchester (above) and Glasgow, facing

the first in Britain to be deliberately blown up. They had been built just 20 years earlier but within two years of completion they were plagued by maintenance problems and vandalism. They became so unpopular as to be unlettable; all over the country hundreds of others suffered the same fate.

Although bad design and bad construction dogged the high rises, they were not bound to become a disaster. Several councils have converted tower blocks into secure housing for elderly people, to the satisfaction of both landlord and tenants. A few, mainly in London, have gentrified into owner-occupied flats, usually after extensive refurbishment, and fetch good prices. Meanwhile the private sector is now building a new generation of luxurious and expensive apartment towers in city centres to add to the 4,000 or so council-built blocks still standing.[11]

High rises became infamous largely because they were the first buildings to be identified with the social problems which were gradually besetting all types of post-war council housing – slab blocks of flats, maisonettes and conventional houses. The fact is that most council flats were not to be found in high-rise towers, and most council homes built in the post-war years took the form of houses rather than flats. Only for a few years during the late 1960s did construction of flats overtake that of houses.[12]

As council homes took the place of demolished slums, municipal housing was moving down market. The urban exodus was concentrating those who lacked options in the big cities; the unskilled and poorly educated, the elderly, the long-term unemployed and people of colour. Poverty brought problems of crime, vandalism and family breakdown in its wake and the new designs only made matters worse. In towers and blocks of flats children could no longer play in the street just outside their front doors, under the loose supervision of their parents and neighbours indoors. Mothers either had to accompany their children to the lawns and playgrounds around the estates or leave their children to their own devices far out of sight. Slum clearance broke up communities and left neighbourhoods blighted and increasingly abandoned while they waited for the demolition crews. Norman Dennis, a sociologist who followed the process closely in Sunderland, wrote 'in some areas, the descent to slumdom can be traced directly to a single cause; the slum clearance proposals themselves'.[13] Demolition utterly eradicated neighbourhoods, leaving almost nothing of the familiar streets, pubs and other local landmarks behind. Between 1955 and 1975 some one and a half million houses were condemned and demolished in Britain, three times the number destroyed in the Blitz.

The way in which this newfangled municipal housing stood out like a sore thumb among the older fabric of towns and cities added to the problems of its tenants. It appeared alien and uninviting to people who did not have to live in it. The new estates with their thousands of flats, their pedestrianised interiors and walkways and their cliff-sides of windows and balconies, were isolated from the streets and the street life of the city. People living on them began to be branded by their addresses and their tenure. Bevan, who had hoped his hundreds of thousands of council homes would be part of mixed communities, who had dreamed of an end to 'East Ends and West Ends, with all the petty snobberies this involves', must have been spinning in his grave.[14] By the early 1970s another leading Labour politician and thinker, Anthony Crosland, had seen that council housing had taken a wrong turn, becoming associated with 'the whiff of welfare, of subsidisation, of huge uniform estates and generally of second-class citizenship'.[15]

Some council estates were much more desirable than others. Tenants might all have paid pretty much the same rent wherever they lived in a borough, but that did not stop hierarchies from forming with estates being graded – in the opinion of both housing officers and residents – according to how well behaved and prosperous their occupants were thought to be. Informal, unspoken allocation policies for tenants and the odd bit of downright corruption helped make this happen; you could occasionally bribe your way into the best council housing. But once an estate had a bad reputation it would often get worse. People who could leave would and those on the council's waiting list would refuse offers of homes there. The estate would then have a higher number of empty properties than usual and a higher rate of tenant turnover making it less settled, less friendly, a less desirable place to live. In the worst cases it would fall into an abyss in which only the most desperate people would take tenancies, the poorest, largest families with uncontrollable children or, worse still, criminally inclined people who felt comfortable living on a shunned estate.

As the 1970s closed, the proportion of Britons living in municipal housing reached its highest ever level – from only a tenth at the end of the Second World War to almost a third. But the proportion of owner-occupiers had grown even more rapidly, from 33 per cent in 1945 to 52 per cent in 1979, and from the mid-1950s onwards most of the new homes being built were for sale, not for rent. By 1976 there were generous tax incentives for owner-occupiers with mortgages and both Conservatives and Labour were in favour of as many people as possible owning their own homes. Renting from private landlords shrank from almost half of all households in 1951 to a tenth at the beginning of the 1980s. This drastic decline had been caused by slum clearance, which turned masses of private-sector tenants into council tenants, and government controls on rents. These controls kept changing as the two main parties alternated in power, the Tories trying to revive the fading private landlords while Labour continued to view the old class enemy with utmost hostility. The net effect was to discourage individuals and companies from investing in the maintenance or construction of rented housing, and to incline them to sell up. Their tenants were increasingly drawn from those who could not get onto a council housing waiting list or who were very low on it – the single, the young and the transient.

The mass conversion of grammar schools into comprehensives during the 1960s and 1970s, under both Tory and Labour governments, was also sending ripples of anxiety through the social fabric of the cities. The grammars, many of them in inner city locations, were largely patronised by the children of middle-class parents whose background gave them a better chance of passing the entrance exam, the infamous eleven plus, than working-class youngsters had. These schools enabled the prosperous to live well inside the city and still obtain a high quality, free education for their children. Once such schools became comprehensive and dominated by an unselected, working-class intake, it was likely that examination results and academic standards would decline – that, at any rate, was the perception. The comprehensive revolution was one more force pushing people with choices and money out of the cities and into the suburbs or beyond. There the intake for the local state schools might be more middle class, with parents having higher expectations.

London was the exception to this great national emigration of aspiration. Its population had been shrinking as citizens moved out – from the inner city to the outer boroughs and clean out of the metropolis – but it could still attract a steady inflow of prosperous, talented people, or people who were destined to become prosperous. From the 1950s onwards waves of gentrification swept around the capital's dilapidated inner city. Young, affluent singles and couples bought and renovated hundreds and thousands of run-down eighteenth- and nineteenth-century houses, turning them into desirable urban homes – often with the help of government grants. This has proved to be an enduring movement, bringing in huge quantities of fresh investment, transforming the look and feel of large areas of the city. It continues to this day. There has been post-war gentrification in some other British cities, such as Edinburgh and Bristol, but nowhere else has it had anything like the same impact as in the capital. Why?

The metropolis is so large that if you want to live close to a city-centre workplace and minimise the time spent commuting you have no choice but to live well within it.* And the capital has always retained tracts of high quality, expensive residential areas near its centre, giving the gentrification process edges from which it could spread into surrounding areas of decay and dilapidation. Estate agents, house buyers and sellers could claim that a bad neighbourhood would assume the identity and branding of the posh one next door. Furthermore, London has always had a huge supply of medium to large, good-looking family houses that had the potential to become beautiful homes after a century or more of neglect and seediness. And, finally, the city exerts a mighty pull as the national capital and a global financial centre. It has a higher than average proportion of well-paid, white-collar jobs, bringing the wealth and the kind of people which underpin gentrification into the city in bulk. London has long had hordes of young, affluent, university-educated

---

*Journeys to work in the capital take much longer than elsewhere in Britain. In the six largest urban conurbations outside of London about 80 per cent of journeys to work took 40 minutes or less. In London, only 50 per cent of all journeys took 40 minutes or less. See *Transport Statistics for Metropolitan Areas 1998*, London, Department of the Environment, Transport and the Regions, 1998, p. 15.

and professional people because it is *the* place for them to begin or to consolidate their careers – in finance, medicine, advertising, journalism, design and many other fields. They do not want to spend hours commuting to distant suburbs and they like being close to the cultural offerings of a great city. Many do eventually move out having risen. London is known as an escalator city for that reason; it has by far the highest rates of emigration and immigration of any British city and the largest net outflow to the rest of the country.[16]

It is time to move on to the turn of the century. We stopped in 1976 because it was a handy quarter century on from the previous halt. It was also the year in which the government at last linked inner city decline to the exodus of talent and money which it had spent decades encouraging. But there is one more reason for picking this date: it was a year of events that made it almost certain that Labour would lose the next general election. Labour in power had once again run into an exchange rate crisis, as it had in the 1920s, 1940s and 1960s. The government had to make sharp cuts in the ambitious public spending plans it had promised the electorate and negotiate an emergency rescue loan with the International Monetary Fund. Once again its hopes of two full terms in office were doomed, and the Conservative's radical new leader, Margaret Thatcher, was set fair to be Britain's next prime minister.

FOR THE CLOSING QUARTER of the twentieth century, our tour guide could choose between two different narratives. One is a tale of the two and a half Thatcher governments letting counter-urbanisation and urban decline rip from 1979 to her fall in 1990. Then a decade of reaction follows as governments recognise how much social and environmental damage this has done. It ends with the election of a Labour government in 1997, which then acts decisively to begin an urban renaissance. This is neat, because the previous chapter ended with the looming defeat of a Labour government that had just begun to realise how desperate the plight of the cities was.

The other narrative is less colourful, less hopeful. It's more of the same. Mobility and electronic communication keep growing, expanding, making it easier for people and businesses to site themselves outside of large towns and cities. Which is what those who can generally do in their pursuit of more space and a better environment. Governments, irrespective of party, come under pressure to slow the process down; they try so to do but it is futile. A steadily increasing share of the population live outside the big conurbations, in something like H.G. Wells' urban regions or town provinces. The conurbations are fated to become increasingly obsolescent, problematic and divided between prosperous and poor. Which of the versions is more truthful? We won't know for a couple of decades. All we can be sure about for now is that neither will be entirely wrong.

Margaret Thatcher wanted to roll back the frontiers of the state and encourage people to take responsibility for themselves. If the natural tendency of things was for families and firms to relocate themselves out of town, then so be it. The winds of counter-urbanisation continued to howl through the 1980s and into the 1990s. Housebuilders were granted planning permission for enormous new suburbs on green fields. Thousands of superstores and retail sheds popped up like mushrooms on urban fringes all over the country, sucking the life out of many market town high streets and suburban shopping centres, sowing them with charity and discount stores and boarded up shop fronts.[3] They were accompanied by gigantic shopping developments such as Merry Hill near Dudley and the Metro Centre at Gateshead. Each of these colossi offered as much shopping space as a medium-sized town's entire centre. These were usually built on old industrial sites rather than greenfield sites, but they were well outside the city centres and the vast majority of their visitors came by car.

The government tinkered and dickered with the planning system in the 1980s, shifting the balance of power away from local government and towards the developers with new legislation and official guidance to councils. The system of Industrial

Margaret Thatcher meets building workers, 1988

Development Certificates, which enabled government to control the location of new industry, was abolished in 1986. Development, any development, represented businesses investing and creating employment. After the deep recession at the beginning of the decade, when millions of jobs were lost, an administration whose instincts were to reduce state intervention became more strongly opposed to council planners obstructing entrepreneurs by refusing planning permission for their schemes. Some councils did try to resist the rush out of town and insist on higher design standards. Often they were over-ruled by central government.

The housebuilders, an industry now dominated by some twenty big regional and nationwide firms, were among the leading beneficiaries. They had long favoured building on the green fields on the edge of towns and cities rather than redeveloping previously used, abandoned and generally more risky acres within them. Building sites were usually cheaper outside, and it was easier to acquire land in large lots there rather than struggling to assemble good-sized sites from more fragmented urban land holdings. The access to roads that a stream of lorries and construction vehicles requires was also easier to secure at virgin sites on the urban fringe. Besides, their customers usually wanted out of the city.

In 1980 central government told councils that from then on they had to ear-mark sufficient vacant land to cover the next five years' worth of demand for new housing. This meant their planners had to keep identifying new sites for potential development to take the place of those which had had houses built on them. Planners found it easier to keep their land banks stocked up by designating large chunks of farmland on the urban fringe for several hundred homes, rather than scouring the built-up area for a dozen smaller sites which could take as much housing. Greenfield land is a known quantity; derelict land, emerging sporadically and unexpectedly as factories and warehouses close, is not. And, in another exodus-encouraging move, ministers over-ruled an attempt by big city councils in the early 1980s to make housebuilders concentrate new developments within their built-up areas rather than building outside. The councils wanted to reverse urban decline; the government said homes should be built where people wanted to buy them. 'If people were to be able to afford houses there must be sufficient amounts of building land available,' wrote Mrs Thatcher.[4] 'Tighter planning meant less devel-opment land and fewer opportunities for home ownership.'

Returns from the 1991 Census, and statistics showing where people were reg-istering with local GPs, demonstrated that every one of the seven largest urban areas in England was experiencing a net outflow of people. This exodus was running at about 90,000 people a year.[5] A team from Newcastle University showed there was a 'counter-urbanisation cascade', with people moving out from larger, denser urban areas to smaller, more dispersed communities – from inner city to outer suburb, from large towns to smaller ones, from towns of all sizes to villages and the coun-tryside. Rates of net out-migration generally were highest among better-off households with professional and managerial occupations.

This cascade was flowing much more strongly than the long-running flow of migrants from the slow-growing North to the more dynamic, wealthier South. And the suburbs of the largest conurbations, especially the outer London boroughs, were contributing more to this outflow than the inner cities. People were still coming into the conurbations from other parts of Britain. But they were consider-ably outnumbered by those headed in the opposite direction.

Employment was also flowing outwards. Researchers from Glasgow University showed that the trends of the 1960s and 1970s had carried on through the 1980s and into the 1990s.[6] This was a period in which the total number of jobs in Britain grew, partly as a result of more women entering the workforce, with especially strong growth in such areas as retail and distribution, tourism, catering, financial and business services, education and health. But employment in manufacturing shrank by 26 per cent between 1971 and 1981 and in the following decade a further quarter of the remaining jobs was lost. These massive sectoral shifts were accompanied by equally large changes in the types and status of jobs; unskilled and semi-skilled manual jobs fell away while the number of professional and managerial jobs rose rapidly.

The decline in manufacturing hit the largest towns and cities much harder than the rest of Britain. They lost more than their *pro rata* share of jobs in this sector; a sign that manufacturers were either doing better outside the big cities than within them, or preferring to move out to or start up in the more rural areas. The largest urban areas also gained proportionately fewer of the new jobs in growth sectors than the rest of the country. The eight great UK conurbations suffered most of all, and their inner areas were harder hit than their more suburban surroundings. Overall, the twenty largest urban areas – all with a population above a quarter of a million – lost some 500,000 jobs between 1981 and 1996 while the rest of Britain gained 1,700,000.

Urban employment

	1981 (thousands)	1996 (thousands)	% change
Seven largest conurbations (excluding London)*	4,497	4,208	−6.4
Greater London	3,560	3,348	−6.0
Free-standing cities**	1,730	1,749	1.1
Rest of Britain – towns, rural areas	11,278	12,953	14.9
Total	21,064	22,258	5.7

Notes:
*West Midlands, Greater Manchester, West Yorkshire, Glasgow, South Yorkshire, Merseyside, Tyneside
**The next twelve largest British towns and cities after the eight conurbations, populations over 250,000, e.g., Cardiff, Hull, Edinburgh.
Source: Turok and Edge (1999)

A growing number of men of working age quit the labour market in the big cities, registering unfit for work and depending on benefits for the remainder of their lives. When jobs were created in the urban cores, they were increasingly likely to be taken by outsiders commuting in.

The exodus of people and jobs required a steady supply of new homes and workplaces outside the largest urban areas. Rising levels of car ownership, economic growth and the workings of the planning system combined to bring them into being. Meanwhile those who used public transport within the cities found it becoming more expensive, and sometimes more crowded and less reliable, driving more of those who could afford to switch to cars to do so. A tight rein on government subsidies for public transport meant that the price of bus, suburban train and Underground fares rose faster than inflation without any corresponding increase in quality or comfort.

Unemployed: Wakefield, West Yorkshire, 1980s

Outside of London the combination of bus service privatisation and deregulation (allowing operating companies freedom to run vehicles where and when they chose) drove fares upwards and total passenger mileage downwards; only on the busiest, highest earning routes did services become more frequent and better patronised. In all seven of England's largest conurbations the proportion of people travelling to work by car grew markedly between 1981 and 1991 while the proportion using buses fell.[7] Journeys to work lengthened, conurbation-fringing motorways built to handle long-distance traffic filled up with car commuters and the numbers of people who lived in the same city as they worked declined sharply. In 1971, just over three-quarters of all the jobs in Birmingham were taken by people who lived in the city. By 1991 the proportion had fallen to less than two-thirds.[8]

In the meantime, improvements in the Intercity rail network made it possible to commute into the urban cores from deep, distant countryside. High-speed trains were introduced on a growing number of routes from the mid-1970s and the number of long-distance rail passengers grew rapidly. Peterborough, 75 miles from London, was now only three-quarters of an hour away by rail.

People living in the solidly Conservative shires, horrified by the threat of new estates engulfing the woods and fields around their homes, campaigned noisily to keep housing away. The term NIMBY, Not In My Back Yard, was imported from the United States and gained common currency among frustrated planners and politicians. It was followed by BANANA – Build Absolutely Nothing Anywhere Near Anything. Nicholas Ridley, Mrs Thatcher's Secretary of State for the Environment in the late 1980s, was burnt in effigy in Tory shires because he was seen as too favourable towards the housebuilders. These protests worked, to some extent, for while hundreds of thousands of new homes were built on greenfield sites in the 1980s and 1990s, the Green Belts fringing the largest towns and cities held firm. Several plans drawn up by major housebuilding firms for private sector new towns in the open countryside of the South East and East Anglia were all eventually refused planning permission.

After Mrs Thatcher's fall in 1990, John Major's administration tried to slow the gadarene rush of shopping and housing development out of town. The government also endeavoured to ensure that as many as possible of the endless planning controversies which arose in a fast-changing world remained local political matters, rather than dragging in central government. New planning legislation enacted the next year made the system of development control 'plan led'; from henceforth the most important determinant of what was granted or refused planning permission in any place was the type of development the local council's plan specified for that location. These locally produced plans, which cover a period of a decade or more and should be regularly updated, divide each council's area into zones according to what type of development is desired there. They tend to be overwhelmingly conservative; green fields are mostly zoned to remain green fields, residential neighbourhoods are particularly sacrosanct.

A government with a minuscule majority clinging to power could not ignore the NIMBY tendency, nor the demands from the rising environmental movement that it must counter the growth in car journeys and pollution which counter-urbanisation was fuelling. By the mid-1990s official planning guidance from government to councils had been rewritten; local planners were instructed to favour development that reduced the need to travel by car. Shopping centres, offices, and leisure facilities that attracted large numbers of people ought to be sited on bus routes and near railway stations. Mixed-use development, in which homes, workplaces and shops are built within a few minutes walking distance of each other, was starting to be encouraged. Proposals for superstores and other retail sheds on the fringes of towns and cities were only to be granted planning permission if there were no suitable sites in or near the centre.

But it took years for these changes in government guidance to have an effect. There are long time lags between land being zoned for particular uses, planning permissions being applied for and granted and, finally, buildings actually being put up. The out-of-town supertanker held its course through the decade, with colossal, car-dependent regional malls still being opened just outside Manchester – the Trafford Centre – and London – Bluewater, near Dartford, Kent – as the 1990s closed. The latter is a shimmering behemoth of silvery metal and glass standing on the bottom of a gigantic, worked-out chalk quarry. There is a great deal of twisting and turning, zigging and zagging before you find your own, empty parking space, even with 13,000 available. But once you are finally on foot and inside the development there is a choice of more than 300 shops and eateries on two floors along with a 12-screen cinema. They are laid out in a triangle and the Holy Trinity of this awesome cathedral of consumption consists of three large department stores, one on each corner.

Labour continued to restrain growth in out-of-town shopping centres after the Conservatives lost power in 1997. Their floor area kept rising, but it was now a slow creep rather than the explosive growth of the 1980s and 1990s. There was greater emphasis on retail development and redevelopment in town centres. By 2000, however, it was estimated that 29 per cent of the UK's 62,000,000 square metres of retail floorspace was to be found in the new, car-dependent shopping locations outside the traditional centres, such as regional malls, superstores and retail parks.[9]

It was the outwards spread of houses, rather than shops, which continued to generate most controversy as the century closed. In 1995 government demographers and planners projected that between 1991 and 2016 the number of households in England would increase by 4,400,000, an increase of just under a quarter.[10] These projections are produced every few years to give council planners guidance on how much land is needed for new housing.* This latest growth forecast was actually a little below the trend for the post-war decades, when baby booms and the shrinking size of the average household had fuelled a strong, steady demand for house construction. But at a time when Britain's population growth had almost levelled off and environmental issues were high on the political agenda, almost two Londons' worth of new homes sounded enormous.

Government-commissioned research produced an estimate that 1,690 square kilometres (nearly 700 square miles) of English countryside would disappear under bricks, mortar and asphalt if the implied demand for new homes were met.[11] The stage was set for unending confrontation. Old campaigners like the Council for the Protection of Rural England marched towards the sound of gunfire while Friends of the Earth – which had previously given the issue little attention – joined the battle, sensing that this could be a fertile new field in which to grow support.

John Gummer, the Tory Cabinet Minister responsible for environment, housing and planning for most of the Major years, responded to this challenge by setting a target for at least 60 per cent of new homes to be built on previously developed, 'brownfield' land, most of it within towns and cities, compared to the 50 per cent or so achieved in the 1980s and 1990s. Soon after he did so the Conservatives were out of office and the new Labour government found itself

*In 1999 a fresh set of projections for England put the increase in the number of households between 1996 and 2021 at a lower figure of 3.8 million.

holding the hot political potato of housing numbers. We pick up its response at the end of this chapter.

As this new Britain outside the old conurbations grew, the reputation of the inner cities plummeted. The recession at the beginning of the decade had added sharply to their already high unemployment levels, making more of their residents dependent on state benefits. A riot in St Paul's, Bristol, in 1980 opened a series of fiery, sometimes lethal explosions of hatred and fury in most of the big English inner cities during that decade. These were not simple uprisings against unemployment and poverty, although both played a part. The riots mostly began with an arrest or a raid by the police, sparking a confrontation with black youths which then spiralled out of control. Today Britain's inner cities remain as deprived and troubled as ever but mayhem on the same scale as that of the 1980s has not recurred (although there were some vicious disturbances in the early 1990s, mostly involving white youths on deprived peripheral estates, and Asian youths rioted in several northern cities in 2001 following racial confrontations and police arrests). Petrol bombs, arson, looting and murder generated masses of press coverage, a measure of national shame and some extra government funding for urban regeneration. But the rioting and its worsening of the inner cities' already poor reputation made residents and businesses more eager than ever to quit these seemingly hopeless areas.

The Thatcher government's first big idea for urban regeneration, which predated the riots, was to bring back private investment in the most blighted areas by doing away with local government. The Labour councils which mostly ran the big cities were seen as irredeemably incompetent and anti-enterprise; they had their planning and economic development powers stripped away in a dozen of the most depressed and abandoned areas. Those responsibilities were handed over to government-appointed Urban Development Corporations. Chaired by captains of industry, these were rather like the state-owned corporations created to build the post-war new towns, with a mission to attract businesses and new owner-occupied housing into their areas. They had broad powers to buy and sell land, to carry out reclamation work on derelict and contaminated sites, to install new infrastructure such as roads and drains and even start businesses. Another regeneration tool of the early Thatcher years was the enterprise zone, where investors and developers were given freedom from the bulk of planning controls along with several years of exemption from various taxes and local council rates.

Neither development corporations nor enterprise zones transformed the prospects of local people. They usually covered areas dominated by vacant industrial land where relatively few people lived. At best, public spending levered in large sums of new private investment and transformed the appearance of some very down-at-heel places. There were new roads, shiny warehouses and office blocks, smart housing estates. But people who commuted into the area took most of the new jobs created. When local people from deprived areas did get work as a result of the new investment, they would often use their extra incomes to move away and become commuters themselves. Furthermore, any gain in the places covered by enterprise zones and urban development corporations could cause pain and jealousy for neighbouring areas, which were usually also in decline. They could see jobs and investment being lured away into the favoured area next door.

What hardly ever happened outside of London was the creation of a sustained virtuous circle in which a large chunk of a city was uplifted and became a desirable place to live, attracting thousands of new residents and the new small businesses which provided them with services. The same was true of City Challenge and the Single Regeneration Budget schemes, the later, more inclusive attempts at urban regeneration made under John Major's one and a half administrations, which tried to forge partnerships between central government, local councils, private sector employers and property developers.

Mrs Thatcher realised that the upsurge in enterprise and personal responsibility that she wanted were not happening deep inside the big cities. On election night in June 1983, as she began her second term, she told her celebrating supporters that their party's next objective must be to take 'those inner cities'. In her memoirs she argues that her traitorous dismissal from high office in the middle of her third term prevented her from implementing her overarching plan for run-down urban areas:

> Once we had solved the problem of the British economy . . . we would need to turn to these deeper and more intractable problems. I did so in my second and third terms with the set of policies for housing, education, local authorities and social security that my advisors, over my objections, wanted to call 'Social Thatcherism'. But we had only begun to make an impact on these by the time I left office.[12]

In fact, some of these policies would only deepen the problems of deprived urban areas. We look at housing later in this chapter. As for education, the big reforms of late Thatcherism which made schools compete for pupils and publish their examination results highlighted the gulf in standards and attainment between state schools in affluent suburbs and those serving inner cities. That made it even harder for the latter to attract pupils whose parents cared about education and demanded high standards.

If central government was failing to turn around the big cities, their own councils were not doing any better. They were overseen by locally elected councillors, they ran crucial local services such as education and council housing and when the Conservatives came into office in 1979 they still raised a very substantial part of their income, 40 per cent, from local property taxes – the rates – paid by local citizens and businesses. They should have been the best placed of organisations to understand, fear and attack urban decay. They had every reason to pull out all the stops to halt the continuing urban exodus because their votes, power and income from the rates were all at stake. But the clash between a centralising, contemptuous Tory government in Westminster and hostile, grandstanding Labour councillors in the town halls made the 1980s the century's most dismal period for local government.

Central government launched a rolling offensive against the councils' autonomy and financial independence. It killed off the top tier of urban government, abolishing the Greater London Council and the metropolitan county councils in 1986. It removed the power town halls had to choose how much to tax local firms by taking over the setting of the business rate. When councils tried to maintain or boost their services by ramping up the rates that local households paid, ministers responded by introducing the capping regime which curtailed their ability to spend more than the centre judged they should. Then Mrs Thatcher steam-rollered through a short-lived replacement for the rates, the regressive poll tax which weighed more heavily on middle- and low-income families than the wealthy. Because the poll tax ended up costing so many families extra, the government moved to cut it by making local councils even more dependent on central government finance. By 2000, local authorities collected only 25 per cent of their budgets from local taxation.[13] They had been reduced to little more than agencies of central government.

All of this made for a decade of ceaseless confrontation, a wretched blame game in which central and local government accused each other of causing high local taxes and poor services in declining areas. Liverpool City Council, run through much of the 1980s by the extreme-left Militant faction within the Labour party, refused to cut its budget and sought a showdown with the government. The revolution might yet begin on Merseyside and if it didn't then a future Labour government could always bail it out of the heavy loans from foreign banks which defiance of Thatcherism entailed. Elsewhere, especially in London, several big city Labour boroughs embraced the new politics outside the old class struggle, baiting the Conservatives with strident campaigns against nuclear weapons and for gay rights. The local tax level they set (first in the rates, then in the poll tax, then in its replacement the council tax) soared way above the national average despite central government capping their spending. Metropolitan councils appeared to be almost boasting about how deprived their people were, how oppressed by Thatcherism. Liverpool's chief manufacturing industry became self-pity. Not surprisingly, the urban exodus continued.

The town halls also came under attack in their role as landlords to that third of the population living in council homes. Here, too, the government doubted the competence of local councils. With the 1980 Housing Act it forced them to sell their houses whenever tenants wanted to buy them and compelled councils to offer mortgages and increasingly generous discounts to the purchasers. Selling municipal houses to tenants was nothing new; a few had been purchased even before the Second World War and many Conservative-run councils had encouraged sales from time to time in the decades after 1945. But now, with the full weight of central government backing, a trickle turned into a flood. About 1.5 million council homes have been sold to tenants since 1980 in the largest privatisation of them all – and predominantly to better-off, employed and middle-aged tenants. It was mostly the more suburban houses that sold rather than the council flats concentrated in the inner cities. Sales tended to run at high levels in prosperous areas where a relatively low proportion of all homes were council-owned. Fewer tenants bought in the big cities where much of the housing is municipal.

Some of the more traditional-looking council estates in prospering areas became largely owner-occupied in just a few years. Today it is hard to tell these apart from private housing built at the same time, for much of the post-war municipal housing was built to high standards with room and garden sizes which were as generous, if not more so, than those the private sector had offered to the bottom end of the market. The best of these largely ex-council estates have lovingly tended gardens, well maintained homes and a multitude of personal flourishes – chunky extensions, diamond-paned windows and York-stone cladding – which try to make each house stand out from the crowd. The right to buy made hundreds of thousands of families more wealthy, independent and self-sufficient.

Elsewhere, however, the government's hard sell of council housing has proved a curse for purchasers. Some tenants who bought at the end of the 1980s – the top of the house price boom – found the values of their homes plummeting a few months later, while in declining areas urban abandonment and decay lowered market values to absurdly low levels. Others who bought flats in badly built council blocks found themselves saddled with huge charges for refurbishing decrepit lifts and other shared structures. On some of the less affluent council estates you can find the odd house looking noticeably worse for wear than its neighbours. These are homes that have been sold to tenants who cannot or will not carry out maintenance work, or who have sold them on to people who neglect them. The local council has managed to keep its stock in good condition, while that which is privately owned has rotted.

The gulf between suburban areas and smaller towns, with generally low levels of council housing, and the inner cities with their high levels, was widened by the great sell off. The most destructive impact of the right to buy was to 'residualise' the council home. The proportions of council tenants who were long-term unemployed, single parents and impoverished elderly folk all rose. The policy of encouraging council house sales was not entirely to blame for these changes; mass unemployment, rising rates of family breakdown (a major cause of poverty) and the general preference for owner occupation all played a part. Those millions of houses and flats that had been a flagship of the welfare state, occupied by an increasingly prosperous working class, became a symbol of failure. Across much of the country renting from a private landlord was no longer the tenure of last resort, the least desirable way to keep a roof over your head. Council housing had now moved into that bottom place, with its estates linked to drug abuse, crime and hopelessness.

The Thatcher government questioned council housing's right to exist, breaking decades of political consensus. Selling homes to tenants was one way of reducing the town halls' role as landlord. Finding other organisations to take over from councils was the other. There were new policies that encouraged transfers of single estates and of councils' entire stock of municipal homes to not-for-profit landlords. At the same time, councils were prevented from building new homes. The task of providing new social housing was given to the housing associations, descendants of those charitable, independent organisations set up in the nineteenth century to provide decent homes for the better-off working classes.

The associations had been in the shadows of council housing for most of the twentieth century. They had concentrated on niche markets, such as housing elderly people or immigrants and converting big, dilapidated old houses in inner cities into flats for rent. From the 1970s onwards housing associations had begun a comeback, thanks to initiatives by both Labour and Tory governments wanting to see an alternative to council housing and private landlords. Now, in the 1980s, the government turned the associations into mainstream providers of new housing for low-income families – and this at a time of mass unemployment, rising rates of crime and drug abuse, and the withering of the two-parent family among the poor. Many expanded rapidly into large, business-like organisations, and that brought a profound change in their culture. 'No more the open-toed sandals, beards and bicycle to the office,' wrote Julian Blake, editor of the housing charity Shelter's *Roof* magazine. 'This new sector was suited and booted, with company cars and mobile phones in tow.'[14]

There were mergers, conferences, expense-account dinners, the odd corruption scandal and more and more chief executives earning salaries of over £100,000 a year for housing the poor. The associations began teaming up to build large estates on edge of town sites, seeking economies of scale. But even though they avoided the discredited flat and maisonette designs of the 1950s and 1960s, many of these new estates soon ran into difficulties. In 1993 a landmark report by a housing consultant, David Page, demonstrated the depths of their troubles.[15] Within a few years of completion housing association estates could look battered and neglected, be plagued by high levels of crime and vandalism and have become highly unpopular among their dispirited tenants.

> The problems are not new but the time-scale is; housing associations are getting there much quicker than local authorities. Run-down council estates are generally the result of two or three decades of decline: housing associations are now meeting similar problems in under five years.[16]

The problems stemmed from concentrating large numbers of children and impoverished families without work in one place, Page concluded. The associations were being encouraged to build family houses in bulk and then fill them with tenants with the greatest need; families off the council waiting list or who were homeless. The result was a high ratio of children and teenagers to adults and high levels of unemployment, single parenthood and dependency on state benefits.

The Conservative government had another reason for favouring housing associations as the builders of new social housing, aside from its gut feeling that councils should not be landlords. This was a way of keeping down the all-important Public Sector Borrowing Requirement (PSBR), the quantity of money the state has to borrow from the capital markets and a key indicator of the government's financial probity. Because the associations are independent, non-governmental organisations their borrowing does not count towards the PSBR; council borrowing does.

The associations meet the costs of constructing new housing partly through government grants and partly by borrowing from financial institutions. The more they borrow from the private sector, the less grant they need from government – but the more they need to charge tenants in the form of higher rents in order to pay off the loans. During the 1990s, the associations – with the encouragement of John Major's government – got a rising share of their house-building funds by borrowing from the private sector (it rose to almost half) and a falling share from government grant. This enabled Britain to keep up the construction of new housing for low-income groups who could not afford to buy their own homes at a time when the government was itself spending less and less on building social housing. But it also meant housing association rents would have to rise well above the rate of inflation. More and more of their tenants could not afford the rents out of their own income, and so the state had to fork out higher and higher quantities of housing benefit – the means-tested rent subsidy available for low-income households. High rents reduce the incentive for tenants to quit unemployment and take paid work, for a small rise in earned income drastically reduces the amount of housing benefit received. And so, for this and other reasons, many housing association estates swiftly joined their council counterparts at the bottom of the social pile.

The century closes with a great gulf between the two dominant forms of tenure – owner occupation (or, rather, taking a large, long-term loan in order to eventually become an owner), on the one hand, and social housing through councils and housing associations, on the other. This is what 'two nations' means today; nothing so divides our state into haves and have-nots. Of the wealthy, developed nations, Britain is one of the most heavily biased towards home ownership. So deep runs this passion that the cost of housing has risen well above the rate of inflation in the long term. The lack of alternatives to home ownership and the fact that spells of sustained incomes growth tend to overheat the housing market, have given us wild booms and busts in house prices. These end up influencing, if not dictating, the interest rate set by the Bank of England. Which, in turn, influences the value of the pound against foreign currencies and the competitiveness of UK industries.

It took Britain most of a decade to recover from the biggest of these upsets, the giddy rise and fall of the late 1980s and early 1990s which, by 1993, had plunged approaching two million households into negative equity – they owed their mortgage lenders more than the value of their homes. The house price collapse that began in 1989 opened, deepened and lengthened the recession of the early 1990s. By 1997 more than 400,000 homes had been repossessed, mainly because unemployed people could not keep up their mortgage repayments.[17] This, it was said, was the largest forced eviction seen in Britain since the Highland Clearances of the 1830s.

The housing slump also halted the long-term decline in the number of private landlords; people who had to move were unable to sell and so rented their homes out instead. Yet, astonishingly, this sickening roller coaster ride does not seem to have caused any fundamental disillusionment with home ownership. The fact that the recovery in house prices was gradual and halting through much of the 1990s suggested buyers had become more cautious and lessons might have been learnt. But as the decade closed, prices were once again rising more rapidly than earnings.

Social housing, meanwhile, became more stigmatised than ever, branded as the place where we keep our losers and readily identifiable because of the way in which it was built in dollops of hundreds or thousands of units. But there has never been any shortage of lobbyists and housing professionals claiming more of it was desperately needed. In the mid-1990s the Major government began to wind down investment in new, state-subsidised, low-income homes and its final spending plans implied less than 40,000 new units a year across Britain. During most of the 1950s, 1960s and 1970s construction of social housing was running at four times that level. There were forecasts that this falling off would produce a rising tide of homelessness and overcrowding.

Yet there was no sign of any housing numbers game being played in the 1997 and 2001 General Election campaigns. Conservative and Labour no longer showed any inclination to compete over how many units a year would be churned out under their governance. Through most of the 1990s, a variety of academic experts and pressure groups using various computer models had declared there was a need for approaching 100,000 new social, subsidised homes to be built *each year* to house the rising number who could not afford to rent or buy in the open market while clearing the backlog of unmet need. But after the 1997 election there was only disappointment for Old Labour stalwarts hoping for a council house revival. The new government began to pump some extra money into refurbishing and repairing council and housing association stock but the number of new affordable homes completed in Great Britain in the four years after the election was just under 100,000, compared to just over 150,000 in the four Tory years before.[18]

In the meantime, the rate at which homes were being transferred from municipalities to housing association landlords actually speeded up. The new government didn't seem to be any fonder of the idea of the council home than its predecessor, not even in Scotland where a higher proportion of the population were living in municipal housing than in England. New Labour ministers at Westminster and in the new Scottish executive generally preferred competing, not-for-profit landlords managed by professionals to provide housing for those who could not afford market prices over council housing departments run by local politicians. A housing Green Paper published in 2000 made this clear[19]

More and more town and city councils saw the writing on the wall. They knew that tens or hundreds of millions of pounds would have to be spent on refurbishing their crumbling council homes if they were not to become a source of increasing shame and embarrassment. They judged that the government was unlikely to allow them to borrow the necessary funds. If transfer deals, backed by private sector loans and some government help, looked like maintaining lowish rents while guaranteeing refurbishment, then councils were interested – and their tenants could usually be persuaded to vote in favour. Birmingham and Glasgow, with around 80,000 homes and some 300 tower blocks each, came down in favour of transfer in 1999, although when this book was completed it remained to be seen whether their tenants would vote for it. The government envisaged that by 2004 – and possibly even sooner – housing associations could overtake councils in England and be the majority providers of social housing.[20] Councils and central government were also looking for other ways for town halls and councillors to step back and bring in private sector capital to refurbish and managers to manage their flats and houses, using subsidiary companies and the government's Private Finance Initiative.

The great national housing shortage that lingered for decades after the Second World War has long disappeared. In large parts of the country, particularly in northern England, surpluses of social and private housing have accumulated. The growth in abandoned and derelict housing in the 1990s has made this pitifully obvious. The 'tinned up' house, its windows and doors covered by steel shutters or plywood, has now become common both in inner city terraces and the semi-detached homes of more recently built council estates. Councils and housing associations had been loath to discuss this problem under a Tory government; signs that their product was no longer wanted would hardly have helped their cause. But with Labour in power the abandonment kept on spreading, from house to house, street to street, like a disease, and it became impossible to deny the phenomenon. Occasionally housing associations demolished properties that had only been built a few years ago because they could find no tenants who wanted to live in those neighbourhoods.

The story moved out of the housing press into the mainstream media. When it became known that the Salford street whose rooftops had graced the opening titles of *Coronation Street* for years was now a near empty husk, the issues of urban abandonment and surplus social housing became part of the political furniture. It has to be seen to be believed, though, and even then it's not quite believable. Great patches of urban fabric, entire streets of terraced houses and blocks of flats, pavements, little corner shops, all built for people and their doings, now deserted, with the process of abandonment sometimes taking less than a couple of years. After 17 years as a journalist I thought I knew something of this country. Then I looked around Glasgow, Liverpool, Newcastle and Manchester and found a new state. I had seen dereliction on that scale only once before and that had been 20 years earlier when, as a student summering in the USA, I had visited the South Bronx.

What on earth is going on here? If a first cause for abandonment can be nailed down, then it must be local and regional economic decline, the shrinkage and clo- sure of manufacturing and transport industries that have erased hundreds of thousands of jobs in places like Clydeside, Tyneside and Merseyside. This creates the high levels of long-term unemployment and poverty that form the background to urban and social decay and drive people with ambition and aspiration to leave. Abandonment first got under way in the big cities that had suffered the worst economic collapse and the highest levels of population loss, such as Liverpool and Glasgow. The latter has been demolishing thousands of corporation homes a year through the 1990s.

An imbalanced housing stock increases a neighbourhood's risks of abandonment. Quarters dominated by council flats and houses which were built for a mass working class 20 or 30 years ago have been left stranded by the combination of rising home ownership, industrial decline and the residualisation of municipal housing described earlier in this chapter.[21] Abandonment has sometimes spread like wildfire amongst cheap owner-occupied homes and privately rented housing but this is usually linked to problems in nearby social housing. One of the links is that absentee private landlords move in tenants who have been evicted from social housing because of crime and anti-social behaviour. These tenants then make life hell for the other residents. Some of these landlords are themselves criminals, who use threats to force owner-occupiers to sell their properties at pitifully low prices.[22]

Crime is an important driver of abandonment. People who burgle, who deal drugs, who get drunk, who are quick to threaten violence, are highly effective evictors of their neighbours. Parents who shout and scream abuse and let their children thieve, smash things up and run wild late into the night can also act like miniature neutron bombs; the buildings are left standing but the occupants vanish. Empty houses then become magnets for more criminals and vandals. The copper piping and central heating boilers are ripped out and sold to the scrap trade. Abandoned dwellings make handy drug dens and fly-by-night brothels.

Even if there is no increase in crime, the mere presence of two or three empty, boarded-up houses is likely to cause more homes in a vulnerable, low-income street or estate to empty. Imagine how you would feel living next to such a house, frightened by any sound you heard coming through the walls. There is no more obvious badge of a failing neighbourhood than empty homes. Which brings us to

Abandoned housing, Beswick, East Manchester, overleaf

another cause of abandonment – stigmatisation. People living in this kind of neigh-bourhood know what the rest of the world thinks of them. They know that their address makes it harder for them to get work, loans and mortgages, a good school for their children, a taxi home, the police to respond to a call.

Those who flee these bad and emptying places need other homes to flee to; without them mass abandonment could not have happened. Many have been able to move into owner occupation. Across much of Britain, a couple in which both partners work in relatively low-paid jobs can afford the mortgage required to buy a small house – even a small new house. And most of the new, for-sale houses built in the post-war years have been built on the edge or outside of the big towns and cities. New estates of boxy, homogeneous houses built on countryside, the show home flags fluttering in the wind, are twinned with derelict inner city terraces. The government's own expert advisers have linked urban decline and disinvestment with this nearby greenfield housing; they have argued that it makes regeneration more difficult.[23]

It is widely supposed that the lure of the prosperous south of England has played a leading part in the exodus from the big conurbations in the North and the Midlands, as well as adding to housing pressures in the Home Counties. In fact, most people leaving these cities remain fairly close to them rather than heading south. The biggest exodus of all is from Greater London, where 48,000 more people were leaving in the 1990s than moving in from other parts of the UK each year. London's outflow was larger than that of the six next largest English conurbations combined. It has fuelled the growth of Greater Greater London, the booming region shot through by high speed railway links and motorways which now extends out to Southampton, Swindon, Northampton and Peterborough.[24]

Urban abandoners do not generally flee straight into new, greenfield homes. Because they are often poor and rely on benefits, they are unable to purchase them. Instead there is a chain with mouldering council estates and by-law terrace streets being deserted for something better nearby at one end, and people with more choices in life making the urban exodus at the other. When people come to hate, fear or despair of their immediate surroundings, they might move back in with their parents, or with other relatives or friends. They might try another council house or housing association property in the area; the fact that others are getting out alto-gether (to suburbia, or smaller towns, or down south) creates the vacancies that make this possible.

As an area's status declines, as more and more people who can move out do so, the residual population faces mounting problems while its capacities to deal with those problems decline. Research in Newcastle's emptying West End found a large proportion of people taking out new tenancies on council homes had come from other council properties nearby, and had moved because of relationship break-downs and problem neighbours or because of electricity and gas debts.[25]

Councils with homes in these emptying quarters find themselves competing with nearby councils or housing associations for a dwindling number of social housing tenants. At least one has resorted to trying to attract tenants from far away, simply to keep their houses occupied. Meanwhile, some of the urban poor prefer to live in private landlords' homes rather than the worst social housing estates – or are forced to do so, having been evicted from social housing for rent arrears or anti-social behaviour. The rent may be higher, the house will often be in poor condition, but the state still pays the rent through housing benefit.

And so it is that the private slum landlord has made something of a comeback. In blighted parts of Britain's big cities, he or she can buy a house for few thousand pounds and readily fill it with a family on housing benefit. The message is 'DSS (Department of Social Security) tenants welcome'; you see it on brightly coloured stickers in the windows of these houses and read it in the classified advertisements in the evening paper. For the landlord it is a sound investment; he spends nothing on maintenance while the taxpayers pay her or him a rent that assumes the house cost a normal price and is not a neglected slum.

Some of these DSS-tenanted, privately owned homes are ex-council, sold under the right-to-buy inducements of the 1980s. The state built them, then sold them off at a discount, and now it pays high rents (considering their quality and location) to house poor people in them. In doing so it helps to accelerate the process of abandonment that is destroying urban homes and neighbourhoods. Thus state payments to slum landlords are helping to trash the state's earlier, heavy investment in building council homes. It is hard to imagine a larger, sorrier mess.

Neighbourhood and city-wide surpluses of social housing, combined with devalued and abandoned private housing, are doing immense harm. The problem seems to be getting worse, with around a quarter of a million homes in the North of England, the Midlands, either abandoned or at risk of being abandoned.[26] Here and there, especially in the north, it is accepted that the decline and abandonment of estates and streets are irreversible. Managed retreat is now on the agenda, with more than 10,000 homes being knocked down each year.* Demolition is an obvious solution to homes that no one will occupy, that look desolate and attract crime. It has become one of the favourite responses of local councils to the problem and the amount of money the state is spending on destroying failed housing is growing.[27] There are euphemisms; people speak of lowering densities, loosening the urban fabric, introducing much needed green spaces – which turn out to be bleak urban prairies or rubble plains. More often than not demolition is followed by redevelopment of social housing, invariably at lower densities (houses replacing flats) and sometimes with low-cost homes for sale. The worry is that redevelopment won't

---

*The idea of managed retreat before the advancing sea, as sea levels rise and the earth's crust sinks along England's eastern and southern coastline, was promoted through the 1990s by experts and environmental bodies as an alternative to building expensive new sea walls. In a few coastal areas managed retreat is now being practised. The term is not, as yet, used to describe what is happening in a growing quantity of residential areas.

work, the houses will empty once more, and the next time they are knocked down there will be nothing to replace them.

But recall that social housing was invented because the private sector could not provide decent housing for people on low incomes. Despite its mass abandonment in parts of Britain, there remain very large shortages of social housing elsewhere and no sign of these being reduced. Millions of people in the more prosperous towns and cities and the wealthier urban areas still find the private market completely out of reach, even if they are in work. In Greater London in the year 2000, the average price paid by individuals and couples buying their first home was around £120,000. Which means that in the absence of any large savings or inheritance, she, he or they would have needed an income of about £40,000 in order to afford the necessary mortgage. In the South East the average first home purchase cost about £87,000, requiring an income of about nearly £30,000.[28] These are way above average (median) incomes, even in the nation's richest conurbation and region.

This level of house prices – and the correspondingly high rents charged by private sector landlords – make it difficult for key service providers, like nurses, teachers, firemen and bus drivers, to live anywhere near their jobs. Many people on middling incomes have to commute long distances from areas where market housing is more affordable. The capital's stock of social housing has been depleted by the right to buy while the number of new households unable to afford market prices anywhere in the region has grown. In London nearly 200,000 households were on council waiting lists and judged to be in housing need in 2000, while the numbers accepted as homeless and lodged in temporary accommodation had risen above 46,000. More than 6,000 were living in bed and breakfast accommodation.[29] Haringey Council began offering £2,000 grants to families in social housing and temporary accommodation to persuade them to move out of the borough, while next door Camden signed a deal with Huddersfield to promote empty council homes up north among London families desperate for improved subsidised housing.

A housing commission set up to advise the capital's new mayor, Ken Livingstone, concluded that London needed 43,000 new houses a year, of which at least half should be subsidised for rent or sale at below market prices. That is more than double the rate at which new homes have been built in the capital through the 1990s, and double the rate at which London planners have estimated sites and surplus buildings can be supplied for housing in the future.[30]

What did the Labour government elected in 1997 make of all of these urban ills? Britain's highly combative political system does a poor job of preparing parties and politicians for power. It encourages a dumbed down, reactive, never-mind-the-morrow style of opposition; do and say everything possible to make the government of the day look bad while developing the bare minimum of your own policies. Keep them short and simple.

Single member constituencies and first-past-the-post elections tend to rule out coalition governments. This has led to wild swings in direction when power changes hands but it also makes coalitions out of the main parties themselves. When it came to power in 1997, Labour was short of detailed policy and long on disparate opinions. Old Labour saw the salvation of its urban heartlands resting in more council houses, more taxpayers' money for physical regeneration and job creation and better local services. But New Labour had grave doubts about the competence of the town halls run for so long, and so badly, by Labour councillors to play a leading part. It intended to reduce poverty and inequality of opportunity by boosting employment nationally, by benefit reforms and job subsidies which shifted people off welfare and into work, and by a mass of local programmes focussed on Britain's most deprived urban areas and aimed at improving people's education, health and employability. Whatever else happened, New Labour was anxious not to offend the suburban and ex-urban Middle England voters whom it depended on to win a second term.

The commitment to stick to the previous government's spending plans for its first couple of years meant that it could make no major new investment in cities for the time being. Most of its time and effort in those first 4 years to 2001 was, therefore, taken up in getting advice and in planning what it might do.

The internationally renowned architect Richard Rogers was asked to head an expert task force to conceive an urban renaissance. Deputy Prime Minister John Prescott's sprawling new superministry, the Department of the Environment, Transport and the Regions, set to work drawing up an Urban White Paper – the first since *Policy in the Inner Cities* back in 1977. Ministers who were initially sceptical or uninformed began to be converted to the 'new urbanism' agenda that an international school of architects, planners and developers had been pushing for a decade. This agenda, the subject of Chapters 12 and 13, calls for new housing to be built at higher densities and within walking distance of shops, schools and public transport hubs, to be more varied, better designed and mingled with workplaces.

Meanwhile the Cabinet Office's Social Exclusion Unit, created by the new administration, was tracking down the most benighted bits of urban Britain, and pondering what to do about them. A variety of surveys had demonstrated that not only had the gap between the prosperous and the poor grown much wider in the previous two decades but that a very substantial proportion of Britain's poor, several million people, are concentrated in a few thousand urban neighbourhoods, each with a few hundred to a few thousand homes.[31] These are the so-called 'worst estates'. They are not, in fact, all council and housing association estates; they also include decrepit streets of inner city private housing. But they are wretched slums where crime and vandalism, drug use, long-term sickness, unemployment, teenage pregnancy and under-education are all far above national averages.

In the autumn of 1998 the Social Exclusion Unit published its first, scoping report on attacking the problems of deprived neighbourhoods and the Prime

Minister, speaking at a rebuilt council estate in Hackney, London, announced what would happen next. In a huge policy development exercise eighteen committees consisting of civil servants from ten Whitehall departments and experts from outside of government were appointed. Each was sponsored by a government minister. These Policy Action Teams covered fields ranging from financial services and information technology to housing management, housing abandonment and anti-social behaviour. Each was asked to come up with workable recommendations for improving impoverished neighbourhoods and bettering the lives of their residents. The teams met, visited and wrote long reports with a total of around 600 recommendations. The government drew on them in setting out a new national strategy for neighbourhood renewal early in 2001.[32]

The assault on urban problems gradually intensified during Labour's first term in office for 18 years. The big departments of government rolled out supplementary spending programmes for deprived urban districts – employment zones, education action zones, health action zones, 'Sure Start' to promote the health and welfare of pre-school children. The Conservative's Single Regeneration Budget was retained and expanded and two new regeneration programmes were launched – first, the New Deal for Communities and then the Neighbourhood Renewal Fund.

The New Deal for Communities was focussed on thirty-nine extremely deprived urban neighbourhoods of 1,000 to 4,000 homes. Each of them would receive around £5 million a year to be spent on programmes that aimed to raise people's ability to get and hold a job, to reduce crime and improve housing, the area's appearance and the delivery of public services.

The Neighbourhood Renewal Fund was spread wider and more thinly among the eighty-eight most deprived local government areas in England, each getting around £1–5 million extra a year. In both of these programmes, the government's intention was to avoid the manifest failures of previous top-down regeneration efforts; renewing housing and other buildings rather than transforming people's prospects, failing to engage locals, failing to improve key public and private sector services such as schooling and shopping. There was now to be much emphasis on obtaining the consent and participation of local communities and on local government entering into partnerships with other public service providers like the police and the NHS and with local businesses.

The government also launched nine regional development agencies in England, tasked primarily with boosting regional economies through workforce training, redevelopment of derelict land and urban regeneration. They had large and growing budgets, but no one seemed quite sure about what they were meant to achieve and precisely how they fitted with existing local and emerging regional government.[33] New Labour's cautious efforts to reform local government are covered in the final chapter.

In the summer of 2000, as the government prepared to face the electorate the following year, it committed itself to huge increases in public expenditure over the next three financial years, the great majority of which would go into public services in Britain's towns and cities and the transport links between them. By 2003/4 an extra £33 billion a year was planned to be spent on education, health, transport, social housing and leisure, culture and sport, a real spending increase of almost a third. Of course that increase was conditional on Labour being re-elected and the economy continuing to grow. The difference that all this taxpayers' money makes to urban life will depend on how wisely it is spent by central and local government and on how much private sector spending it levers in. After all, £33 billion a year works out at less than £600 per citizen.

In the meantime, the urban renaissance project seemed in danger of being overwhelmed by the complexity of partnerships, zones, initiatives and strategies with bureaucrats only emerging from endless meetings to write yet another progress report or consultation document. Ministers and senior civil servants wanted strong local leadership but they also wanted community consultation and participation. They wanted local government to be bold, imaginative and independent but they weren't at all sure most town halls had it in them and felt they needed close watching. They fretted endlessly about integration, and feared that people's lives might not be very different once all the money had been spent.

At the close of 2000 the Urban White Paper was belatedly published.[34] As with most White Papers, much of the content of *Our Towns and Cities: The Future* turned out to be a compilation of existing, albeit fairly recent, government tactics and strategies knitted into something resembling a grand plan. Lord Rogers, who had been making worried noises about the government's lack of response to his urban task force's mighty report, was under-whelmed. The White Paper, he said, was 'an important step along the road. But on crucial issues this white paper falls short of what is going to be required to engender a real urban renaissance.'

More than a dozen of the task force's 105 recommendations were rejected or ignored. Among them were three that were particularly controversial and radical: a new tax on derelict urban land, council tax to be paid in full on empty homes and tax breaks to encourage institutional investment in homes for rent at market rates. The government's response to several other key recommendations was partial, or left to policy reviews that might eventually bear fruit. Gordon Brown, Chancellor of the Exchequer, agreed a modest package of tax changes which, he claimed, would increase investment in towns and cities by some £200 million a year. But on the other side of the urban coin is the demand for new homes and the consumption of countryside by sprawl. And here the government really did make radical changes.

The post-war system for planning where new homes should be built runs to a timetable; every 5 years the government's statisticians produce national and regional projections of household numbers which update the previous set and cast them further into the future. Then councils from each English region, working together in regional planning bodies, must take these projections into account when drawing up their Regional Planning Guidance. This guidance, RPG, looks ahead over the next couple of decades and sets down an outline of what type of development and infrastructure is desired in the region and where. It tells individual county councils and the larger towns and city councils how many new homes they must plan for within their boundaries. But before it becomes official guidance, there is a public inquiry into the draft RPG, conducted by a panel of planning experts. The panel writes a report giving its critique of the guidance.

There is then further consultation and the government eventually approves a final version, usually taking some of the panel's criticisms on board while inserting some amendments of its own.

It sounds most unexciting but the process stirs great passions. The biggest clashes between planning for new homes and countryside conservation have long been in England's South East, the wealthiest, fastest-growing region of Britain which is also among the most densely populated and built up. In the mid-1990s the government's forecasters estimated the demand for extra homes in this region, the twelve counties surrounding London, as being 1.2 million between 1991 and 2016. This implied that more than a quarter of all the new housing required in England had to be fitted into this single region around the capital.

SERPLAN, the umbrella planning organisation for all of the region's councils, proposed a very much lower target of 718,000 new homes between 1996 and 2016 in its draft Regional Planning Guidance which it submitted to government at the close of 1998.[35] It called for policies to constrain new development in the zones experiencing the most intense growth pressures, to the west and south-west of London. It advocated that 40 per cent of the new homes in the region should be affordable – either to rent or buy – for people on below average incomes. And it said 60 per cent of new homes should be built on recycled urban land rather than green fields (a little less than half had been in the 1980s and 1990s). This was in line with the target put forward by the previous government and then adopted, with minor modifications and after a pause for thought, by the new Labour administration.

And then came the Crow Report. Professor Stephen Crow, formerly the government's chief planning inspector, conducted the public examination of SERPLAN's draft RPG in 1999 and then produced his own response. So savagely critical was his 140-page report that what would usually have been among the dullest of documents became quite an interesting read – at least in parts.[36] He tended to agree, he wrote, with those who had found the strategy over-long, confusing, inconsistent and incoherent, evasive, specious and 'irresponsible in relation to providing for legitimate planning needs'. 'Sufficient housing should be available for all who wish to live in the region,' but SERPLAN had found 'one excuse after another' to avoid planning for the number of new homes required. Crow advocated 1.1 million between 1996 and 2016, almost two-thirds more than SERPLAN's figure for the same period. And only 50 per cent of these new homes, not 60 per cent, should go on recycled urban land – provided enough of it could be found. Massive new greenfield development should be planned in four areas: next to

Gatwick and Crawley in West Sussex, Milton Keynes in Buckinghamshire, Ashford in Kent and near to Stansted Airport in Essex.

Behind the SERPLAN strategy, wrote Professor Crow, lurked the notion that the South East's dynamism was harming other parts of Britain and that restraining growth in this region would divert it to less favoured regions. But this was a false analysis. Investment discouraged from settling in the South East of England was just as likely to go to other parts of the European Union. The South East was 'an engine of growth in the national economy'; further expansion was in the national interest. The three sentences below sum up the Crow Report and give a flavour of its exasperated, 'stop being silly and listen' tone.

> In the final analysis, if expansion of towns into the countryside is to cease, then this can only come about through the South East ceasing to be an area of economic growth. It is a mere pretence, indeed a cruel delusion, to suppose otherwise. Choices have to be made, but all must recognise that the economic stagnation of the South East has never been on the agenda of national policy. There is a price to be paid for every material benefit in this world.[37]

The document made several approving nods towards the new urbanism agenda, for higher densities, mixed uses and less car dependence. But it was, in essence, a vigorous defence of the post-war status quo – of counter-urbanisation, greenfield growth and the drift of people and jobs out of the conurbations and from north to south. The Crow Report's most telling single sentence reads: 'The essence of planning lies in taking a view of what is likely to happen in the future and planning to meet it.' Indeed. The best bet for what happens in the future is that the trends of the immediate past continue. Assuming, of course, that you don't want, or don't try, to change those trends.

This document provoked a horrified reaction from the environmental groups, the Conservative Party and the *Daily Telegraph*. But it was strongly defended by the Confederation of British Industry, the housebuilders and by professional planners, in the shape of their Royal Town Planning Institute and the Town and Country Planning Association. Some newspapers ran articles and editorials urging the government to back Crow and make a stand against the NIMBYists; there was still masses of countryside left in the South East and the new homes needed to go where the new jobs were.

The never-ending national debate about development in the countryside boiled up in confusion and became inextricably, confusingly entwined in the long-running debate about the South (a goose laying golden eggs which might be killed by planning constraints) sucking life from the North (a wasteland of abandoned homes and industry). Suddenly it seemed as if all those new greenfield homes were being taken by northerners in search of jobs, or better-paid jobs. But this was not so; the new housing in the South East was mostly needed for the children and grandchildren of people who had moved out of London decades earlier, and for the continuing exodus of people out of the capital.

The government had to respond to SERPLAN and Crow with its own final, authorised version of Regional Planning Guidance for the South East. John Prescott, Secretary of State for the Environment, eventually opted for a compromise. He ordained figures for the number of new homes to be built in the region lying nearer to SERPLAN's than to Crow's. Even so, they still implied a speeding up of the rate of housebuilding in the South East compared to what had taken place in the 1980s and 1990s. Prescott stuck to the target of constructing 60 per cent on recycled urban land. And he avoided attaching his name to a colossal total figure for 20 years' worth of new homes (780,000) by putting down annual numbers for the next five; after half a decade the situation would have to be reviewed and the rate of land supply and housebuilding probably shifted upwards. Major urban expansion should, he said, be seriously contemplated at Ashford, Milton Keynes and alongside the lower half of the M11 London to Cambridge motorway.

This response, and the way the government handled parallel Regional Planning Guidance for East Anglia – where the growth pressures were just as intense as in the South East – showed that the central government was moving away from its old 'predict and provide' approach to land for new housing. It could contemplate providing slightly fewer new homes than projections based on trends called for – at least until politically painful shortages started appearing.[*]

A slim document published in 2000 with the innocuous title *Planning Policy Guidance Note 3: Housing* (known in the planning and housing trade as PPG 3) was altogether more radical.[38] This was the latest version of the government's official guidance to all local English councils, not just those in the South East, on the planning of new homes. It tells them what to take into account when deciding whether to grant or refuse planning permission for proposed housing developments. What the new PPG 3 aimed to subvert, Prescott told the House of Commons, was the practice of 'wasteful, badly located and poorly designed housebuilding which has gone on over the last 20 years'. (It had, in fact, being going on for rather longer, under Labour as well Conservative governments.)

The new guidance turned much of the new urbanism agenda into official policy. It said new housing developments should be more varied, offering homes to a wider variety of income groups and household sizes rather than reinforcing social distinctions. Councils should 'encourage the development of mixed and balanced communities'. Housing should be built in ways and places that minimised car use and maximised people's ability to walk, cycle or take public transport to where they needed to go. The first priority should be to build new homes on disused urban land such as old industrial sites; only when that was exhausted should green fields be covered.

[*]But in the Regional Planning Guidance for the South West of England published at the close of 2000 the government aimed for a rate of housebuilding 10 per cent higher than the local councils in the regional planning body had proposed.

Throughout the century, planners had tended to push housing densities down. Unwin and the Tudor Walters report had proposed thirty homes per hectare as a *maximum* 80 years earlier (see Chapter 3). The average in the 1990s had been twenty-five per hectare with more than half of new housing being built at less than twenty.[39] Now government was pushing things in the other direction, putting thirty per hectare as a *minimum* density for new housing developments and encouraging up to fifty. For decades, council planners had been trying to raise the number of off-street car parking spaces in garages and on front drives in residential areas by specifying a minimum standard. Their aim was to tidy a rising number of cars off the streets. Now the guidance set a *maximum* of 1.5 off-street car parking spaces per home to allow higher densities and discourage multiple car ownership.

PPG 3 applied with immediate effect, superseding any housing guidance and plans which councils had devised on their own. And just to make it absolutely clear that the government was determined to change things, Prescott ordered that any proposal to develop housing on greenfield sites larger than 5 hectares should be notified to him. This enabled government to remove the decision on whether to grant or refuse planning permission from the local council and take it itself.

This new guidance had the essence of real planning. Instead of merely trying, *à la* Crow, to accommodate existing trends, it took a view about what a better world would be like and tried to make it happen. Will PPG 3 work? As we shall see in later chapters, 'mixed and balanced communities' go against the grain; people with choices appear to want more space and privacy, implying lower densities. They want plenty of off-street car parking spaces and they usually opt to live among people with similar wealth to their own. Many housebuilders will do all they can to resist the government telling them how to build their product.

So, over the next decade or two, it might all go wrong. Perhaps the punters won't like the new, denser estates stuck within towns and cities and built on what used to be wasteland. There may be persistent scares about this land being contaminated with toxic chemicals from former industrial use; a fortnight after PPG 3 was published the BBC's *Panorama* devoted a documentary to the subject. Low demand for new homes built in unpopular locations might lead to low supply, causing housing shortage and a general surge in overall house prices. The result would be a nation even more polarised between suburb and inner city, between conurbation and prosperous hinterland. Once again, the best of intentions would produce the worst of outcomes in our cities.

PPG 3 is a big step in the right direction. But it cannot achieve an urban renaissance on its own. If the other things that need to happen do not, this brave and bold initiative could fail.

Chapter 6

TEN OPPORTUNITIES

The patterns of development characteristic of most of the last century cannot continue. They have been: socially unstable, concentrating the poor in inner city areas; environmentally damaging, destroying the countryside and creating a car-dependent society; economically harmful, since they have undermined our core urban areas which remain the essential centres of the English economy; and wasteful because schools, shops, even houses, lie waste in urban areas while new infrastructure is provided at great expense outside.

House of Commons Environment Committee[1]

We cannot find a single case brought forward by a member country of a successful turnaround of a deprived urban area . . . policy has been reactive, at best it has prevented a bad situation from getting worse.

Josef Konvitz, Head of Urban Affairs,
Organisation of Economic Cooperation and Development[2]

WE'VE REACHED THE PRESENT, the end of our historical tour. The twenty-first century opens with about four-tenths of Britons living in large towns and conurbations with populations of a quarter of a million or higher.[3] We are now mainly a nation of small and medium-sized towns, many of which are growing fast.

But while most Britons live outside the big cities, most of the poor are found within them.* In some places poverty stretches back for several generations. In others it has been widened and deepened by more recent industrial collapse or the arrival of poor immigrants and refugees. And in others, poverty has been lifted lock, stock and barrel from inner city slums to new peripheral estates miles from the centre and has then persisted for generations. But it is an abidingly big city phenomenon.

There are 354 local authority, or council, districts in England. Of the thirty-five districts suffering the highest levels of overall deprivation – the worst tenth – twenty-five are part of the largest towns and cities with populations above a quarter of a million people.[4] At the other end of the scale, of the thirty-five English districts with the lowest levels of deprivation, only one is part of a large conurbation; the outer London borough of Richmond on Thames. The remainder are small town, rural and suburban places, mostly in the Home Counties.

This poverty is prevalent in council and housing association households, but it is not confined to them. There are many impoverished owner-occupiers, mainly pensioners or from ethnic minorities, in cheap, dilapidated little terraced homes which they cannot afford to maintain. Hundreds of thousands of poor people also live in privately rented housing in the inner cities.

Population loss appears to have slowed in the big cities during the 1990s. Of the eight largest British conurbations, three had their populations stabilise and two – London and West Yorkshire – had slight increases while the Merseyside, Tyne and Wear and Glasgow conurbations saw significant further population losses.[5] But even where there has been population growth, it was driven by migration from overseas and by local births exceeding deaths. The urban exodus continued with the number of people leaving the conurbations exceeding the number moving in from other parts of Britain. If this exodus continues to leave people without choices in life in the big cities, then their problems can only worsen.

Some conurbations – particularly London, Manchester, Leeds and Birmingham – saw a large increase in employment in the growth areas of tourism,

---

*In England 42 per cent of the population lives in towns, cities and conurbations of above 250,000 people. But 54 per cent of England's households that are so poor as to rely on means-tested state benefits live in these urban areas. Author's estimate from 1998-based Short-Term Sub-national Population Projections for Local Authority

Tower block, Stepney, East London

culture, finance and information technology in the late 1990s. This growth in employment, fuelled by a long boom, appeared to reach into the more deprived areas of towns and cities.[6] Yet their unemployment levels remained well above the national average, and the fact that many residents of working age have effectively dropped out of the jobs market makes the position worse. According to the usual measures used to assess encouragement for business start-ups and growth, Britain's big cities compete poorly with the rest of the nation. And, as we shall see in the next chapter, our attitudes remain anti-urban.

If we could resurrect one of the Victorian urban reformers she or he might conclude – at least on first impressions – that our age had eliminated slums. Our poorest city neighbourhoods offer broadly the same amenities as prosperous suburbs; streetlights and trees, indoor lavatories and central heating. But poverty is relative

Areas in England, ONS, and *Indices of Deprivation 2000*, Regeneration Research Summary 31, London, DETR, 2000.

and so are the slums where it is concentrated. Our society, our governance, our housing markets are as effective as their Victorian antecedents in segregating the poor, concentrating them and making their prospects worse in the process. Throughout much of the twentieth century we have had governments which promised to abolish slums and civilise our cities. Despite the vast increase in wealth, the astonishing advances in technology, life expectancy and standards of living, they – or we – have failed.

Yet there is hope. I can think of ten reasons for optimism, ten possibilities that have opened up within a generation:

1 Change is in the air. You could smell it even before May 1997, when the Conservatives lost power after 18 years. Urban councils are willing to own up to the problems of abandonment and to recognise that the flight of people with choices and aspirations is among their biggest problems. Since the change of government there has been delay and hesitation, more rhetoric than action, some signs of retreat and confusion. But no one in high office can get away with ignoring Britain's urban plight for very long.

2 Overcrowding is no longer the great problem of the inner city; it has become much roomier over the past century. The abandonment of old industrial sites and the failure and desertion of some housing have created large areas that can be redeveloped. This emptiness and dereliction are all too obvious in the big northern cities but even in London, which is more densely populated and economically buoyant, a steady stream of former industrial sites provides space for thousands of new homes each year. Lots of urban land will continue to become available, and some of it can be used for new parks and woodland, bringing the countryside into town as Ebenezer Howard envisaged.

3 Of the developed nations, Britain has a high level of home ownership with two-thirds of households being owner occupiers. This proportion may yet rise a little higher.[7] The great majority of new houses are built for sale. Owning a home, with or without a mortgage, gives people a strong incentive to maintain it and to care about the quality of their neighbourhood.

4 But not everyone wants to be a home-owner. A small but significant proportion of the population earns too much to qualify for social housing or rent subsidies but does not want the tie of a mortgage. Britain's cities need professional landlords to compete for tenants by offering good, well maintained housing, promoting and protecting urban neighbourhoods in the process.

Conditions are now beginning to favour increasing investment by landlords of the non-slummish variety in towns and cities. In his 1999 Budget the Chancellor announced the abolition of the last remnants of income tax relief on mortgage interest payments, shifting the balance from owning towards renting a little further. In recent years landlords have been given more freedom to evict bad tenants and to set rents at levels they choose.

A growing number of small investors with a few thousand pounds to spare have been taking out mortgages in order to buy and rent out houses, and several mortgage companies have offered 'buy to let' products. By 2001, 110,000 individuals had taken out mortgages worth more than £8 billion in order to purchase then rent out properties.[8] But the view of many people in the housing world, the Rogers task force and the government is that major investors such as pension funds need to become involved if the desirability and scale of open market renting are really to expand. This would improve the reputation of private renting – for both investors and tenants, accelerate development and refurbishment in towns and cities and introduce economies of scale, particularly in managing housing.

The government should do more to encourage this tendency. It made a start in the urban White Paper by offering 100 per cent capital allowances for converting redundant space above shops into flats for letting. But it turned down the task force's recommendation for a new kind of investment trust that would encourage individuals to invest in letting homes by limiting their risk and tax liability.[9]

5   There is also a recognition that the third form of tenure, state-subsidised renting for low-income households, is in grave trouble. The worst estates are a world apart. But the fact that councils and housing associations rather than myriad private sector landlords or owner occupiers own these troubled areas gives hope. It pushes responsibility and shame onto publicly accountable, regulated bodies. And their ownership makes it easier for housing sites to be redeveloped or refurbished in large packages.

6   The physical environment of the big cities is better than ever and the improvement seems likely to continue. The rivers that run through them are the cleanest they have been for 200 years; salmon now swim up the Thames. People might not believe it, but their air is gradually becoming cleaner too. This is due to the switch from coal to gas as the predominant household fuel and to European Union laws which are imposing tougher and tougher restrictions on pollutants in vehicle exhaust fumes.

The concentration of polluting industry in cities was one of the most powerful drivers of suburbanisation and a great cause of anti-urbanism. But in an economy increasingly dominated by services and knowledge-based industries, homes can be much closer to workplaces. They can even share the same building, as was the norm before the nineteenth century. And that proximity can, in itself, offers further environmental benefits; less need to travel, less traffic congestion and pollution.

7   In the past an influx of prosperous newcomers into a neighbourhood and the resulting rise in property values could displace poor people from cheap housing. Private landlords would be tempted to evict their tenants in order to charge higher rents to wealthier tenants or to sell their property at the higher

values created by gentrification. Today, however, gentrification should generally be welcomed. Only one in ten of the urban population rents from private landlords. The remainder – home-owners and tenants of council and housing association property – stands to gain from any uplift in their neighbourhood. The former will see the value of their homes rise, while social housing tenants could benefit from the wealth and improvement in status which gentrification brings to their neighbourhood without any risk of losing their home.

8   More and more women want to raise children while remaining in full-time or part-time work – partly because their families need the money, partly because work can provide company, stimulation and other life-enhancing things. Working mothers will generally want – or be under pressure – to work as close as possible to their homes and their children's schools. That proximity can best be obtained by living in a city or a town.

Distant, dispersed suburbs were the creations of a male-dominated society; they were meant to be pleasant, safe places where women and children could be left while men went to work in town. Men can afford to waste hours in commuting when they take the lesser role in childcare and housework. Women can't, and nor can men who provide their fair share of family support.

9   Every large town and city offers some superb environments, some places where things are manifestly working well. In Manchester, the city centre's resident population grew twenty-fold from some 300 in 1995 to around 6,000 in 2000 thanks to a boom in apartment building initiated by the public sector.[10] Newcastle-upon-Tyne has suffered industrial collapse, massive population loss and some of the worst housing abandonment in England; even so, its inner city still has Jesmond where a four-bedroom terrace house with a small garden can sell for around quarter of a million pounds. The same size of house and garden in pleasant countryside or a pretty village outside the city would fetch less. Here and elsewhere the market is telling us that the environment people are willing to pay most to live in is a high quality urban one – even if the city as a whole has colossal problems.

Other UK cities have demonstrated that they can revive their centres, see off the competition from out-of-town development, and be places of beauty and excitement. Even Birmingham can; if you do not believe me, suspend your prejudices and spend a day there. The question now is whether the revival can spread outwards from the centre.

10 The tenth, and last, opportunity has often been portrayed as a threat to city living; this is the rapid growth in information technology and electronic communication. The suggestion is that in the near future hordes of office workers will escape the agony of commuting and retreat to rural telecottages. There they will labour using Internet, fax and phone, either as salarymen and women or as e-lancers. More and more people will shop, be entertained and work in their homes using personal computers, cable links and digital television. About 1.3

million Britons, one in twenty of the working population, were doing at least some of their work in this way in 1999, according to one survey.[11] People will be able to spend more of their lives at home, and have less need to live close to the facilities and workplaces traditionally concentrated in cities. A large American study has found that as people's Internet use grows, they spend less time with friends and family and in the world outdoors. They also tend to spend more time working from home without cutting back their hours in the office.[12] 'The Internet could be the ultimate isolating technology that further reduces our participation in communities even more than television did before it,' said one of its authors, Stanford Professor Norman Nie.

But it does not follow from this that people will actually want to spend more time at home or aspire to live in less urban environments. Rising levels of wealth and education enable and encourage us to travel more, to eat out more, to seek entertainment outside the home. And most of us crave company beyond our immediate family. The friendships, the gossip, the acts of creation, negotiation, and competition that we love all depend on face-to-face encounters. Which is why industry has invested heavily, and perhaps prematurely, in Internet access outside the home, via portable, wireless devices – the mobile phone and its descendants.

Even if the revolution in information processing and communication weakens the need for us to live in cities, it may still incline us towards urban living. Workers who spend long hours on line will want to swap screens and e-mail for bustle and real conversation. Cities, with their variety, activity and dense populations, provide both in abundance. If you are working on a computer alone, from home, you may want to live in surroundings that are livelier than those that a village or low-density suburb can provide.

Some proponents of telecommuting have suggested that many people who could work from home using electronic communication will actually prefer to work in small, local centres with a few other people in similar circumstances just to have some human company. Some have already been set up. But it will surely be easier to found or to join one of these centres within a short distance of your home if you live in a town rather than deep in the countryside. In the city the density of people in the market for such a place will be much higher.

The long, strong tendency of companies making the same kind of product to cluster in one place is, some say, another reason to believe the expanding knowledge economy will favour cities. Clustering gives employers a greater choice of the skilled, trained workers their particular industry needs whilst giving those workers more employers to choose from. Clustering also fosters networks of suppliers and specialised services and it aids the exchange of information and ideas. In the knowledge economy highly skilled employees and the swapping of ideas matter more than ever; hence a stronger tendency to cluster. And the weightlessness of the product favours this tendency still further. High costs of transporting finished goods to market will tend to oppose clustering, but in the knowledge economy these shipping costs are very low indeed with much of the goods (information, entertainment, advice) flowing, as electrons and photons, down copper wires and fibre–optic cables.[13]

Clustering might, therefore, be expected to increase. But in the absence of pro–urban, anti–sprawl policies, the clusters may well form along greenfield sites beside motorways rather than within cities. Silicon Valley in California and Route 128 in Massachusetts, the USA's two leading areas for computing and software production, are not within big, traditional cities.* Nor are the cluster of IT-based industries dotted along Britain's M4 motorway (although, that said, the small, new, Internet-based enterprises springing up at the end of the 1990s appeared to favour more urban locations).

There may come a time when digital electronics can satisfy our need for face-to-face encounters. We will no longer need to be in the same room as someone else in order to impress, moan, gossip, plot or feel useful. When that happens, one of the main reasons for having cities will disappear. One way computers will do this is by becoming extremely good mimics of humans. Another is through advances in communications that will make two people feel they are together – observing each other's faces and body language in the closest detail – when they are many miles apart. I'm assuming these advances will not come onto the mass market for at least 20 years. Mere videophones won't do.

Finally, digital electronics may make life much more difficult for criminals, as we shall see in Chapter 11. But the spaces in which such security systems work best will be urban and suburban ones.

To sum up, information technology and the knowledge economy may favour cities. The arguments that they will are about as persuasive as the arguments that these momentous changes will promote further dispersal. And that, for any optimist, is an opportunity.

So much for the new opportunities which could help to bring an urban renaissance. If they are to be seized, then other things will have to happen. Before we come to what these might be, we need to look more closely at why most people with choices choose to shun cities.

Chapter 7

PUSHES AND PULLS

Lower class of people moving in when other people moved out . . . We wanted a better education for our children. We wanted a better area to bring up our children.

The area had deteriorated with people who did not have the same standards as myself.

Totally dissatisfied – dirty, dilapidated, unsafe, uncaring, no respect.

Residents of Hayes, an outer London suburb,
commenting on the neighbourhoods they had lived in previously

THE MESSAGE FOR THE PRESENT from our potted urban history is that powerful, entrenched forces are at work. Without big shifts in people's attitudes and further changes in government policies it is likely that cities and city living will continue their decline. There may be new opportunities, things may be moving in the opposite direction in places, but the main trend is still for a destructive exodus. What changes might halt this? To attempt an answer, we need to know why most people dislike, even loathe, cities today.

There are two ways of approaching the question. The first involves universal, hand-waving explanations that embrace culture, psychology and evolution. These are interesting and endlessly debatable but of limited use in telling us what we can do now to improve urban life. The second, more prosaic and useful approach, involves breaking down people's objections to cities into smaller, more manageable components.

Let's have the big, hand-waving explanations first, beginning with 'Size matters'. People are troubled by the idea of hundreds of thousands of strangers living in one place, never knowing each other, not caring for each other. These themes of loneliness and anonymity in the urban crowd, the inspiration for so much great art and literature in the past two centuries, lie behind the recurring image of the city as an ant heap, a place of uncountable numbers of inhuman beings forever scurrying about their own business, oblivious of the observer.

The notion of cities being inherently too large to be comfortable can be made to fit with the speculations of evolutionary psychology, which treat the human mind as a computer produced by Darwinian evolution. *Homo sapiens*, the thinking goes, is adapted for living in hunter–gatherer bands of a few dozen people, many of whose members will be kin. In these tight, nomadic societies the great majority of interactions involved people who already knew each other well and were often blood relations. Strangers were treated with great caution if not outright hostility.* Much the same could be said of the village societies which replaced hunter–gatherer groups after the agricultural revolution.

Around the globe, half of humanity has now abandoned those ways of life and crowded together into cities because the benefits of urban living outweigh its high costs. Urbanisation was an essential part of building an industrial civilisation which

*But we might also expect people to be adapted to dealing peaceably and cooperatively with strangers too, for the purposes of trade (which goes deep into prehistory) and gaining mates from other groups.

**How does the population split between urban and rural living? It depends on definitions. If 'rural' means only small villages or isolated homes in the countryside, then around 90 per cent of England's population is

gave people longer lives, wider choices and more education than hunting, gathering and farming ever could – even if huge numbers of the urban poor were denied these improvements for generations and still are. Nonetheless, the fundamental unease that comes from living among a huge mass of strangers remains.

Many people will find the proposition that we are somehow anti-urban by nature daft. They rejoice at the way in which cities bring together huge numbers of strangers, contrasting it with the social claustrophobia of the village where everyone knows everyone else's business. They welcome the spectrums and fusions of cultures and cuisines that result from waves of immigration into large cities. They like being able to choose friends and collaborators from among the large number of people they encounter while keeping the option of having very little to do with their immediate neighbours.

But I believe most people do find something disconcerting about living in huge, built-up areas among hundreds of thousands of strangers. Suburbanisation and counter-urbanisation point to this, and so does plenty of opinion polling. One survey, carried out for the government's Countryside Commission, found that while less than a quarter of the population resides in the countryside or a village, 54 per cent of all adults wanted to live there while only 6 per cent would prefer a home in the inner city.[1] Focusing on the majority of the population who now live in towns and cities, the poll found that 45 per cent of them wanted to live in the countryside. Among people living in suburbia, 51 per cent would prefer to reside in rural surroundings or a smaller town. The commission's researchers divided their sample into people living in the countryside, in smaller towns, suburbia and inner cities; only in the first of these four groups did a majority prefer to live in the environment it was already in. In the other three, the majority wanted out, and it was mostly out of town.[**]

The would-be urban emigrants tend to believe that the smaller towns and rural communities they have set their hearts on are kinder, less stressful and more closely-knit places with a stronger sense of community.[2] They are looking for a change in the social, as well as in the physical, environment. The village is idealised by millions as a place where everyone is known and where the class struggle is suspended; differences in wealth do not cause the same envy and tension as they do in cities. Real villages may not be at all like that but the dream lives on.

Another big, hand-waving explanation for our dislike of cities invokes national culture and national identity. Continentals may be happy living close together in cities; we British cannot (although you could make an exception for the Scots, whose tenements appear so strikingly European to the southerner). The cleverest, most widely read explanation of why this should be so comes in *English Culture and the Decline of the Industrial Spirit 1850 to 1980* by Martin Wiener, an American

urban. If any town with a population less than 10,000 is considered rural, then 80 per cent of the population is urban. See Urban Task Force, *Towards an Urban Renaissance*, London, E. & F.N. Spon, 1999, p. 29.

**Left behind:
Drumchapel, Glasgow,
overleaf**

historian.[3] He was primarily concerned with explaining Britain's relative economic and industrial decline through the twentieth century, but he believed our rejection of the urban has the same root.

In the nineteenth century, he argued, the industrial revolution created a class of capitalist entrepreneurs who, with their increasing wealth and confidence, challenged the ruling class of the landed aristocracy. But conflict and crisis between the old and the new elites were avoided because the rising bourgeoisie absorbed the gentry values of those it might otherwise have ousted. Instead of concentrating on further industrial expansion, many of the sons of the Victorian capitalists aped landed gentlemen, getting out of trade and into politics or the professions, buying rural estates, taking up country sports. The aristocracy succeeded in making a large part of the bourgeoisie distance itself from, even disdain, cities, factories and wealth creation whilst infecting them with an attachment to the countryside and rural pastimes. Revolution was avoided. The *ancien régime* retained power and influence while partially merging with the wealthy entrepreneurs.

This settlement shaped Britain's twentieth century. 'The nation that had been the mother of the industrial revolution was now uneasy with its offspring,' wrote Wiener. Green Belts, new towns, our love of gardening and 'a wariness of most modern architecture' all stem from an anti-industrial culture spreading down the social scale. 'Having pioneered urbanisation, the English ignored or disparaged cities.' The Victorian public schools were mostly sited well outside the big conurbations and the same was true of the plate-glass universities of the 1960s' higher education expansion. Today rich men, from the former deputy prime minister Michael Heseltine all the way down to the pornographer David Sullivan, still signal their wealth by acquiring or building great country mansions, while the most senior members of Cabinet are loaned them as a job perk. Among people of more ordinary means this city-shunning instinct manifests itself in a desire to own a cottage in a village or the countryside.

Britain certainly does have an anti-urban culture with strong and deep roots but there is really no need to explain it as a matter of attitudes and fashions trickling down from the upper classes to the masses. People take cities as they find them and, across all classes, generation after generation has found them pretty unpleasant (although it has to be said that elites have always been rather good at coping with city life, with many of them able to afford a town house in a pleasant neighbourhood as well as a place in the country).

But whatever the explanation, we cannot say we are culturally or genetically inclined to hate cities and leave it at that. If our culture is anti-urban, then it has to change because we are among the world's most urbanised nations and will continue to be. We have to make cities work.

To argue that there is a particularly British – or Anglo-Saxon – anti-urban prejudice also misses a more interesting and important point. Britain is more like than unlike other advanced industrialised countries in respect of the main urban trends.

While some mainland European cities retain much higher densities of people living in their cores than their UK counterparts, all Western nations have seen extensive suburbanisation and most have experienced counter-urbanisation. Most have also faced problems of inner city decline linked to poorer people being left behind by the exodus of wealth and talent. And most married couples with children in rich Western nations now live in a suburban, owner-occupied house rather than rented flats.[4] Which suggests that our British urban distaste is universal; big cities don't satisfy most people, for most of their lives.

What should we do? Continue to leave them, with those who have wealth and choices leading the charge? But that, as we have seen, is a debilitating and wasteful process that increases traffic pollution and congestion, obliterates countryside and makes our society more unequal and unjust. Imagine, too, what would happen if the majority of the population which says it would like to live in a village or the countryside achieved its ambition. It would destroy the things it loved. The hamlets would all have to grow into towns or the entire countryside become a new kind of suburbia; endless clusters of housing with narrow, token ribbons of countryside separating them, criss-crossed by crowded roads.

There remains our original solution to the urban crisis, the suburb. Could we not just keep on decreasing housing densities so that everyone lives in a suburban environment? We are already in sight of that goal, for only 15 per cent of Britons now reside in places with inner city levels of population density (grossing out at 50 people per hectare or above).[5] We are an overwhelmingly urban society mostly living at suburban densities; the average Englishman or woman dwells in surroundings of 20 people per hectare.[6] We would just have to thin out the remaining inner city housing stock, replacing terraces and council blocks with semis and detached houses while continuing to build the millions of extra homes needed on greenfield sites at the lowish densities favoured by the bulk of developers. Huge areas of woods and fields would disappear and we would have to make longer journeys for work, play and shopping within settlements which were much more diffuse. Even so, Britain would be completely suburbanised and content.

Would it? Where would the process end? People usually aspire to more space than the amount they currently have. If they can afford it, they generally purchase more and more room on their journey through life, moving from flat or terraced home to larger and larger properties in a quest for privacy, autonomy and status. Only retirement, or the children leaving home, or divorce, or the death of a partner, incline people to reduce their claim on space and land and move into smaller homes.* Families that become seriously affluent usually buy large houses with ample grounds, or farms, or second homes in Britain and abroad. Some 400,000 English households own second homes for their own use, as opposed to renting

*Although divorce often enlarges a broken family's total requirement for space and land – as when the mother and children remain in the old family home, while the father takes a flat or house with several bedrooms so that the children can visit for weekends.

them out, and most of these are not abroad. According to the Council of Mortgage Lenders, owning two properties is no longer the preserve of the rich.[7] As we grow wealthier we want more room; even if we were able to suburbanise all of Britain's cities, millions of households would still be left wanting more land and space. Ultimately, there is only enough land in all England for each family to have one and a half acres, and that would leave no room for farming or anything else.*

But living room is not the only issue. The dissatisfaction with large cities reaches out from the high-density inner areas into their suburbs; these, too, can become tired and dilapidated as people and their money move on.[8] A hierarchy emerges. The inner city of terraced streets and council blocks near the heart of a large city is, generally, the least desirable place to live. The most prized environments are completely outside the conurbations, in medium-sized and small towns, villages, or the countryside. In between lies the varied, complex landscape of twentieth-century suburbia around the larger towns and cities. Of course this is an over-simplification. Prosperous small towns often have decaying, stigmatised council estates where no one would choose to live; indeed, there are some small towns such as Peterlee in Durham and Barrow-in-Furness in Cumbria devastated by industrial decline where deprivation is deeper and wider than in most cities.[9] And big cities usually retain some fine Georgian and Victorian quarters near their centres which remain upmarket, while London is a special case. But these are exceptions; the urban to rural hierarchy exists across most of Britain. We sense it intuitively as we look around towns and cities, searching for clues on whether a neighbourhood is good, bad or indifferent. We pick it up from friends and colleagues and imbibe it from the media.

Two research projects have illuminated this hierarchy. Robert Rogerson, a geographer at Glasgow's University of Strathclyde, has been investigating the quality of life around Britain for more than a decade.[10] He has tried to establish what the average British adult wants from her or his community and has also attempted an objective assessment of how good 189 towns and cities are at delivering those attributes. The smallest places and the five largest conurbations – London, Birmingham, Manchester, Liverpool and Glasgow – were excluded from his analysis. Rogerson began by asking a representative cross-section of more than 2,000 adults what was most important for their quality of life. This opinion survey gave him a list of sixteen key attributes which included such things as crime rates, local health and school provision, housing costs, employment prospects, shopping facilities and travel to work times. He also ranked these attributes according to how important people thought they were. The rate of violent crime was given the top priority with 72 per

---

*The more land there is available, the bigger houses become. Average income family homes in Australia and the USA are one and a half times larger than those in Japan, despite all three nations having roughly equivalent per capita GDP. Japan's population density is more than ten times higher than the other two. See *Is the UK Different? International Comparisons of Tenure Patterns*, London, Council of Mortgage Lenders, 1996, p. 30.

cent rating it 'very important' while leisure opportunities were given the lowest.

The next step was to score the 189 towns and cities for how they performed on each of these attributes. Rogerson and his colleagues used the best available data from local councils and other official and commercial sources. The police, for instance, supplied crime statistics while estate agents gave information on house prices. Then the score for each attribute was adjusted to take account of how much importance the average person attached to it; the mark for violent crime rates was given the heaviest weight, that for leisure opportunities the least. Finally, the sixteen marks were aggregated to give each of the 189 towns and cities a grand total score representing the quality of life it offered.

The upshot of all this wrestling with data and indicators was great woe and shame for Nottingham which came bottom. Hull was in second worst place while the little market town of Dumfries in south-west Scotland notched up the highest score. The largest cities – industrial centres and seaports which had mushroomed in the nineteenth century – were almost all to be found in the bottom half of the league. The lower you go in the league, the more concentrated they become. The only really sizeable places found in the top quarter are Edinburgh and Aberdeen, both of them rather untypical.** The upper echelon is dominated by small to medium-sized market and county towns that have grown rapidly during the post-war decades and are within easy reach of large cities and towns.

The second investigation which illuminates the hierarchy of towns, cities and residential areas used a fearsomely clever statistical technique to estimate how discontented the English were with their surroundings. Roger Burrows and David Rhodes of the University of York have devised a method for predicting the degree of local dissatisfaction in every council area in England and in almost every electoral ward (there are over 8,600 of these wards and they each contain around 6,000 residents).[11] They wanted to contribute to an intense debate buzzing around academia, pressure groups and government; how should the most deprived areas of the country be identified in order to attack poverty most effectively? The usual practice is to create a deprivation index which combines such variables as the local unemployment rate, what proportion of households have no car, are outside owner occupation, are headed by lone parents. The places that score the highest on these indices are then highlighted on maps, usually in strong shades of red.

Burrows and Rhodes set out to complement the growing number of poverty maps with information about the places where people were, on average, most discontented with their surroundings. Since surveying a representative sample of people in every ward was prohibitively expensive, they devised a way of making credible predictions. Let me briefly explain how.

**One a capital city and cultural centre with a large number of professionals, senior administrators and financial service workers, the other benefiting from two decades of oil wealth.

Their starting point was a very large housing survey carried out by the government's Department of the Environment in 1994 and 1995. Among the questions asked of 20,000 households were more than a dozen which probed how satisfied or dissatisfied people were with their surroundings. Were there problems with graffiti, litter, crime, vandalism and noise? How good or bad was your area for schools and public transport? Rhodes and Burrows decided that any householder answering that she or he was seriously dissatisfied to four or more questions was severely discontented with their neighbourhood. About one in ten English households fell into this category.

Next, they found out how all of this unhappiness was distributed around the regions of England and among different categories of people. They could do that because the survey recorded the social class, tenure, employment status and race of the householders along with several other factors. The lowest levels of dissatisfaction were to be found among home-owners (7 per cent), people of the higher social classes (4.5 per cent for professionals and senior managers), those living in rural areas (7 per cent), in detached houses (4 per cent) and in full-time employment (8 per cent). The most discontent existed among those in the lower social classes (14 per cent for unskilled manual workers), single parents (22 per cent), unemployed people (21 per cent), those living in council or housing association properties (18 per cent), in flats (15 per cent) and in urban areas (10 per cent).

There was, then, a strong link between prosperity and satisfaction with one's surroundings. Why should this be? Is it because poorer people are more likely to notice and complain about problems in their area than the better off? That seems unlikely. A more plausible explanation for the difference is that the former are living in housing and in areas which really do have more problems than the latter.

Burrows and Rhodes analysed the survey data to pick out which of all the differences between householders best explained their widely varying levels of dissatisfaction. It turned out that discontent was most strongly linked with whether you owned your own home or lived in social housing or private rented property. Using a statistical technique called segmentation modelling, the University of York researchers sorted the householders into eighteen distinct categories which provided the best purchase on the varying levels of discontent.

Each of these groups had its own average level of discontent. The largest of these, accounting for just over a fifth of English households, consisted of employed or retired owner-occupiers living in semi-detached homes. They were relatively happy; only 4 per cent expressed serious discontent with where they lived. The unhappiest group of all consisted of unemployed people living in social housing in North-East England, with close to half of them – 42 per cent – seriously dissatisfied with the area they lived in.

Now comes the ingenious finale; the construction of what Burrows and Rhodes call a 'geography of misery' covering all England. They used the returns for the 1991 Census to divide the residents of almost every ward in England among their eighteen mini-groups. The Census could tell them, for example, how many retired and employed people there are living in semi-detached houses in any one ward (with the inevitable caveat that things may have changed a little since Census night). Since they knew the average level of discontent for each of these eighteen groups, they could estimate what proportion of people in each place – down to the level of the individual ward of a few thousand residents and up to the level of cities and counties with more than a million – was unhappy with its surroundings.

Taking the forty-seven larger, county level areas of England, the most satisfied place of all was affluent Surrey, closely followed by Dorset. In both, less than 6 per cent of households were projected to be unhappy with their surroundings. Unhappiest of all was Inner London followed by the Newcastle upon Tyne conurbation (just over and under 15 per cent, respectively). Raise the scale of magnification for the map of misery and look at England's 366 local authority districts; you find that four inner London boroughs – Tower Hamlets, Islington, Hackney and Southwark, in that order – are top of the unhappiness league. Indeed, all but three of the thirty-one inner and outer London boroughs are near the top of the table, with levels of dissatisfaction above the national average of 9.7 per cent. So are all but five of the thirty-five districts which compose the cores and suburbs of England's seven largest conurbations.

East Dorset, Wokingham, Surrey Heath and Christchurch lie at the other end of the league, with only 5 per cent of their residents discontented. The greatest dissatisfaction festers in the inner areas of the largest cities and in smaller, separate cities that have suffered from industrial decline, such as those in coalfield and textile areas. The most content places ooze prosperity and have had rapid population growth – the small and medium-sized towns near the outer edge of London's Green Belt are typical specimens. Suburbia lies in between these extremes.

Burrows and Rhodes found a close match between their estimates of misery in individual wards and the local poverty scores produced by three different indices of local deprivation. One could be forgiven, then, for seeing their league table, for all the number crunching and cleverness which went into its production, as a statement of the painfully obvious. Poor people are less happy. But that, I think, would be a mistake. If, as I argued above, we accept that people are reasonably impartial detectors of and complainers about problems in their area, whatever their level of income, then the University of York researchers have produced a rough and ready guide to the quality and attractiveness of all of England's urban environments.*

The two league tables – quality of life from Strathclyde and levels of discontent from York – have a great deal in common. Although Rogerson's analysis excluded Britain's largest cities and smallest towns, while Burrows' and Rhodes' looked only at English places, they share 160 local council districts in common. There are one

*In fact, this approach may underestimate the gap in quality between the best and worst areas. Wealthier households might well be inclined to voice greater discontent about neighbourhood problems than poorer ones.

or two glaring disagreements over rank (Strathclyde puts Kendal, Hereford and York very close to the top of its table; York University places them in its bottom half) but by and large the two leagues tally reasonably well. Their sumps are crowded with big towns associated with manufacturing, often bygone manufacturing, and ports. Both the top twenty and bottom twenty places of the two leagues have six names in common:

In the top twenty of both leagues	In the bottom twenty
Wokingham	Leeds
Fareham	Barnsley
New Forest	Sheffield
Eastleigh	Stockton on Tees
Horsham	Hull
Reigate and Bansted	Middlesbrough

Note: based on 160 districts common to both.

If you wanted to depict Britain as a nation deeply divided – between poverty and wealth, manufacturing decline and service industry growth, north and south – then these places would give you as strong a set of contrasts as you could ever hope for. The league toppers all have populations that are growing as fast as or faster than the national average growth (just under 3 per cent per decade). But three of the six from the bottom of the table have shrinking populations, while the remainder are growing at well below the national growth rate.[12]

There is, then, a strong, clear hierarchy of urban desirability and success. Size does matter (small seems generally more beautiful) and so does poverty. Concentrations of urban deprivation and discontent drive those with choices away, apart from in the special case of inner London, which combines high levels of discontent and social polarisation with the highest house prices. We can't do much about the size of cities apart from carrying on with the sorry business of allowing them to fail and depopulate. As for urban poverty, it seems unlikely – as I argue in the next chapter – that it could ever be abolished simply by redistribution, by spending more on job creation, education, training and social housing in deprived areas.

The solutions to urban problems will lie in breaking people's discontents and fears about cities into lumps and seeing what can be done about each of them. What are the things they object to most and what are they looking for when they consider moving? There has been a mass of opinion research in this area. 'For anyone seeking to promote urban repopulation, this work makes depressing reading since it suggests that anti-urban sentiments in the British public remain as strong as ever,' said a report to the Rogers Urban Task Force.[13] The key urban hates, expressed in survey after survey, are high levels of crime, especially violent crime, bad schools, heavy road traffic and the noise and pollution which attend it, and a lack of greenery and open space. People want to escape from dilapidation and incivilities – graffiti, litter, rowdiness, out-in-the-open drunkenness and drug dealing. They do not like living close to places that are often noisy and crowded.

Two other anti-urban prejudices need to be added to this list, expressed by fewer people but not few enough to be ignored. Car drivers want to park their vehicles close to their homes, preferably right outside or off-street and on their own property. In high density, inner urban areas that becomes difficult or impossible. And then there is racism. A small proportion say they want to leave or have left inner city areas because there are too many people from ethnic minorities.

But people are not only pushed into moving house by objections to their current circumstances and surroundings; they are pulled towards better places. There has been much debate about whether the urban push or the rural pull matters more in the exodus from the cities, with some pessimists concluding that, however much effort is poured into improving the cities, the British will always prefer to live in the countryside. Opinion surveys show, however, that towns and city living exert their own strong pulls. People put living near frequently served bus stops and train stations, being close to schools, shops, their workplaces and leisure facilities such as cinemas high on their list of priorities. Which is why people are moving into cities all the time, even if there is a faster flow in the other direction.

Like many journalists, I'm somewhat sceptical about opinion polls (while fully satisfied, naturally, with the integrity of all of the ones referred to in this book). We are no experts on sociological research but we know that people's answers depend on how you put the questions. We have written too many flimsy stories hooked from the river of opinion polls and surveys flowing into newsrooms. Most of them are commissioned by firms and pressure groups wishing to plug their own preconceptions, products and campaigns. I needed to be convinced about the depths of our dislike of cities, so I carried out my own opinion research around our (then) home in Hayes, Bromley, a chunk of archetypal inter-war suburbia on the furthest edge of the metropolis, using a questionnaire which tried to avoid putting my own preconceptions into people's answers.[14]

My little poll (more than 200 questionnaires returned and a gratifyingly high response rate of 60 per cent) mirrored the findings of other larger, more professional soundings. I focused on those households that had moved into Hayes from elsewhere in Bromley and Greater London during the past 20 years; they make up about half the neighbourhood's population. People had come mainly because of its greenery, open spaces and nearness to the countryside and the high reputation of the local state schools. A substantial proportion claimed Hayes was like a village (indeed, it once was, but that all finished in the 1930s once Henry Boot and other housebuilders had put up hundreds and hundreds of semis in the adjacent country-side). My neighbours wanted to get away from places that they described as 'overcrowded' or 'too built up'. They had left areas which, they complained, had too much crime and congestion and where they lacked confidence in the local school. All these fears were expressed more frequently among those who had moved out from inner London boroughs, but they were also to be found among those who had moved from other parts of Bromley which they perceived as being in decline.

A childminder who had come from Southwark wrote:

> Next door neighbour mugged at end of road, policeman stabbed in local sweet shop . . . on the day I received this survey I had been for a ride on my bike on the lanes, played badminton in the village hall, popped into Bromley for the shops . . . my children can walk to friends and school.

That seemed to sum up the attitudes and contentment – or, if you're unkind, the smugness – of many fugitives from the inner city. But this suburban bliss was not universal. A small proportion of households, most of them elderly, feared urban ills were invading Hayes. They complained about the growth in road traffic, the decline of its own little shopping centre (hit by the opening of superstores nearby) and, most of all, about groups of teenagers hanging around there in the evening. 'We would not dream of going out these days late in the evening because of the risk,' wrote a retired couple.

But the great majority of my sample had no plans to leave Hayes for the fore-seeable future. Of those who were contemplating a move in the next few years, three out of five intended to head out of London for the countryside, a village or small town. My survey encountered two households that were considering moving closer to central London. One of them was my own.

Hayes cannot stand for the entire nation and I will soon stop telling you what it thinks and wants. Dominated as it is by families with children and retired couples, lacking in non-white faces (just 4 per cent of its inhabitants were from ethnic minorities at the 1991 Census; for London as a whole the proportion was more than a fifth) and rock-solidly Conservative, it is hardly typical of modern Britain. And yet, when I considered what I found alongside all of the other opinion research I was reading, I had to conclude that it gives most people pretty much what they want from their home surroundings for a large part of their life. It left me in no doubt about the national distaste for large cities.

Urban revivalists, myself included, want to believe that large numbers of people are disinclined to live in low-density residential areas like Hayes. Many people in their twenties, for instance, would probably not find life in such a place satisfying. In fact 16 to 24 year-olds are one of the few groups that do defy the 'counter-urbanisation cascade' mentioned in Chapter 5; they generally move from smaller to larger places for work and to go to college.[15] And, in the special case of London, the upper middle classes would usually prefer to live in older property in a prosperous street of an inner borough, closer to the shops, workplaces, and high culture of the city centre, rather than a place like Hayes.

But there is a much larger, fast-growing group which has been portrayed as naturally urban: single, childless households – the never-married, the divorced and the widowed. This group is expanding because of our lengthening lifespans, high divorce rate and a low marriage rate. About 80 per cent of the growth in the number of households up to 2020 is accounted for by growth in the number of single person households. Could this group spearhead a return to the big cities? The lack of companionship and family meals at home might incline singles to living in the city where it is easier to socialise, to get to restaurants and pubs, to shop for one at the end of a working day. Nor do they need suburban houses with several bedrooms. If the rising tide of singles could be accommodated in compact flats, they could be fitted inside the towns and cities, bringing in life and money and sparing the woods and fields from endless new estates.

Alas, most singles are no more naturally urban than couples.[16] They prefer houses with two or more bedrooms because they like the extra space and being able to put up friends and relatives. They like gardens, too. And most of them have the same negative image of city living – crime, incivility, congestion, over-development, and so on – as families do. Nonetheless, the fact that they don't have children (or, in the case of divorced fathers, no children living with them permanently), may incline many of them towards urban living. There could be a very substantial and under-exploited market here (see text below).

SINGLES IN THE CITY

Two kinds of people tend to live in city centres and inner cities – many poor and much fewer of the prosperous. The Joseph Rowntree Foundation believes these urban areas badly need middle incomers to come in. It has backed its convictions to the tune of £6 million by building two striking, stylish but highly economical apartment blocks for rent at market prices.

The Foundation considers that there is a large and largely untapped demand from working people earning £15,000 to £30,000 a year for rented housing in or near city centres. They earn too much to qualify for social housing (and would probably reject it, in any case) but they wouldn't, or couldn't, pay the high prices which obtain in the owner-occupied market for fancy city centre lofts and apartments.

The Foundation has unmarried, childless people in their twenties and thirties in mind and also divorced fathers. The fact that these households have no children living with them permanently means that shortages of good state schools, private gardens, greenery, and safe play areas deep inside the big city do not put them off living there.

They either don't want a mortgage, preferring to spend their money on other things, or would struggle to afford one for the kind of home they want. They value being close to their workplaces and the after-work attractions such as pubs, restaurants and cinemas.

The Foundation commissioned research into what men and women in these categories wanted from a city centre home, then set out to find the land, the architects and builders who could deliver it at the right price. The aim was to have rents that middle incomers would be willing to pay while providing a net yield on the investment of more than 6 per cent; enough, hopefully, to attract big investors for future projects along the same lines. The two CASPAR (City-centre Apartments for Single People at Affordable Rents) blocks, just off the centres of Birmingham and Leeds, were completed and occupied in 2000 after construction consortia had been chosen through a design and build competition. The average rents for the flats in both blocks were set at about £110 a week (those with the best views pay a little more), enough to provide a yield of just over 6 per cent.

The flats — 44 in Birmingham, 45 in Leeds — mostly have their own balconies and come with carpets, curtains and appliances. They are roomy enough for a couple to cohabit in, or for a single to have a friend or a couple of children stay the night. Both blocks offer secure communal entrances as well as secure car parking on site — one space per flat. There is a charge for this which tenants without cars avoid.

The two buildings have several other things in common; total cost close to £3 million, five stories, both built on compact, undistinguished and not particularly appealing sites that used to be used for car parking. The West Midlands block was the first to be completed; it sits beside the Birmingham and Fazeley Canal in the regenerating Jewellery Quarter. The Leeds building, a large semicircle with the flats facing into its core, sits next to the city's inner ring road. It has a timber frame, timber-clad walls and a roof of corrugated sheet steel. The apartments were built in a factory and lifted into place by crane, either as entire modules or flat pack room dividers. As well as saving time this advanced construction system allowed for high levels of insulation including triple-glazed windows, making for very low energy bills for the tenants.

What next? The Foundation hopes that its new buildings will help to regenerate their immediate surroundings and that their value will therefore rise, making them an even better investment. Its larger ambition is to encourage the big pension funds and insurance companies, which are already huge investors in commercial property such as shopping centres and office blocks, to move into residential property and finance hundreds or thousands of CASPAR-style apartment blocks for rent.

In Britain (but not in continental Europe) the big property players hardly touch the rented housing sector and there has been much debate about what inducements and changes in taxation might encourage them to do so. By treating the land purchase, design, construction and letting of the two blocks as a totally commercial venture, the Joseph Rowntree Foundation is hoping to demonstrate that a sound investment opportunity already exists right under their noses in Britain's big cities. But the real point, says the Foundation's director Richard Best, is to bring in tens of thousands of people who would otherwise have rejected urban living, thereby giving the inner cities a nourishing, regenerating spread of incomes and skills.

Cities do have their own assets and pulling power. And recall that their pushes, their failures, can be broken down into individual components. Most people do not suddenly decide that they are finished with city life and move out. They give the matter a great deal of thought, weigh up the pros and cons, before they decide when and where to go. The main motive behind many changes of address is not to move to a better area but to live in more suitable housing. If the right kind of property is available and affordable near to their existing home, people will often take it because they like to be in familiar surroundings and want to stay close to relatives and friends. Which is why most house moves only cover a few miles.* Familiarity and family are powerful factors helping cities to retain people.

So anything which noticeably improves some aspect of neighbourhood – or an entire city – will make a difference to the sum of those location decisions. Some people will postpone their departure for years, or indefinitely, or move to another house nearby. Any improvement is likely to bring in some newcomers who would otherwise have shunned the area. The counter-urbanisation cascade is a two-way street, with more people moving out than in. If the outward flow from the conurbations was reduced by a fifth and the inward flow from smaller towns and the countryside increased by a fifth, the net outflow would cease.[17]

If a secondary school improves its GCSE examination results that should have an effect. Any measures that make people feel more safe from crime would help. Better bus, train, tube and tram services within the cities will retain some people who would otherwise leave. These are covered in Chapters 10, 11 and 14.

But an urban renaissance will require cities to offer some of the pulls of small towns and the countryside. They need to be scattered with swathes of meadows, woods, lakes and streams where, within a minute's walk, you can escape from right angles, hard surfaces, crowds and traffic. They need more parkland, and it would be a colossal blunder to lose any useable public greenery for the sake of providing more housing in the city.

Cities have to give their residents the option of feeling they are part of a place and a community. The best way they can do this is to provide more mixed use developments in which homes, workplaces, shops, schools and leisure facilities are all within a few minutes walk of each other and identified as belonging to a neigh-

*Half of all household moves covered a distance of less than three miles, according to one survey (see *How Far Do People Move House?*, Jarlath Costell, Housing Finance No. 5, London, Building Societies Association/Council of Mortgage Lenders, February 1990). Three in five households move less than six miles, according to the 1991 Census.

bourhood. Finally, cities have to give their citizens the things that only they can and that small towns and villages cannot – memorable, magnificent buildings and animated, atmospheric public spaces. Somehow all of these urban improvements must be tied in with increasing the supply of new homes within towns and cities without over-crowding them.

It can be done, but we need to be sure it's worth the bother. It will require fresh interventions by the state but, as previous chapters have shown, the state's good intentions have often had dismal outcomes. People must be persuaded to live in places they are not choosing to live in now. There are risks in this endeavour. Freedom of choice could be restricted. The springs of enterprise could be blocked.

So perhaps we should shrug our shoulders, turn away and tolerate urban decay, as the USA has tended to do. As cities and economies grow, a few people and neighbourhoods are bound to get left behind. It is one aspect of the creative destruction and unending obsolescence that underlie economic progress. Besides, urban decline can itself create opportunity and wealth. In the short term, it provides a convenient market place for some of our most obnoxious industries, such as sex and drugs, tucked away from where better-off folks live and work. In the longer term, as housing and workplaces spread outwards, inner city land becomes cheaper, allowing it to be reused or redeveloped in ways that the original builders might never have conceived of. Decay and blight enable cities to be flexible. They are, for instance, a prerequisite to gentrification which can bring in billions of pounds of investment and create many new jobs.

Fair enough. But surely the free play of the free market has to be constrained if the actions of self-interested actors impose grave costs and damages on society as a whole. The housebuilding industry meets the demand for more space and larger homes with continued suburban sprawl. That implies more roads, more travel and more pollution as well as the loss of countryside, all costs imposed on society. Disinvestment and decay within the cities add massively to the bill.

We can argue about whether suburbanisation and counter-urbanisation cause this urban rot or whether it is the other way round. I am convinced it is the former but the sequence of eggs and chickens come to mind; we are never going to prove which came first. They belong to a cycle and as long as we continue to invest in huge numbers of new homes and other developments in the countryside rather than spending the money within the cities the wheel will keep turning.

The strongest justification for intervention is that it will reduce poverty and increase equality of opportunity. But can we be sure it will?

Chapter 8

THE MILTON KEYNES EFFECT

The increasing isolation of the lower class is a problem, to be sure, but it is hard to see what can be done about it. The upper classes will continue to want to separate themselves physically from the lower, and in a free country they probably cannot be prevented from doing so.

Edward C. Banfield, *The Unheavenly City*[1]

For the poor always ye have with you . . .

St John's Gospel, Chapter 12, Verse 8

The next time someone leans back in their armchair and tells you that the poor are always with us, simply punch them.

Nick Davies, article in the *New Statesman*, 6 November, 1998

Milton Keynes is of enormous importance, not only for the London region but for the entire civilised world.

Steen Eiler Rasmussen, *London: The Unique City*[2]

UNTIL THE 1970S, the debate about declining cities concentrated on their physical environment. Sort the traffic problems out with new roads and public transport, clear slums and build decent housing to replace them, persuade industry and commerce to stay with new sites and facilities and all should be well. The cities would become civilised and attractive to all.

Now, however, it is clear that the social environment is as much, if not more, to blame as the physical for the unpopularity and disinvestment that plague cities. They have concentrated and thereby nourished deprivation. People with a choice about where to live don't wish to be troubled with poverty's companions – crime, drug abuse, low aspirations in schools and vandalism. Urban salvation rests in finding the most effective methods of reducing poverty. But before we can talk about methods, we need some agreement about its causes.

People compete for limited resources; there is a spectrum of outcomes between winning and losing. The political right views winners as winning because of their superior talents and energies. Those who end up in middling positions in the struggle are comfortable. Those who lose suffer relative, if not absolute, poverty. That may be because they are idle through choice or it may be misfortune – a matter of bad parenting or genes or circumstances – but whatever the cause, it is tolerable, in the right's sometimes covert opinion. For as well as winners and losers, the ceaseless competition and self-interested collaboration of individuals produce a flow of improvements, new enterprises and inventions. It expands markets, raises efficiencies and creates wealth. The rising tide of money lifts all the boats, including the smallest and poorest craft.

The political left sees those with wealth and power retaining them, from generation to generation, by using their inherited or acquired advantages to exploit those with less. Redistribution, public services and maybe even some common ownership are needed to create a just and sustainable society – one that will offer equality of opportunity, if not equality itself.

The right emphasises that people can move between social classes and points to the huge expansion of the middle class in the previous two centuries. It warns that the price of redistribution and common ownership is stifled enterprise and economic stagnation, while revolutionary attempts to establish equality invariably fail. Inequalities are quickly re-established. The left, for its part, stresses the rigidities of class, the continuing gross inequalities of wealth and opportunity and the waste of human potential among the poor.

After a century of political experiments and extremes it's clear that both sides are right about each other's wrongs. And so our democracy muddles along, somewhere in the middle. We try to harness and tame globalised capitalism with state education, state health care and a meagre income for the incapable and unemployed. Living standards rise fairly steadily yet we don't feel we are anywhere near to eradicating poverty. Some might claim we have, pointing out that the Victorian underclass had no televisions, but they are ignoring the relativities.

The notion of positional goods gives us a better grip on what poverty means today. Once people can satisfy their basic needs for food, shelter and warmth, they move on to acquire more and more goods and services. They do so partly in order to signal their wealth and status to society and to themselves. A couple of centuries ago only a small elite had the resource to play this game seriously, using such things as houses, carriages and servants. As wealth increased and spread, a growing proportion of the population could afford to join in. More and more items fall into this category – cars, gadgets, clothes, holidays, restaurant meals and, perhaps most important of all, addresses – precisely where you can afford to live.

But there has to be a finite supply of positional goods, as Frederick Hirsch pointed out in *The Social Limits to Growth*; if there weren't, they would fail to position people.[3] If most British households owned two cars and a detached home and took two foreign holidays a year (things which are now the preserve of a substantial minority), then these items would no longer be positional and the better-off half of the population would find alternatives that the poorer half could not afford. Commerce – in the shape of advertisers, marketeers, and entrepreneurs – would spot new markets and encourage them to do so. There will always be novel symbols of wealth and those with the least money will always be kept aware of their lowly position, even if overall living standards are rising.

Another important idea about long term, deep-seated poverty – one with a history, but enjoying a renaissance – is that its most important aspect is exclusion from the mainstream of society. Poor, unemployed people who live in deprived areas and depend on state benefits lack useful connections and networks that the rest of us take for granted. Because they are short of both work experience and contacts inside the world of work, they find it much harder to get jobs. Because their income is low and their address has a bad reputation, it is difficult for them to get credit. Potential employers may also reject them because they come from a bad neighbourhood. They face prejudice and feel powerless, they have low self-esteem and little stake in society. Lives are lived from day to day. Crime is an obvious way of gaining wealth, networks and esteem in the world of the socially excluded. Drugs and drink are means of escaping a hopeless reality.

But how do we separate causes and effects? Who is to blame for social exclusion – the excluded, or the rest of us? The American sociologist Edward C. Banfield argued that the main reason why people were poor was that they did not wish to work and had no ability to plan ahead; any income was spent immediately.

This is, of course, an old, old song but he gave a rather fine rendition; it was the bleakest analysis of urban poverty and urban crisis I came across in researching this

Town centre, Milton Keynes, Buckinghamshire

book. Banfield made an eloquent, careful case for *laissez-faire*, for coolly ignoring the social and physical rotting of great cities.[4] His time was the late 1960s, his place the United States of America, where many of the largest conurbations had suffered huge race riots. By then inner city abandonment, social collapse and 'white flight' to new suburbs were taking place on an awesome scale following the migration of millions of African Americans to the big cities from small towns and the countryside during the century's middle decades.

Banfield, a Harvard professor who advised the Nixon administration, argued that the most severe, intractable problems of the cities were due to the way they concentrated masses of 'lower class' people. For him, the chief definer of the social classes – upper, middle, working and lower – was their differing abilities to envision the future and their degree of willingness to make present sacrifices now to secure some future gain. The lower class, which provided most of the urban poor, made up somewhere between 10 and 20 per cent of the population and was 'radically improvident'. Its members lived from moment to moment, found a work routine difficult to settle into, suffered low self-esteem and felt they had no control over their chaotic and often violent lives.

Banfield considered this condition to be 'pathological' but he did not insist that it was an inherited disease, passed down through the genes; it was just as likely to be transmitted from generation to generation culturally through bad parenting and bad example. Whatever the reason for the persistence of a large lower class, it was almost impossible to change its deviant culture, which became fixed in its children at a very early age:

> So long as the city contains a sizeable lower class, nothing basic can be done about its most serious problems. Good jobs may be offered to all, but some will remain chronically unemployed. Slums may be demolished, but if the housing that replaces them is occupied by the lower class it will shortly be turned into new slums. Welfare payments may be doubled or even tripled . . . but some persons will continue to live in squalor and misery. New schools may be built, new curricula devised, and the teacher pupil ratio may be cut in half, but if the children who attend those schools come from lower class homes, they will be turned into blackboard jungles, and those who graduate or drop out from them will, in most cases, be functionally illiterate. The streets may be filled with armies of policemen, but violent crime and disorder will decrease very little.[5]

A mass movement of the affluent to new suburbs was inevitable at a time when both household numbers and incomes were rising and technology was giving people greater freedom to travel, he argued. By leaving their inner city homes for pastures new, the prosperous were helping to make affordable housing available for low-income incomers. The African Americans were only the latest in a long line of poor migrants moving into the cities; like their predecessors – the Irish, the Sicilians, the

Eastern Europeans – they had to start at the bottom, in the cheapest, oldest housing near the city's core. The professor conceded that they were, to some extent, the victims of prejudice, although 'much of what appears as race prejudice is really class prejudice or, at any rate, class antipathy'.[6] But prejudice alone could not explain why their neighbourhoods were plagued by high unemployment, poverty, disorder and crime; these things happened because a very high proportion of residents were lower class.

Banfield offered his sympathy to the 'normal' black working-class and middle-class people who found themselves living in such squalid and criminal quarters. They deserved better and many would eventually make their escape into more respectable, congenial neighbourhoods. He ended his book by considering what could realistically be done to tackle the urban crisis. Very little, he concluded. There were several feasible solutions; he listed a dozen including 'intensive birth control guidance to the incompetent poor', greater police powers to stop and search, and increased incentives for employers to take on the poor and unskilled (such as the removal of minimum wages). Regrettably, he wrote, most of these measures would prove politically unacceptable.

Today plenty of ordinary Britons look upon the residents of the worst neighbourhoods of their towns and cities in much the same way that Banfield saw his 'lower class'. Driving me to Bulwell, a deprived, declining and white Nottingham suburb, a pleasant, friendly taxi driver told me: 'They're a different breed of people up there.' That is the harsh reality of social exclusion; to be instantly judged not on the content of your character but by your address.

There is small proportion of people who would fit into the professor's category of 'lower class' and, frankly, most of us would not wish to live anywhere near them. But it is monstrously unjust and plain wrong to claim that the majority of people living in a poor neighbourhood are irredeemably feckless, criminally inclined, incapable of thinking ahead and making sacrifices for the future. It is hardly surprising that many lack ambition and live their lives from day to day. If you know your prospects are bleak, if you know your neighbours are all in the same boat (which seems to be sinking rather than being lifted by any rising tide), it is difficult to live any other sort of life. There is little in the way of reasons or role models to encourage you. And how can you save for the future when you have to spend all your income feeding your family and keeping the cold at bay in the present?

Poverty defines a bad neighbourhood, a sink estate, a slum. But how are these places created? Are we giving everyone a fair chance to succeed in life, then efficiently sorting out the losers   a fraction which has the least to contribute, the lowest earning power, the most anti-social habits – and herding them together away from the rest of us? Or is it the other way round? Are slums creating and perpetuating poverty that would not otherwise have existed? The people who grow up in them are denied their fair share of opportunity not only from the moment of their birth but from the moment of their conception. (Babies from the poorest households are disadvantaged throughout their time in the womb; they have highest incidence of low birth weight, which can set back childhood development and affect their health.)

It is surely more a matter of perpetuating poverty in slums rather than sorting through society and pushing the poor into them. If the poorest, most excluded people were 'diluted and dispersed', their opportunities might be increased and their stigmatisation decreased because they would no longer be tarred by their address. Even Banfield, an archpriest of urban non-intervention, appeared to accept the arguments for dispersion while deeming it politically impossible. A society, he wrote, that wished to protect itself from its lower-class people and to protect lower-class people from each other, should scatter them through its cities 'in such a way that they will not constitute a "critical mass" anywhere'.[7]

We have to try to stop cities segregating people by incomes. We need to break up concentrations of poverty and achieve a finer social mix instead. In an imperfect world of wide inequalities the best we can hope for is a spectrum of housing – from large houses for the wealthiest down to small, cheap homes for the lowest incomes – in every street. That would create a varied, vibrant townscape that would help to maximise equality of opportunity. Housing would still function as a positional good; wealth and status would be signalled by living in a posh place in an ordinary-looking neighbourhood (instead of the present system of living in an ordinary-looking place in a posh neighbourhood). Yet this line of thinking goes against most of the housing policies pursued for the past 80 years. The state generally provided homes for lower income families in bulk, in towers and big blocks and estates, because that is the way of states.

The socially mixed street or neighbourhood I am advocating has often been described as 'balanced'. Generations of sociologists, planners and politicians have yearned for this balance, have sung its praises and sighed at society's inability to accomplish it. It is a notion with broad appeal. Most people like the idea of mingling with people from different backgrounds and classes to their own, of learning from and getting along with them, even if they don't actually do any of it.*

But experience tells us that the socially mixed street or neighbourhood is actually out of balance. It appears to be unstable which explains why it is uncommon. Most people end up using what wealth they have to obtain homes alongside people of broadly similar circumstances.[8] Housebuilders develop estates with this in mind. Hierarchies form with rank registered precisely by house price. Addresses are positional goods. There are several companies that exist simply to gather and sell data on the income status and social composition of every British neighbourhood to marketing departments in other companies.

*Snobs and Marxists would be the exception. The latter have tended to argue that efforts to create fraternity between social classes, including attempts to maintain 'balanced' communities, represent an attempt to suppress working-class solidarity and struggle.

Talking to the academics and professionals who have pondered and studied this question of social mix, I found weariness, wariness and even a little bitterness. They knew all about the ruthless self-interest of the middle class, its formidable abilities to put some space between itself and the lower orders. 'People have been wringing their hands over this one for decades, and I've spent more years on it than I'd care to remember,' Professor Ray Pahl, a wise old sociologist at the University of Essex, told me. 'You just have to keep on trying.'

A thought experiment, a fantasy we can think through but never enact, illuminates the benefits of socially mixed neighbourhoods. Imagine millions of households and families being moved, overnight, so that every street and housing estate in the land has the same mix of classes and income groups as the entire surrounding region. Dual income, professional households are abruptly dumped in grim council maisonettes. Single mothers with several children conceived with several fathers find themselves whisked off sink estates into detached houses in prosperous new suburbs. Those are the extremes; in most cases the change in housing circumstances are far less drastic and, across the board, about half of the households do not have to move at all. There is no need for people to travel very far to achieve the mix – a mile or two at most – so they all keep their old jobs (and incomes) along with their possessions (homes excepted) while their children remain at the same schools.

A few more initial conditions have to be set for our thought experiment. After the great swap, each household faces the same housing costs in rent or mortgage repayments as it used to, however much better or worse its new housing is, and it retains the same type of tenure. Exclusive suburbs now have a scattering of impoverished, unemployed social housing tenants whose rents are paid by the state while the bleakest, worst-maintained council estates are dominated, like the rest of Britain, by owner occupiers. For an indefinite period which may stretch into several years nobody is allowed to move, sell or rent out property. And so the cities wake up one morning with a fine social grain. But while every neighbourhood is now well mixed, there are still some familiar faces around because a proportion of households has stayed put. What happens next?

A great wail goes up at the unfairness of it all but nothing can be done. People realise that they will just have to cope with their new circumstances for the time being. Most folks introduce themselves warily to their new neighbours with strained smiles, wondering whom this cruel national lottery of housing redistribution has brought next door. They want a quiet life without any hassle and unpleasantness; they want to get along.

The tiny minority of anti-social and criminally inclined families who had been concentrated on sink estates find life in owner-occupied suburbia is much harder. Their neighbours seem to be watching them all the time and are less easily intimidated. Loud noise and abuse, vandalism and drug dealing, are no longer tolerated.

The police and the local council are bombarded with demands to take action against these neighbours from hell and there is even some unpleasant vigilante activity. The problem families find themselves faced with a choice of endless run-ins with the surrounding community, their landlords and the law, or of changing their ways and becoming less anti-social. Most of them opt for the latter.

But the great majority of low income, social housing tenants – who are as decent and sociable as any other group of people – enjoy their new surroundings. True, some of their owner-occupier neighbours seem snobbish and aloof, but most judge them by their character rather than their income or tenure. When they apply for a job they no longer have to feel ashamed of their address. They are living in an environment in which more money circulates, so there is more casual work available in childcare, gardening and decorating.

Meanwhile, the stunned, demoralised owner-occupiers who have found themselves in possession of most of the homes on run-down council estates pull themselves together and begin work on a transformation. They want an improved environment and the highest possible value for their homes; they are desperate to remove the stigma that attaches to their address. A frenzy of DIY and improvement work begins and they also open noisy, impassioned campaigns for cash help from the government and local councils to renovate the housing. Their lobbying cannot be resisted and improvement grants are soon made available.

Within a couple of years the estates are transformed into what they were always intended to be – collections of decent, compact homes, reclad, spruced up and surrounded by well-tended greenery. The anguish and anxiety caused by the great overnight move give way to a blossoming of neighbourliness. People have to co-operate more in these strange new circumstances. The spirit of the Blitz is abroad; the nation feels more at ease with itself. Entire cities are better looked after and their environment and schools improve while crime drops.

And then comes the great day when the government announces that the moratorium on house sales and moves is to be lifted. At first there is no rush to buy or sell. The market is extremely cautious. But soon a trickle turns into a flood. The size and state of repair of each property play a big part in determining its price, but those estate agents' three old favourites – location, location, location – become increasingly important. The well-to-do gravitate towards their former haunts, concentrations of large houses in outer suburbs or grand pre-twentieth-century terraces in the inner city. The councils and housing associations, which had found themselves owning large numbers of these expensive homes after the great overnight shuffle, start to sell them off. They can easily justify these sales; the receipt from each pays for the purchase or construction of several smaller homes, enabling them to house more of the low-income families they exist to serve. The social landlords try to buy or build these cheaper homes in bulk; that way it is easier to manage and maintain them. Just ten years after the moratorium is lifted, the cities are well on their way back to their bad

old ways. Authentic sink estates, plagued by crime and unemployment, are starting to reappear, while the best streets are entirely in the hands of wealthy
owner-occupiers.

Let us quickly run another thought experiment; one with an opposite starting
point but roughly the same conclusion. A city springs into being in which all the
homes and streets appear identical. The houses are equally spacious and, while
their occupants are allowed to maintain them, they cannot alter them in any way.
Every household is given its home, which it can sell. The city is completely
mixed with respect to income and social class – every neighbourhood is equally
heterogeneous.

But almost as soon as houses start selling, the perfect mix begins to break
down. In some of the more desirable neighbourhoods house prices start to rise,
even though the houses themselves are no different from those anywhere else in
town. These areas are desirable because they are next to the countryside, parks, the
river, or they are on a hill with fine views, or they are conveniently close to the
offices and shops of the central business district. In some neighbourhoods, however,
house prices fall because they are next to busy roads or large, noisy factories, or they
are furthest from shops and workplaces.

The proportion of higher-income families in the more expensive, sought-after
neighbourhoods starts to rise. Only they can afford to buy into these areas, while
some of the poorer residents sell up, cashing in on the rise in prices. The changing
social complexion and rising prestige make these places still more desirable and
expensive. Meanwhile, in those quarters where house prices are falling, the proportion of poorer families rises steadily; high-income families flow out faster and
faster, to be replaced by those who can only afford below-average house prices. The
city begins to turn its back on these places; their schools perform poorly, crime rates
rise, prices plummet further.

After a decade, the city has become thoroughly unmixed. There are exclusive,
spick and span neighbourhoods, sad, tatty slums and a range of areas in between,
each dominated by a particular income group or social category. Houses in the best
places now cost several times as much as those in the worst, even though they are
identical.

It may be difficult to create and maintain mixed neighbourhoods but it has not
stopped local councils, housing associations and central government from trying.
Government regeneration money has been used to lure housebuilders into building
homes for sale in areas they would otherwise have run a mile from. And government agencies which fund low-income housing (the Housing Corporation in
England, Scottish Homes north of the border, Ty Cymru and now the National
Assembly in Wales) spent hundreds of millions of pounds in the 1990s subsidising
the construction of cheap shared ownership homes next to homes for rent; this was
done as part of a wider policy to spread home ownership among lower-income

households during the 18 years of Conservative rule. The occupants own (or take a mortgage out on) a quarter to three-quarters of the value of their home, while a housing association owns the remainder on which the occupant pays a rent. The monthly outgoings are cheaper than for outright ownership, and the costs to the taxpayer of building the house are cheaper than they would be if it were entirely for rent.

People opting for shared ownership were generally couples earning an average wage or less, typically young and in clerical or manual jobs. Often they had grown up on council estates. In Scotland many of them wanted to live next to the council estate they grew up on and where they still had close relatives, but wished to be owners, not tenants. Subsidies for low-cost home ownership was at first run down by the new Labour government, then increased with a new emphasis on providing housing for key workers such as nurses, teachers and police officers in booming cities like London where open market prices were out of their reach.

Three main arguments are used to justify interventions and subsidies to create mixed neighbourhoods. The first of these – let's call it 'defending the neighbourhood' – goes like this.

DEFENDING THE NEIGHBOURHOOD

Middle-class people are good at defending the things that an entire neighbourhood shares, from schools and libraries to clean pavements and safe, well-maintained parks. They may also be more effective resistors against crime. They are articulate and confident and they know how to lobby authorities such as the local council and the police in order to defend their interests. Less wealthy people who live in the area benefit from their defence.

Put this way, the argument is wincingly laden with assumptions about class superiority. Classes and class definitions are contested and 'middle class' means different things to different people, so let us make the same assertion in a way that avoids those two awkward words.* A neighbourhood needs to retain a sizeable proportion of self-reliant people with earned incomes who value education for themselves and their children. These are people who hate crime, who want to raise their quality of life and believe they can do so by their own efforts, who save and plan for a better future. These middle-class values have actually been held by generations of working-class people. When they see their neighbourhood threatened they act, raising petitions, writing letters, holding public meetings. Keep them in a district in sufficient numbers and it will remain civilised, hopeful and able to look after itself, to the benefit of all of its inhabitants.

---

*The American political class seems very comfortable with the term, but 'middle class' in the USA refers to people on roughly average (meaning median, or modal rather than mean) incomes while in the UK 'middle class' usually refers to a higher income bracket. Perhaps 'Middle England' comes closest here.

The government says that poor quality public sector services such as education and health care are among the most important problems facing deprived neighbourhoods, and gives the highest priority to improving them.[9] But it appears to ignore the reasons for the low quality of services; that there is a lack of local people able to demand or campaign for higher standards, and that most of the professionals providing the services prefer to work in places where conditions are less challenging, less dispiriting.

The flaw in the defending the neighbourhood argument is that prosperous communities are ready and willing to keep out the poor and the socially excluded, which further harms the prospects for those poor communities where society's rejects are dumped. Anyone working in the field of supported housing will tell you that middle-class areas are brilliant at resisting new housing for 'problem people' such as young people leaving care, those with learning difficulties or drug and alcohol addiction. They can exclude by price, too; once a neighbourhood becomes predominantly occupied by the prosperous no one on a low income can afford to move in.

Nonetheless, the findings of a study of four new West London housing estates with mixed tenure give some support to the 'defending the neighbourhood' argument. Consultants David Page (mentioned in Chapter 5) and Rosie Boughton looked at housing association estates built in the 1990s, mostly for rent but with some shared ownership homes.[10] Page and Boughton quizzed 200 residents, both owners and tenants. They looked for wear and tear and litter and vandalism, they spoke to the estates' landlords and went through their complaint files.

The findings were broadly similar for all four estates. Stark differences emerged between the tenants and the owners. Well over half of the heads of households from the rented homes were either unemployed or 'economically inactive' (of working age, but not working). Of those tenants with children, almost twice as many were single parent families as were headed by adult couples. But as for the owners, very few were not working and there was not a single parent family among them.

A clear majority of both tenants and owners thought the balance of numbers between them was about right. But of those dissenting from this view, there was overwhelming support for the notion that there were too many tenants and too few owners. Interestingly, that applied to both owners (no one thought there were too many of them) *and* the tenants (for every tenant who thought their estate had too many owners, there were seven who thought there were too few). Nearly nine-tenths of the tenants preferred the estate to have a mixture and only 6 per cent wanted to live in a place where there were only tenants.

Page and Boughton's study suggests low-income renters want owner-occupiers in their neighbourhood. They recognise that their estates, and they themselves, are in danger of being stigmatised if only social housing tenants live there. The West London research also found that owner-occupiers were less tolerant of litter, dumped rubbish and vandalism than the tenants and more likely to complain to the estate's landlords about these problems.

NETWORKS AND CONTACTS

The second argument in favour of the socially mixed neighbourhoods is that they promote contacts between different income groups. This may foster social cohesion, for when better-off people see and know more of the poor, they may be more willing to tolerate wealth redistribution. Well, maybe. More importantly, contact with the better off brings poor people out of a hopeless world in which all those they meet are also poor. It might provide them with role models. They see that it is normal to work and be self-reliant, to have ambitions and aspirations and for children to have two parents living together. Having a wider network of acquaintances also gives them a better chance of hearing about job vacancies and making contact with potential employers. A paucity of 'weak ties' to others, a lack of acquaintanceships gained through work, leisure and education (as opposed to the strong ties which bind close friends and family members) are both a symptom and a cause of chronic poverty.[11]

The flaw in this argument is that the prosperous, the comfortable and the middling can isolate themselves very efficiently from those at the bottom of society, even if they live close to them. They find ways of blanking them out. In inner London, in boroughs like Islington, affluent professionals and City people inhabit the same street as jobless, low-income families or live next to their council estates while having no connection with them whatsoever. Their children go to different schools.

It's not just a matter of differences in wealth. We have become a less neighbourly people. 'People round here keep themselves to themselves': how many times have we heard or read that phrase, or some variation of it, in a news interview with neighbours after something horrific happens to the couple living a few doors down the street? In an increasingly free-footed world, our contacts, friendships and sense of belonging to a community are less and less rooted in the places where we live and more and more likely to come from the places where we work and play. People jabber endlessly on mobile phones to friends spread across the city and never meet the neighbours living a couple of dozen yards from them.

Keith Kintrea, a researcher at the University of Glasgow, studied the social lives of people living on three deprived council estates around Glasgow and Edinburgh.[12] All three had undergone extensive regeneration work which included building subsidised homes for sale in order to introduce a social mix. Kintrea persuaded forty-nine households to keep diaries in which they detailed where they went and whom they met during one week. He and his co-researcher Rowland Atkinson also interviewed the owner-occupiers and tenants to gauge their views about each other.

They found that on all three estates the two groups had very little to do with each other. More than three-quarters of all the owners' activities outside the home (shopping, work and play, visiting friends and family) happened away from the estate. The owners were much more likely to have a car than the renters and much less likely to use the local shops. Often they would buy the sort of things one could purchase from the local corner shop – milk, cigarettes, a newspaper – from a petrol station or supermarket during a car journey. The tenants, many of whom had no work, spent much more of their time on the estate and had far more contact with relatives and friends living on it. Only 40 per cent of their activities took place off the estate and they had far fewer contacts with outsiders than the owners did.

Home ownership has been introduced onto such estates in the hope that it would reduce social exclusion, but Kintrea found it was usually the owners who ended up feeling excluded. They tended to have little involvement in local community life, often because it went on during the day when they would be at work. Kintrea found some tenants resenting the owners, believing they were snobs, while the owners detected this hostility. The hoped-for blossoming of contacts between owners and tenants had failed to materialise and there had been no transformation in prospects for the workless tenants. While around half of the owners had purchased on a social housing estate because they had a family connection to it, many of the remainder had only bought a home there because the housing was relatively cheap. They were likely to move on after a few years once they could afford to live in a better neighbourhood or needed a larger property for their growing family.

But while Kintrea's study suggested the tenants and owners remained divided and the contacts argument therefore failed, the defending the neighbourhood argument succeeded. A majority of both groups felt the introduction of owner-occupied housing onto the estate had improved it. Residents thought crime and vandalism had reduced, the appearance of their neighbourhood had been improved and their address had become more respectable.

Kintrea concluded that instead of building only small 'starter' homes for owner occupation on such estates, there was a case for building some larger units to give owner-occupier families who wanted to stay in the area greater choice. He decried the tendency to build the new housing for sale in blocks that were well separated from the homes for tenants; if the two groups were expected to interact, then their housing ought to be mingled. And he judged that the owners' heavy reliance on their cars was 'inimical to social contact' on and around the estate.

Both his study and Page and Boughton's in West London suggested that the local primary school could bring owners and tenants with children into contact. Children lacked the class sensitivities of their parents and played with whomsoever they wanted to, something which several owners commented on favourably. On the four West London estates, Page and Boughton found that households with children were twice as likely to know the names of their next-door neighbours.

Ben Jupp, of the think tank Demos, has carried out the largest UK study of social dynamics on mixed income, mixed tenure estates.[13] He and his associates questioned 1,000 residents living on ten such estates, most of which were large and had a high proportion of subsidised housing (housing association homes for rent and shared ownership properties). The team from Demos was trying to find out whether mixing did secure the benefits which its proponents claim or, on the contrary, if it caused extra tension and hostility between classes, as its opponents have suggested.

'Both the aspirations and the worries about social relations on mixed estates are often exaggerated,' Jupp concluded. Most people did not even know anyone on the estate with a different type of tenure than their own. But this dearth of society between the different groups reflected the design of the estates they studied and today's dispersed, fragmented patterns of social contact rather than any class antipathy and prejudice. The private and social housing were usually in separate blocks rather than intimately mingled in the same streets, and people usually only knew their immediate neighbours. Two-thirds of people interviewed in the survey spent little or none of their social time with other people from their estate.

The majority of both owners and renters were indifferent to the mix. They thought it neither caused any problems nor brought particular benefits. The minorities among both groups who believed it caused problems were roughly equalled by those who thought it brought benefits. On a couple of the estates rather larger proportions of residents felt there were problems in the mix. Some owners looked down on renters and blamed them or their children for noise, vandalism and unruly behaviour. Some renters were aware of this.

Jupp argued that the case for mingling different income groups and tenures was quite undented by his findings. Even if people failed to socialise, other strong arguments for a social mix still applied. What mattered, he maintained, was the apparent absence of general and severe tensions between different groups; had these existed they would have posed a real challenge to the growth of mixed tenure estates. He concluded that there ought to be a closer mingling of the different types of tenure within individual streets and that large estates needed a code of acceptable behaviour, policed by someone in an official, or semi-official position with a brief to prevent or resolve disputes between residents.

The Demos study's finding that people are relaxed about living among people with lower incomes and status than their own goes against the strong residential hierarchies and social sorting that we find in towns and cities in Britain and elsewhere. This may have something to do with the fact that most of the estates examined had a lot of subsidised housing and much of their private, owner-occupied housing was in the bottom half of the market. It would be interesting to see if these owner-occupiers would move to an exclusively owner-occupied area if they could afford to.

To sum up, this large investigation did not support the 'social contacts' argument. Nor did it back the 'defending the neighbourhood' argument. But the authors remained strongly in favour of mixed residential areas as a means of combating poverty and stigmatisation.

THE MONEY-GO-ROUND

The third argument in favour of the social mix is concerned with job opportunities and the local economy. It is my favourite; let us call it the money-go-round. If people with low skills, low earning potential and a heavy reliance on state benefits are lumped together in hundreds or thousands of homes, these neighbourhoods can only have the most vestigial and sickly of economies. There is not enough money to support a local service sector of any size or vigour so there are fewer jobs in the neighbourhood. People are more likely to have to leave the area to buy the things they need, so more of what little money is coming in as wages and benefits leaks out instead of circulating. Hardly any money is being invested in property because only a minority of the housing is in the hands of owner-occupiers, and most of them are badly off. The money-go-round isn't turning.

Contrast this situation with a mixed neighbourhood in which most households have an earned income and most homes are owner-occupied. Their presence will create and sustain numerous jobs within a few minutes walk of their houses. There will be shops, restaurants and pubs, leisure centres, home maintenance and improvement firms.

And there will also be a second means of job creation in the immediate area. Some of the earners living in those houses will use their accumulated skills and knowledge to strike off on their own and found small businesses which trade over a much wider area. These may begin as one-person outfits based in homes but a proportion will start off – or grow into – undertakings employing several people. Some of their founders will site these new enterprises close to their homes to cut commuting.

Mixed neighbourhoods have a varied, flexible labour force – less well off, unskilled people who will take lower paid jobs, more prosperous types with the capital and skills needed to found new businesses. It may be that such places are inherently better than the uniformly wealthy suburb at creating local employment. Wealthy families are likely to have very high levels of car ownership, so they will neither need nor expect their shops, jobs and places of pleasure and leisure to be near their home. Besides, tranquil, exclusive, very low-density suburbia is the kind of place that fights to keep out the bustle of shopping and other business premises.

Two centuries ago people lived and worked in the same building. A century ago, most employees walked a mile or two to work every day. But after the Second World War we lost sight of the link between employment and housing, thanks to high car ownership and zone-based planning which split cities into residential, commercial and industrial areas. The notion that people might prefer or need to live close to their jobs was forgotten by local and central government. For decades both sought to promote economic and employment growth by providing big sites for businesses near the edge of the city or outside it, well served by A roads and motorways. How close those workplaces were to the homes of people who could take those jobs was often deemed irrelevant. The employees would live on new estates being built on other greenfield sites miles away. People could always get in their car or on the bus.

Yet vibrant neighbourhood economies have survived the onslaughts of the planners and the car, bringing life and colour to the streets. You find them at their most vigorous in the denser, more prosperous inner areas of towns and cities where people can easily walk – rather than drive – to get the things they need. Almost a third of all employment in Britain is in small businesses or public sector establishments with fewer than ten staff.[14] Most of these little firms are of the type that want or need to be knitted into the urban fabric, sited in streets close to the houses where customers and workers live.

Another important aspect of the money-go-round is the black economy, in which money circulates out of sight of the Inland Revenue and the VATman. If you put a notice in a shop window in a prosperous residential area offering your services for child care, cleaning, gardening and so on, your telephone will ring. In my suburb you can command £5 an hour, comfortably above the minimum wage, and all of it cash-in-hand, untaxed and un-deducted from any state benefits you may be receiving. But your telephone would not ring in a deprived area (there's a good chance you won't have one – approaching half of households in the poorest parts of Newcastle aren't connected).[15]

Despite the enormous and continuing growth in mobility and long-distance, electronic communication, there are such things as local economies in which the demands of local people create and sustain local jobs. But how many jobs? If we look at a new town, we can get some picture of the local job creation powers of the money-go-round.

Milton Keynes, in Buckinghamshire, is Britain's fastest-growing urban area. Its population has more than trebled since 1967, the year when the government designated 35 square miles of Buckinghamshire countryside, embracing four small towns and a dozen villages, as the last of the English New Towns.* Each year about 1,500 new homes are built there. The town is now being slated for massive further expansion by government; within a decade or so it will have more than a quarter of a million inhabitants and overtake big places like Swansea, Southampton and Aberdeen.[16]

*One of which, I found out, gave the town its name. Which is a shame. I liked the idea that there had been some long, inter-departmental battle within Whitehall over whether to name the place after the poet or the economist and that Milton Keynes represented an exhausted compromise.

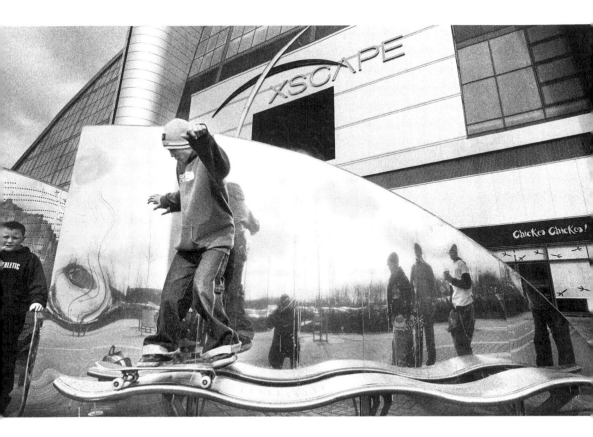

Leisure complex, Milton Keynes

The aim of the town's master builders, the government's Milton Keynes Development Corporation, then the New Towns Commission and now English Partnerships, has been to balance the growths in employment and housing so that as many people as possible could both live and work there. At first they had to spend taxpayers' money wooing new employers and housebuilders to the area, paying in advance for an infrastructure of roads, drains and so on. Established companies were persuaded to come in, moving their factories and offices lock, stock and barrel from other towns or setting up new branches in Milton Keynes. Sometimes the firms would bring part of their workforce with them and the development corporation would provide them with public sector-built homes.

But after a decade of building Milton Keynes gained an economic momentum of its own, blessed as it was with a site in the Greater Greater London region, excellent road and rail links, plenty of building land and attractive countryside all around. The money-go-round effects I described above went to work. The demands of a growing population created more and more local service jobs. The rising number of entrepreneurs – or potential entrepreneurs – living in the area led to more and more new firms being founded. As time passed, Milton Keynes found it easier to attract big, established firms to locate in the area (recessions excepted) because it had a larger, more diverse local labour force, more housing, more facilities to offer them. The growths in the number of residents and employees sustained each other.

We can get some idea of the numbers involved because the Milton Keynes Economic Partnership carries out an annual census of employment in both the private and public sector. Its 1998 survey found that just under 70 per cent of the employers in the area had been founded in the new town, rather than moving in from elsewhere or being established as new branches of existing companies. They provided just over half of the area's total jobs.[17] These local firms were nearly all new town creations, for less than a tenth of them existed before Milton Keynes got its special status in 1967.

The census found that the area's biggest single employer was the retail trade, accounting for one in seven of the local jobs. Most of these jobs exist in order to sell things to local residents. True, a proportion of the retail employees work in headquarters and in administration, dealing with shops across Britain. And some retail jobs have been created by Milton Keynes' rise as a regional shopping centre, sucking in customers from other towns nearby.* Even so, approaching 10,000 jobs have been created in Milton Keynes simply in order to sell things to its 200,000 inhabitants.

I estimate that roughly a third, and perhaps as many as half, of all the jobs in the town exist in order to serve people living there – in their homes and in shops, in hospitals and schools, at work and at play (see table). And since the population is twice the size of the number of jobs in town, that suggests every dozen residents creates and sustains two or three local service sector jobs. Here, then, is a thriving local economy which can be drawn on a map; it's all within a square with sides about 7 miles long. Its people are busy trading with each other as in any old-fashioned, compact town.

This is a pleasant irony since Milton Keynes is a kind of anti-town. With its sub-suburban densities, huge quantities of open space and dispersed grid of fast roads, it is the only town in Britain planned from the beginning on the assumption of mass car ownership. Indeed, a design criterion for these criss-crossing roads was that drivers should not even be aware they were passing through a built-up area. It is a polite, small and well-ordered British version of Los Angeles.

Milton Keynes represents the exact opposite of the compact urban form I argue for later in this book But its new town circumstances, and the existence of the comprehensive employment census, compelled me to use it as a brilliant example of the local money-go-round in action. Its planners are now looking to bring housing right into its heart and to build at higher densities in the new town's future suburbs.

*Milton Keynes had one of the first truly gigantic, car-reliant malls to be built in Britain. It may look like an early classic of the out-of-town shopping movement but, being sited at the core of the town, it actually forms Milton Keynes' centre.

Employment in Milton Keynes – local homes and local jobs

Sector	Total number of employees in establishments sized			Total employees in sector
	1 to 50	50 to 500	500 plus	
Retail	8,349	6,073	2,904	17,326
Business services and real estate	7,869	7,381	1,998	17,248
Education	3,565	4,168	2,730	10,463
Wholesale distribution	3,432	4,011	1,230	8,673
Medical/health services	3,132	1,276	3,911	8,319
Banking, finance and insurance	1,189	1,768	2,770	5,727
Catering and hotels	3,344	1,889	—	5,233
Total employees for the 7 largest sectors above	30,880	26,566	15,543	72,989
Total for all service sectors (there are 15)	37,417	36,738	16,283	90,438
Construction	2,658	258	1,387	4,303
Total in all other non-service sectors	7,560	13,643	3,025	24,228
Total – all sectors	47,635	50,639	20,695	118,969

Source: Milton Keynes Economic Partnership 1998 Employment Survey

Notes: Just over 75 per cent of all jobs in Milton Keynes and its immediate surroundings are in the services sector. *The bulk of these service jobs exist to serve people living in the town,* either at home or in their work-places. The evidence lies in the very nature of the services provided and the fairly small size of most of the companies and public sector establishments involved – 41 per cent of the service jobs are in small businesses or establishments with less than 50 staff. The Milton Keynes employment census split services sector employment into fifteen categories; 81 per cent of these jobs were accounted for by seven categories shown in the table, in descending order of size. Of these seven categories, five (retail, business services and real estate, education, medical, catering) depend *primarily* on serving the local population – in places ranging from shops and pubs to the town's hospital.

The largest category is retailing. Some 60 per cent of the people visiting the town's shopping centre come from within Milton Keynes, so most of the 17,326 retail jobs also exist to serve the local population.

The second largest category is business services and real estate. Because Milton Keynes is a fairly self-contained new town, most of its local workforce live there or in its immediate vicinity. Which means that most of the 17,248 jobs in this category also exist to serve local residents (or their employers) in their workplaces.

In only two of these top seven service sector categories (wholesale distribution and banking, finance and administration) is the bulk of the employment tied to trading across the wider region or nation.

Outside the service sector the bulk of employment will depend on manufacturing goods for national and international markets. But two-fifths of construction jobs are in firms with under 20 employees; these will have most of their business in the immediate area.

This analysis suggests that at least a third, and possibly more than half, of all jobs in Milton Keynes depend on providing services and construction for the town's 200,000 residents.

costs. The UK government estimated the Commission's ruling would cause gap funding expenditure by the public and private sectors to decline by at least 80 per cent.
By the spring of 2001, it had secured the Commission's approval for schemes which
replaced only a fraction of the former gap funding regime. The extra cash the government planned to pump into urban regeneration in the coming years would not
make up for the shortfall. MPs from all parties condemned the European
Commission's ruling as perverse, bizarre and disastrous for urban regeneration.
They judged gap funding to be an effective partnership between the public and private sectors which had enabled hundreds of millions of pounds to be spent a year.[19]

Another objection to placing new private housing at the forefront of urban
regeneration is that there is insufficient space in the cities; their parks would have to
be concreted over and densities raised to intolerable levels. This is simply not the
case, but I want to remit the crucial question of urban density to later chapters.

There is another objection; if hundreds of households were to move into a
declining area they could harm the places they moved from. This is a danger, but
urban regeneration is surely not a zero sum game with one neighbourhood's rise
invariably causing another's decline. The total wealth and resources available for
towns and cities grow along with the economy. New households are continually
being formed and the people who move in to reviving neighbourhoods might otherwise have left the city altogether or never come into it in the first place.

And so on to the most powerful objection of all. Surely it is the locals and their
needs that should have top priority in urban regeneration? If the emphasis is on creating a local housing market and bringing in newcomers, the original residents may
gain little or nothing. They cannot afford the new houses. The incomers may fail
to create any new local jobs if they use their cars to shop and play miles away. And
even if their arrival does create substantial new, service sector employment locally
it will consist largely of low-paid, hamburger-flipping, 'Macjobs'. To add injury to
insult, those jobs might well be taken either by outsiders commuting into the area
or by the new inhabitants themselves.

So while the kind of regeneration I have advocated may bring in more affluent
people and prettify the neighbourhood, the original locals could remain mired in
poverty, feeling increasingly alienated by the new wealth around them. Their needs
were greatest but only the needs of others have been met.

The story of London Docklands is often used to make these points. This was
Thatcherite, property-led regeneration with a vengeance; getting in new private
capital mattered more than anything else, even if it took a great deal of taxpayers'
money to attract it. Between 1981, when the London Docklands Development
Corporation was set up, and 1998 £7.7 billion of private capital flowed in, along
with more than 24,000 homes and 58,000 jobs. It poured into the 8 square miles
around the city's redundant docks besides the Thames.[20] Most of these new jobs

went to outsiders and many of them were not new at all; they were relocations of existing jobs. For most of its 17-year lifespan the Docklands Corporation, which led the regeneration, was accused of under-investing in the education, training and housing of people living in the deprived local boroughs of Newham, Southwark and Tower Hamlets, allowing the employment bonanza to pass them by. Take London City Airport, one of the jewels in the regeneration crown. Having got off to a shaky start in 1987, it now employs some 700 people in the Royal Docks, but only one in eight of them come from the local borough of Newham. Most of the new private housing also went to outsiders.

The Docklands regeneration story is, nonetheless, far more about success than failure. Any endeavour which can double the population of a huge inner city area, enabling 44,000 more people to live deep inside London, must be a good thing. No one ever counted how many of the jobs created across the area during the regeneration actually were taken by people who had been living in the three boroughs before 1981 but the figure certainly runs into thousands; there has been local job creation. As for housing, more than 2,000 of the new homes were reserved for sale at subsidised prices to local council tenants or their children in the corporation's first 7 years.* This demonstrated that there was a demand from local people for owner-occupied housing which had long gone unmet. In 1981, when the corporation began its work, 83 per cent of homes in the area were council-rented and only 5 per cent owner-occupied. By the time it was closed down in 1998, that had risen to 45 per cent. Docklands has become much more of a mixed area, far less of a poor one.

The area's regeneration is far from complete and there is plenty to criticise. I have no love for Docklands, having worked for five years in the area's chief symbol and landmark, the Canary Wharf skyscraper named One Canada Square. It has all been a boom, bust and boom, make-it-up-as-you-go-along juggernaut which produced some appalling architecture. But it is probably the world's greatest feat of urban regeneration, and was a lot better than nothing. And nothing was pretty much what the residents of Docklands had had, under both Labour and Conservative governments, for years and years as one by one the capital's docks closed through the sixties and seventies.

Something is better than nothing, some jobs and homes for the locals are better than none at all, and a Macjob beats being on the dole. Can that justify giving the needs of new residents and housebuilders top priority in urban regeneration? Let's

---

*That enlightened policy was halted by a combination of corruption and soaring house prices. Some council tenants sold their rent books to speculators who then used them to buy the new reserved homes at the special discount price and then sold them on the open market, making big profits.

London Docklands, regenerated Royal Victoria Dock, overleaf

look at a couple of alternative approaches that attempt to put the needs of local people first.

One way in which councils have tried to do this is to reserve derelict industrial sites for new industry or commerce which could employ local people, instead of allowing private sector homes to be built on this land. But the site often stays empty for years, blighting the neighbourhood, because industry and commerce prefer greenfield sites nearer the urban fringe, the ring road and dual carriageways. Then, when an employer finally does come along to occupy the site, it hires nearly all of its workforce from outside the area. And those few locals it does employ use their improved incomes to move out.

Another approach to regeneration – the core, in fact, of government efforts – is to help the people living in deprived neighbourhoods to help themselves. Standards can be raised in local schools, unemployed adults and teenagers granted the training and pre-training required to get work elsewhere in town. Local people can be given the loans and skills needed to set up their own small businesses.

They can also be trained, encouraged and financed to become more self-sufficient outside a mainstream economy that has turned its back on them. So they can build and refurbish their own homes, grow their own vegetables, set up community banks and trade without money through LETS schemes.* This is the expanding world of the Third Sector: not public and part of the state, not private and profit seeking, but of and for local communities. It is a place where government is starting to dabble, and where Greens collaborate with those on the left who want to work from the bottom up, having given up believing that national governments can ever deliver any kind of socialism. If, the thinking goes, deprived communities have been excluded from society, left relying on meagre welfare which no government will ever raise to a decent level, then they must be helped to help and empower themselves. By working together, by internal trade and voluntary effort, communities blighted by poverty can raise their quality of life and self-esteem from within.

This Third Sector, community development approach should be a component of urban regeneration, supported by local councils, businesses and central government. So, too, should the more orthodox efforts to improve education, to give people skills they can market and to encourage small business start-ups. But, on their own, they will often fail to bring depressed, excluded neighbourhoods back into the mainstream of urban life. New small businesses will struggle to survive because of the local shortage of money. LETS schemes may fail to flourish in deprived areas.[21] Attracting private capital and new residents into declining areas, creating mixed neighbourhoods, making the money-go-round, are just as important. The two

*Local exchange and trading schemes, in which people exchange things – mostly services – among themselves using IOUs which can circulate among the members of a scheme. For instance: he mows her lawn, she gives him an IOU for one LETS unit, he returns it to her when she cuts his hair – or he passes it on to me when I paint his front gate, and I give it to her when she cuts my hair. And so on.

approaches, sometimes portrayed as in conflict, need to become compatible and complementary.

More and more urban councils have come to accept this. Some chief executives (the most senior civil servant in a council) and Labour council leaders are now open about wanting to change the social mix in their boroughs. Wendy Thompson, who was chief executive of the London Borough of Newham and a member of the government's Urban Task Force when I interviewed her, told me: 'More ambitious, more able people move out and we want to keep them. We need a more balanced community here. The place is never going to regenerate itself with public money thrown at it.' She noted that the numbers of cafés, restaurants and cinemas in the borough have risen and found that 'a solid performance indicator' for regeneration.

Yet many of the new jobs created by regeneration will be low paid and involve dealing direct with customers in the services sector. In the largest, most thriving cities local people may struggle to win these jobs in the face of competition from students and visitors from overseas. A middle-class college student or an Antipodean backpacker stand a better chance of getting one of these 'front of house' jobs than an unemployed, inarticulate teenager from a run-down council estate. So local employers need to be given incentives to hire and train local people and they should make commitments to do so; we return to this issue in Chapter 11. The area's schools and training opportunities have to be boosted so that local people can take a broader range of jobs. Community development must be part of any regeneration scheme; if it isn't then local people won't expect to gain anything and won't give it their support.

To sum up, the strongest argument for policies favouring socially mixed neighbourhoods is that they can help to prevent concentrations of poverty from widening, deepening and eventually damaging entire cities. But strong, too, is the tendency of neighbourhoods to be socially unmixed. How, then, can we prevent people on lower incomes from being pushed into exclusively poor areas?

Chapter 9

HOW TO MINGLE

In setting a plan of large streets for the dwellings of the rich, it will be found necessary to allot smaller spaces, contiguous, for the habitations of useful and laborious people, whose dependence on their superiors requires such a distribution . . . This intercourse stimulates their industry, improves their morals by examples, and prevents any particular part from being the habitation of the indigent alone, to the great detriment of private property.

John Gwynn, London and Westminster Improved
(to which is prefixed a discourse on Publick Magnificence), 1766[1]

In the interests of social variety, and in order to enable families of different backgrounds and experiences to continue to mix in peace time as they are now mixing during war, each neighbourhood unit should be socially balanced, containing houses of different types and size inhabited by families belonging to different income groups.

National Council of Social Service, The Size and Structure of a Town[2]

My old man's a dustman
He lives in a council flat
He wears Armani trousers
and he shops at Habitat.

Advertising slogan for council housing, devised by **Duckworth Finn Grubb Waters**
for the Institute of Public Policy Research and The Observer newspaper, 2000

IN THE TWENTIETH CENTURY Britain succeeded in giving a very substantial majority of its citizens a decent home – a place that was weatherproof and centrally heated with at least one indoor lavatory and bathroom. But Britain gave rather fewer of its people a decent address. Millions of people who subsist on state benefits or poorly paid jobs live in bad places which damage their lives and reduce their and their children's opportunities.

How, and where, should people who cannot afford the house prices and rents of the market be housed? We will come to the where; as to the how, that must usually be in state-subsidised housing run by not-for-profit landlords. Recall that social housing began a rapid expansion early in the twentieth century because an increasingly prosperous nation was ashamed by the appalling conditions in which so many of its citizens had to live. Philanthropy alone appeared unable to deliver adequate improvement. Today, even though we are a vastly wealthier nation in which most people can afford to buy their home (a thing which would have seemed quite unthinkable 100 years ago), we still need subsidised housing. If we relied solely on the market, then a large proportion of the population who depend on benefits or low paid jobs would once again be crowded into single and double rooms in crumbling hovels.

Even with a state-guaranteed minimum income of more than £10,000 a year for families with children and a working adult (under the Working Families Tax Credit scheme) and a guaranteed minimum wage of around £4 an hour, people in the lowest income groups cannot afford to rent or buy on the open market across most of the country. The only places where there is widespread affordability are the most blighted, desperate areas of urban Britain, where the bottom has dropped out of the housing market and where few people want to live.

Housebuilders complain that planners' restrictions on the supply of greenfield land for new homes push up the general price of all housing, new and old. It does, but even if we ended all planning restraints (a political impossibility), it would not bring the price down sufficiently to enable much more of the poor to rent or buy without subsidy.[3] Many better-off people would, however, use the relaxation to obtain second homes in a rapidly disappearing countryside.

So we continue to need subsidised housing owned by local councils and housing associations. (We should not forget, however, that private housing is also subsidised. While tax relief on mortgage interest payments has been abolished, owner-occupiers remain exempt from capital gains tax.) But let us learn another big lesson from the last 100 years; that, instead of being a mass public service, social housing should be a safety net for a low income minority. Health and education established themselves as state services for the great majority of the population but public housing never came close to being an option for all. Owner occupation has emerged as the nation's indisputable favourite; most people achieve it, and most people who don't, aspire to it.*

There was perhaps a time, back in the 1950s and 1960s, when it looked as if things could turn out otherwise. The state's task was to provide decent homes for a large, stable working class. In Scotland more than half the population lived in council houses by the mid-1960s.[4] Since then there has been a steady decline in the proportion of people living in social housing, north of the border and elsewhere. Some people working in social housing still envisage it playing a leading role in housing large chunks of the population. As recently as 1997 the director of Britain's best-known housing pressure group, Shelter, could advocate 'a wider range of households in an expanded social rented sector'.[5]

He was wrong. If independent housing is what the great majority of people aspire to, if the market can provide most of the population with adequate homes, then we should be advocating a wider range of households in an expanded independent sector. Subsidised housing's task should be to use taxpayers' money wisely and well to provide for that substantial minority of people who cannot afford market prices. It is, hopefully, a shrinking role but it is hard to think of a more important one. A civilised country ought to be judged by the way it treats its weakest members.

It seems that this is what we have now accepted, we meaning the voters and governments of both main parties, even if we're unwilling to spell out the role of social housing in quite such bleak terms. It is a job which needs to be carried out by not-for-profit, social landlords made accountable to society as a whole as well as the people they house. Auditing, inspection and a degree of competition can keep them up to the mark and try to ensure that new social housing meets standards of decency while the existing stock is decently maintained.

There are, essentially, two ways of keeping housing costs low for poor people. First, the state can grant a large part of the costs of building and maintaining a new home (or buying and maintaining an existing one) to a social landlord charged with housing people in need. This 'bricks and mortar' subsidy is attached to buildings and it pushes rents down. Alternatively the rents can be higher – nearer to open market levels – and a rent subsidy be paid directly to the tenant or his or her landlord.

*Given a free choice, 87 per cent would buy their home and only 12 per cent would rent, while 72 per cent of those currently renting would prefer to buy. This according to 1,000 members of the 5,000 strong People's Panel, a large survey group set up by MORI and Birmingham University for the Cabinet Office.

Another MORI opinion poll, commissioned by the Institute for Public Policy Research in 1999, found that 89 per cent hoped their children and grandchildren would own their own home and 47 per cent would prefer to live in an area where most people own their own.

It's possible to envisage a free-market system in which there was no bricks and mortar subsidy whatsoever and no social landlords charged with housing the poor. Instead, the state would provide enough subsidy to ensure that every low-income household could afford to rent a home from a private sector landlord. The main drawbacks with such a scheme would be that the costs to the Exchequer would be vast and uncontainable while some of the housing involved would still fail to meet minimum standards of decency.

In the real world there are both types of subsidy: bricks and mortar and the means-tested rent subsidy known as housing benefit which is claimed by one in five UK households. There is a complex interplay between them. For some years the received wisdom has been that Britain moved too far away from bricks and mortar subsidies under recent Conservative governments. This pushed up social housing rents, trapping a growing number of people into lives of unemployment and state dependency in social housing ghettos. For when an unemployed tenant with housing benefit took up a job, each pound she or he earned resulted in an almost equivalent sum of housing benefit being withdrawn. This meant it was not worth taking low-paid jobs, the kind of jobs people on housing benefit are most likely to be offered.

But there are also problems in shifting towards larger brick and mortar subsidies in order to exert downward pressure on social housing rents, which is what New Labour has decided to do. They are, after all, a kind of covert wage subsidy; the more the state provides cheap housing with artificially low rents, the less employers will need to pay their lowest paid workers. If we have a society in which low pay fails to cover employees' housing costs and therefore does not amount to a proper living wage nor an adequate encouragement to come off benefits and go to work, then the case for raising minimum wage levels or benefits that supplement low pay is as strong as the case for cutting social housing rents.

If, furthermore, we accept that the market ought to provide for as many people as possible, with social housing reserved for a low-earning or benefit-dependent minority who really need it, then a policy of driving down rents could work against this. Social housing tenants who find themselves earning good money would be reluctant to move into unsubsidised housing as owner-occupiers or private sector tenants because they would face a big jump in their housing costs. Instead they would stay put, denying their subsidised home to someone who has no hope of affording market prices and really needs it. Some would argue that this is no bad thing; unemployment-dominated estates need more working tenants to act as exemplars and resistors against social decay. But a better solution, set out below, is not to have social housing agglomerated into large estates in the first place.

New Labour wrestled with these difficult issues of housing subsidy for three years after winning power before producing a cautious housing Green Paper.[6] The challenge the government faced was to devise a personal rent subsidies regime that did not act as a disincentive to taking a job, that maximised housing choices for those unable to afford open market prices, was more simple to administer and better proofed against fraud. At the same time, any reform could not leave large numbers of people worse off, nor make public expenditure soar. This was not at all easy and in the end the government opted for evolution with no prospect of fundamental reform for several years. Its intention was to break with the Tory past by keeping social housing rents at broadly their existing level, well below open market rents.

Where it was at one with its Conservative predecessor was in wanting the maximum number of households to house themselves without any need for state support. Maximising the affordability of open market housing generally enlarges choices while reducing the demand pressures on social housing. Sound stewardship of the economy with lashings of prudence and low interest rates can contribute to this goal by keeping down the costs of paying off a mortgage. But it is also necessary to ensure that the total number of homes keeps pace with the growing number of aspirant or actual households.

One way of doing this is to supply sufficient housebuilding land in and outside of towns and cities; the strategy for town extensions set out in Chapter 16 would contribute to that. Another is to build at higher densities without worsening urban environments; achieving that is the subject of Chapters 12 and 13. But we also have to ensure that housebuilders provide enough homes for the bottom half of the market, even if they find the upmarket sector more profitable. And we have to persuade conservative, stuck-in-the-mud firms to pursue the radical reductions in cost which are commonplace in other manufacturing industry.

For is it not strange that British houses are still built by hand-gluing small lumps of baked clay together, one by one? That they are mostly constructed on site, amid wind, rain and mud, instead of being largely made up of prefabricated parts in factories where workers can be more comfortable and where higher standards of quality control, efficiency and safety are much easier to enforce? That they leak warmth like sieves compared to new homes in Scandinavia, where much higher standards of energy efficiency are imposed? And that they look pretty much the same as houses built 80 years ago? Think how much the appearance of cars, aeroplanes and other large, complex products have changed over that period.

Housebuilders argue that punters insist on traditional-looking, brick-built homes; anything else would not sell. In fact, there are three other reasons why most of them have not tried to offer anything radically different and improved for decades. First, innovation, pre-fabrication and anything looking remotely untraditional are still associated, in some minds, with the failures of the mass social housing built from the 1950s to the 1970s, much of it hurriedly and shoddily. Second, the industry has been spared from having to compete with more innovative foreign construction firms who make more use of pre-fabrication and could provide better value for money; housing remains an almost entirely domestic industry. And, third, investment in workforce training is insufficient to allow many builders to build anything very new or different.

In any case, the builders' argument that consumer taste is generally more conservative for housing than for other types of product is becoming increasingly questionable. A growing proportion of purchasers are opening themselves to something quite different from the brick box with pitched, tiled roofs. Consider the enthusiasm for loft living and the uncompromisingly modern designs of many new, private sector blocks of flats which fetch high prices.

The government recently set up a task force to advise ministers on how to improve the entire construction industry's performance. Chaired by a leading industrialist, Sir John Egan, it looked at practice and results in the UK and overseas and concluded that UK construction was seriously underperforming compared to other major industries, most of which had faced more foreign competition.[7] Profitability, capital investment, research and development and standards of training were all lacking. It called on the entire industry, including the housebuilders, to set itself a target of reducing construction costs and construction times by 10 per cent a year. Even if the industry achieved only half of this improvement, the costs of a new home (excluding the land) would be slashed by 40 per cent in a decade, enough to make a real difference to affordability *if* it worked its way through into price cuts for new homes rather than increases in land prices. But the task force decided that the radical improvements in housebuilding it believed possible should, initially, be focussed on social housing 'for the simple reason that most social housing is commissioned by a few major clients'.

This is surely a mistake. Innovation will once again be in danger of being associated only with housing for poor people. Social housing tenants will, furthermore, lack the influence on producers that consumers with choices would have; if things go wrong with their new, cheaper housing their complaints will not have the same force as those of owner-occupiers. Housebuilders will be less likely to improve subsequent designs and construction techniques as a result.

What we need, for both the market *and* the social housing sector, is the equivalent of the Mini and the original Volkswagen. Being small and cheap does not have to mean being nasty or second rate. We need homes for the bottom end of the market which are appreciated as being ingenious, distinctive, innovative and therefore stylish because their design and construction embody all of those virtues. We need a British volume housebuilder to start thinking and acting like Ikea, which has been selling budget new homes as well as furniture in its native Sweden.[8]

Now to the where. Where should people who cannot afford unsubsidised housing live? In the previous chapter I explained why it would be better for them, and better for everyone, if they were mingled with the rest of society rather than cut off in shunned estates and neighbourhoods. The averagely prosperous and the wealthy tend to exclude and exploit poor people; they do them little conscious good and much unconscious harm. But when the poor are compelled to live together in their thousands and tens of thousands, they end up doing great harm to each other. Centuries of working-class solidarity and collective resistance against oppression

Housebuilding on greenfield site, Essex

won great gains, but that noble history seems irrelevant for today's socially excluded masses. You don't find much in the way of effective, life-changing solidarity or resistance in today's underclass estates, unless you consider shoplifting and the fencing of stolen goods to constitute such resistance.

If we take the need to mingle seriously, then three policies follow. First, wherever there is a shortage of affordable homes, any new, market housing built in the area should have some subsidised units among it. Second, whenever subsidised housing is built, or rebuilt, some open market housing must be included. Third – and this is the most difficult thing – every council and housing association estate, apart from the smallest, ought eventually to be somewhere where most residents own their own homes and where people with choices are content to live.

The first and second of these policies are now talked about all the time but are often ignored, or poorly implemented on the ground. The third remains something of a pipe dream. Let us summarise how far the current system for supplying and locating new social housing departs from them before setting out how the three essential policies for mingling could become a reality.

Under existing legislation, local councils are meant to act as strategic housing bodies, helping to ensure that the separate market and social housing systems serve every household in need of a home. They should carry out regular surveys into the needs for housing in their areas and the extent to which those needs are not being met. To do their job properly, they need to gather and regather information not only on the local stock of housing but also on its price, on rent levels and on the make-up and incomes of local households. Some councils fail to do adequate surveys of housing need and affordability and they all carry them out in different ways, making comparisons between them difficult.[9]

Such surveys, along with the length of waiting lists and the number of households presenting themselves as homeless, reveal a need for more social housing across most of Britain's towns and cities. A recent estimate by a Cambridge University team put the total number of English households in real housing need at 700,000.[10] One way of getting these homes built is through government grants, directed through the government's Housing Corporation (in England, and its equivalents in Wales, Scotland and Northern Ireland) and through local councils. Since town halls are no longer in the business of building council houses, nearly all of this money ends up going to housing associations who carry out new development. When this book went to the printers, these grants were starting to rise again after several years of decline. The government's intention was to focus more of the grant on the booming regions where the need for affordable housing was growing fastest, such as the South East.

The other means of getting new social homes built is for local council planners to negotiate 'planning gain' deals with housebuilding firms: the developer gains planning permission for open market homes in return for guaranteeing to help to provide homes for low-income families. The government actively encourages councils to make such deals and in 2000 they were responsible for more than a third of the new social housing provided.[11] If a housebuilder seeks planning permission for a development of twenty-five or more homes (fifteen in London and rural areas), the local council can insist that a proportion of them are 'affordable' in return for granting this permission.[12] There are several ways in which the developer can deliver this affordability. The most common is for it to give some of their land to a housing association at a discount or for free; the association then builds homes for rent there. Alternatively, the affordable element can consist partly or entirely of relatively cheap homes for sale on the open market. Sometimes the developer simply gives the council a sum of money, which it then spends on refurbishing or building social housing elsewhere.

Despite this planning gain system's growing importance in getting new social homes built, it routinely fails to mingle subsidised and open market housing. The low-income stuff is often put on its own in one corner or along one edge of the site.

It may well be the nastiest corner, the one next to the industrial estate, the railway line or busy road.[13]

Much new social housing is, however, built on the site of old social housing which is so dilapidated or shunned that it has to be demolished and rebuilt or vacated and refurbished. Densities are often lowered in the process. High rises and slab blocks are replaced by little terrace houses with gardens and streets. And when a council or housing association estate is rebuilt, it is now common for some sub- sidised housing for low-cost home ownership to be introduced.

If the estate in question is close to the city centre, then clever councils and housing associations can exploit the higher land values to do property deals which finance much of the rebuilding costs. New supermarkets, retail sheds, private housing and other commercial development will go up on part of the site. When all is said and done, however, the fact is that the great majority of social housing refurbishment and rehabilitation projects still produce estates which are largely devoted to low-income households.

We can do better than this. We need to conflate the two separate housebuilding systems, open market and social housing, into one. And we need to change the planning system from one which attempts to stipulate how many homes should be built into one which ensures that a spread of types and prices of homes are built, including homes for the growing number of single people and households who cannot afford market prices.

As a starting point, it should not be left to the political whims of local councils to decide whether new private sector developments should include some social housing. The fact is that local authorities serving prosperous, privileged areas will often be inclined to under-provide affordable housing. They and their electorate would rather poor people lived in someone else's patch. When it comes to housing needy families, their inclination will be to do the bare minimum. The outer London borough of Bromley (my own patch), which has usually been run by the political right, has a history of resisting attempts to ship the poor of inner London into new council housing within its borders.[14] At the very end of the 1990s, when Labour and Liberal Democrat councillors were able to combine to form a very narrow majority and take political control, a policy of including affordable housing for low-income families in all private sector housing developments of ten or more units was introduced. This policy was intensely opposed by the Conservatives, caused one Liberal Democrat to defect to their side, and was likely to be dropped if the former won back control at the next borough elections.

If, instead of that kind of political uncertainty, we are to have a stable, nationwide policy of incorporating affordable housing within new developments, it needs to be founded on more and better information about local housing needs. The surveys councils carry out must be done regularly, to a standard set by the government, in order to assess how many local individuals, couples and families with children cannot afford to buy or rent in the open market.* Government needs to inspect councils to ensure the assessments are being done thoroughly and fairly. Councils must put their findings in the public domain and be prepared to explain and defend them. The government made a start in 2000 by publishing a good practice guide on assessments.[15]

Estimates of housing need will tend to depend on assumptions about who should be entitled to social housing. Should pregnant teenagers? Or families on low incomes arriving in a borough from overseas, or from other parts of Britain? Or single people who want a place of their own, but cannot afford it in the open market? Councils and central government are required to make difficult and controversial judgements about entitlement. If a local estimate of the need for affordable housing is very high, and if the extra housing is then provided, then a council could be in danger of having too many low-income, benefit-dependent households. Many already do.

Where housing need surveys demonstrate a significant, persistent shortage of affordable housing, then the local council must start filling the gap by insisting – through its control over land use planning – that a proportion of all new homes built within its borders is for people who cannot afford market prices. The way this could be done would be to insist that some social housing would be included on all developments of *nine or more* homes. Call it the nine plus rule. Nine does not have to be the minimum number; it could be anything between, say, eight and twelve. It is simply a much more ambitious figure than the government's current limit of twenty-five. It would ensure that new market housing is usually mingled with some social housing whenever the latter is required.

Setting the limit around nine units would, however, also allow small sites in towns and cities with space for only a few homes to be developed solely for the open market. This would help to attract housebuilders to the many small, infill sites scattered through urban areas, hoping to get a premium price for the all-private developments they could build on them. Such pocket urban developments, marketed as being quite literally 'exclusive', would serve to bring more people with choices into urban living.**

As for the proportion of subsidised, social housing to be included in private developments with nine or more units, that would depend on the scale of the local need for it. In areas where the shortage is modest, it might be only one in nine. But what about places such as London where even ensuring that half of all new homes built were subsidised would not be enough to meet demand for many decades?

There is a good case for having a maximum of one subsidised home for every two open market ones on any housebuilding site. This maximum would prevent

---

*Bromley's most recent survey had found a need for several thousand new subsidised homes for low-income households living in the borough.

**As well as a minimum for the number of houses covered by the obligation, there would also have to minimum area for any development site; above this

private sector developers from being overburdened by obligations to donate their land to social housing. It would also prevent the creation of new low-income ghettoes. Places where there is a major problem of affordability either have a shortage of all types of housing, not just subsidised housing, or an excess of low income households. Either way, they need lots more open market housing.

The government's reworking of planning policies for housing (PPG 3, discussed in Chapter 5) has shifted the system a little nearer to something like the nine plus rule, but not far enough.[16] It should not be left to the discretion of local councils to decide whether or not to use the planning system to provide the affordable housing local people need. Instead it should be a clear duty, so that they can be held to account on how they discharge it. At the same time, there needs to be more clarity about what constitutes affordable housing. It need not all be for rent. It can include some shared ownership properties. It must all be affordable for households on well below average incomes.

If housebuilders are compelled to give up some of their development land for social housing by the nine plus rule, or something like it, then someone has to pay. But it will be neither the developer nor the purchaser of new market housing. Developers have to preserve their own margins to stay in business. And the price of the new houses they sell can be pushed up hardly at all, for they have to compete with the huge stock of second-hand homes that are for sale at any one time. The only price that can readily shift in response to demands to hand over land for social housing is the speculative price of land prior to development. It will be landowners who suffer, so no great harm will be done.

How would such a system work in practice? Imagine you are a housebuilder who has an option to buy a site suited for housing and large enough for 100 units. You know that the council's rule is that one in five of all new homes in its area are to be in the affordable, low-income bracket. You find a housing association prepared to build twenty social homes to be scattered through your private housing. You want the association's homes to look as much like ordinary, for sale homes as possible, and you seek whatever guarantees you can that they will be well maintained and the tenants well behaved. Then you apply for planning permission, confident of it being granted. The nine plus rule, or some version of it, would increase the supply of subsidised homes in the places where they were needed while mingling them with private homes. But it could only provide the land. Government would have to supply sufficient funds to ensure those subsidised homes actually got built.

What about areas where there is either no shortage of social housing or an excess? Places where abandonment has set in because of economic decline, stigmatisation

hectarage the obligation to build some social housing would come into force. Without such an area minimum developers would have an incentive to build at low densities in order to avoid the obligation. For nine plus, the appropriate area limit would be about 0.35 hectares.

abolition of ghetto estates will require a great deal of will and expense but, in the long run, it need not involve colossal *extra* expenditure. Virtually all of this housing is going to need extensive refurbishment or rebuilding over the next three decades in any case. But it must be done carefully, gradually, and with the consent and support of the majority of tenants, avoiding the mistakes of previous rounds of slum clearance.

There are three plausible objections to the major reform in planning and social housing arrangements advocated here. First, it could be argued that the comfortable or prosperous majority of the population don't want to live with poorer people mingled among them. If local councils and central government try to insist that new private sector housing is mixed with subsidised housing, it won't get built because it would not sell. The outcome will be a decline in the output of both subsidised and open market housing, leading to higher house prices and greater shortages.

This objection might be sustainable if *most* of the new mingled housing was subsidised and *most* of its tenants were neighbours from hell – thugs, drug dealers and vandals with armies of noisy, foul-mouthed, out-of-control children. But neither of these things would be true. Open market housing would always comprise the majority of any new scheme. And the majority of people who need subsidised housing are as considerate and neighbourly – or as inconsiderate and un-neighbourly – as owner-occupiers. If new housing is mingled, the occupants of the two different types of housing will not feud. They will just get on with their lives.

Of course, there would, under this new scheme of things, be occasions when developers would find themselves mightily inconvenienced. If you obtained some land in a prestigious urban area, such as beside the Thames in London, the thing you would most want to do is build the largest, highest block of luxury flats possible. The thing you would least want to do is fit some subsidised housing onto the site. (The developers of the twenty-storey Monteventro apartment building at Battersea – architect Richard Rogers – avoided any.) But clever developers and their architects would find ingenious, elegant ways of dealing with an obligation to mingle. True, the value of a socially mixed housing scheme would be less than a scheme aimed entirely at the top end of the market, but that would not make it unviable. It would merely lower the price of the undeveloped land. That, in turn, might make landowners less willing to sell, or it might favour other types of development – such as offices – over housing, but there are ways of dealing with that. We come to them in Chapters 13 and 17.

A senior executive of one of Britain's major housebuilders told me that if his industry wasn't allowed to carry on doing what it knew how to do and what it wanted to do, it would find profitable alternative investments. The Japanese whaling fleet was his wry suggestion. I think he was wrong. If we give UK housebuilders an agenda for contributing to an urban renaissance, as outlined above, and they don't get on with it, then competitors from overseas might start to come in and take over. For there is money to be made.

The second objection to reform is this. If you declare social housing estates to be a mistake and say they must all eventually go, you may only worsen the circumstances of the people living in them. You condemn them to live in doomed, hopeless urban landscapes in which there will be more and more rubble and demolition-cleared wasteland and fewer and fewer people.

But the cleared space where a house once stood is, ultimately, a more hopeful place than a house which has been empty, vandalised and boarded up for several years. Nothing else makes a street or estate look more sad and lost. And, while the ideal might be to transform council and housing association estates into mingled neighbourhoods, this does not mean they should all be abandoned to decline in the meantime. Some should; they have gone past a point of no return.[17] But so long as millions of people live on deprived social housing estates, it must be right to carry on pouring effort and resources into trying to improve most of them. It will often appear as if time and money are being wasted, as if there is no hope of turning the tide of exclusion and private sector disinvestment, but some things will work for some of the time in some places.

It will be difficult to strike the right balance between transforming social housing estates into mixed-tenure, mixed-income communities in the long term and improving the lives and prospects of their residents while they remain as mono-tenure, low-income neighbourhoods in the short term. If we are ingenious and lucky, the two things might occasionally blend into one, with estate regeneration being so successful that it pulls in private housebuilders to build new, for-sale homes while existing owner-occupied houses in the neighbourhood are gentrified.

There is no shortage of ideas of what can be done to improve bad neighbourhoods and estates without resorting to mass demolition. A list of all the actions and programmes going on in three inner city areas of Manchester and Newcastle found eighty separate initiatives.[18] The Social Exclusion Unit and its myriad expert advisers came up with thirty key ideas and more than 100 recommendations.[19] Better policing and lower tolerance of crime and disorder make up one strand of improvement. Raising the quality of local schools and providing training and employment schemes form another. Improving the management of social housing, including the way tenants are allocated to it, is a third strand (see text below).

The third objection to the reform of housing and planning advocated in this chapter is that it does not go far enough. It demands change for all new housing and for existing social housing but leaves established private housing areas alone. It asks nothing of comfortable suburbia and prosperous inner city quarters; these would remain enclaves of owner-occupation excluding the poor. If house purchasers prefer these unmingled neighbourhoods, then the more we try to mingle the remainder of our cities, the more desirable, expensive and therefore exclusive these places will become.

ESTATE RESCUE

How can the reputation of run-down social housing estates and the quality of life of their residents be improved in the short term, without the major social and physical restructuring which this chapter advocates as long term solutions? The leading ideas, all of which have been tried with varying degrees of success and are now entering the policy mainstream, are:

- **Change policies for allocating tenants to vacant homes** in order to bring in more tenants with jobs. Prevent the formation of estates in which hardly anyone works and where there are high proportions of children, single parents and 'problem families' (they have problems and need help; they give their neighbours problems and need rules). Allow only a proportion of vacancies to go to the type of tenant which, once they become dominant, are likely to stigmatise the estate.

   This is not easy. Once an estate has become stigmatised, it becomes difficult to get the right kind of tenant to come in. And making value judgements about people – who is a desirable tenant, who is unwanted – can be an ugly, controversial business which in itself constitutes a form of social exclusion. Such allocation policies may increase the hardship of the most vulnerable and needy people. If a single mother with several children is denied a home on an estate because there are too many of her kind, she will have to crowd in with friends or relatives or take her chances in the worst kind of temporary accommodation, such as bed and breakfast hostels.

Local lettings policies that attempt to create more sustainable communities need to be tied in with the following ideas:

- **Advertising social housing.** When a particular estate becomes unpopular and hard to let, or when there is a shortage of employed tenants on the waiting list, vacancies should be advertised. As well as detailing homes and their rents, the adverts should say what facilities and public transport services are nearby. Several big councils have used advertising to refill unpopular high rise flats, having decided not to have families with children in them. The local evening newspaper and local radio will be the best bet. One council set up a market stall advertising tenancies. By giving all tenants and would-be tenants more information about what is available out of the total stock, it may be possible to increase their choices and their satisfaction with where they live.
- **Involve tenants as much as is possible** in running and improving their estate, and even in deciding who can move onto it when homes become vacant. **Give some of them paid work** in maintaining, refurbishing and managing it. Encourage them to decide, collectively, what kind of behaviour is tolerable and

what offences against neighbourliness and decency should lead to formal warnings and eviction.

Apathy and in-fighting often disable tenants' associations. And giving tenants control over allocations could enable a group of them to keep out people of different colours and creeds. Tenant involvement is no panacea, but there is no alternative.

- **Take better care of estates and streets** by cleaning litter, responding quickly to tenants' complaints and repairing wear, tear and vandalism. Concierges and 'super-caretakers' employed by the landlord and living on the estate may be the best way of doing this. Everyone's first impression of a place – and of its residents – is based on the amount of neglect and rubbish they can see.
- **Encourage diversity in social housing provision.** In every town and city, you need several landlords competing to provide the best value for money in building, managing and maintaining homes for people on low incomes.

Here are some more radical ideas:

- **Rent abandoned social housing to any takers, free of charge or at extremely low rents.** Better still, give them the title deeds if they live there for a minimum time (say, a year or two) and maintain or improve the property to an agreed minimum standard (weathertight, fit and safe). It is a way of getting value into housing which has become almost worthless. The value accrues to a new owner but that's no matter; some value is better than none.
- **Oblige all tenants to sign a 'mutual aid compact'** – a declaration of intent to be well behaved, to help neighbours in need and do their bit to look after the estate. The landlord would encourage and enable mutual aid by compiling a register of people's needs – help with shopping, gardening, babysitting and so on – and keeping it up to date.[20]

And here's one idea that's completely daft:

- **Run a national advertising campaign to improve the image of social housing,** in order to raise the self-esteem of the people living in it and to attract a broader, more prosperous range of applicants for tenancies. This was the task which the IPPR, a leading left think tank, gave to an advertising agency. They thought long and hard, did some work with focus groups and concluded that there was no point in trying to pull the wool over anyone's eyes. The only really positive proposition which a mass audience would find credible about this particular product was that council housing was relatively cheap to live in, leaving money to be spent on other things. Their copywriter duly devised the ditty set down at the beginning of this chapter.

One way to counter this objection is to say: 'If it ain't broke, don't fix it'. If we find urban neighbourhoods which are working well and where people with choices are happy to live, then we ought to let them get on with it. We could also simply accept that it will be extremely difficult to change the social composition of these neighbourhoods against the will of their residents and not attempt to do so. It is not fair that they have excluded the poor, but that's life. You cannot build much in the way of social housing in suburban Acacia Avenue, less still in Mayfair, because the land is too expensive and the locals will kick up too much of a fuss. In future, however, much owner-occupied suburbia will gradually need to be redeveloped as its houses age, become outdated and less worth refurbishing.[21] That may allow the chance for some subsidised housing to be introduced.

Changing the social mix in deprived neighbourhoods and entire cities will depend on curbing crime and the fear of crime and in lifting the achievements of urban schools. But changing the social mix will, in itself, improve the chances of success in fighting crime and raising educational standards. We turn to these issues in the next two chapters.

Chapter 10

EDUCATION, EDUCATION, REGENERATION

Above all, inner cities are often characterised by expectations of pupils that are too low, by parental and pupil anxiety, by a culture of under-achievement and by a perception that failure is endemic. But while these factors may go some way towards explaining why standards in inner cities are low, they do not justify or excuse them . . . and they certainly do not exempt all involved with inner city education from striving to do better.

Excellence in Cities, Department for Education and Employment Green Paper[1]

What Ofsted and the government will not face up to is the simple truth that most of the factors which affect school achievement are beyond the control of schools . . . if the government really wants to improve educational standards, it will take effective measures to reduce poverty by redistributing wealth.

Phil Taylor, Head Teacher of South Manchester High School,
Manchester Evening News, 25 June 1998

ALDERMAN DERBYSHIRE HAD BEEN doing so badly, for so long, that in January 1999 Nottingham City Council decided to close it and make a new beginning under the government's 'fresh start' procedures. The comprehensive school was only the third in England to be chosen for this most drastic treatment.

Fresh Start really is the last resort for a school which the Office of Standards in Education's dreaded inspectors have judged to be failing and which proves incapable of making substantial improvements; the only alternative is complete closure. A new head is selected, all the teachers have to reapply for the jobs and the expectation is that many will choose to leave or fail to be re-appointed. A new board of governors is drawn up and even the school's name is changed.

I arrived at Alderman Derbyshire at 8.30 a.m. a couple of days after the school got the news, intending to write a long piece for *The Independent*. I asked to see the outgoing head, John Dryden, having had no time to make an appointment (the newspaper wanted the article in a hurry). He appeared after ten minutes and told me he wouldn't be able to give me more than a few minutes because he had much to do. But we got talking and in the end he and a senior teacher gave me over an hour and plenty of coffee. Afterwards I spoke to a few more teachers and some pupils.

The reason this school on Nottingham's northern fringe was so helpful, why its harassed head was willing to drop everything and talk, was that it was desperate to explain itself to anyone who looked and sounded sympathetic. He and his staff were fed up with the negative, abusive coverage Alderman Derbyshire had received from the local and national press. It was a tough school to teach in because it serves a deprived area; they were weary of the extra burden of being branded as failures. They leapt at any chance to defend their youngsters and colleagues against the onslaught of criticism.

One of the school's biggest problems was in hiring and retaining teachers. When I visited there were five key vacancies among thirty-one teaching posts. The head of English whom Dryden had appointed recently – there was only one applicant – had resigned after only six weeks in post. A few months before my visit seven of his staff had been trained in a highly regarded American literacy skills programme by instructors who had flown in from the USA. This training was needed because so many of the 11 year olds arriving at Alderman Derbyshire had reading ages which were many years younger. Now there were only two of those teachers still on the staff. It was the departure of the previous head and two of his deputies – all on long-term sickness leave – which had brought Dryden to the school three years earlier as acting head. He had been parachuted in from his job as a schools inspector.

He insisted that the staff who remained were highly dedicated and I was inclined to believe him. You surely had to be to teach there amid the frequent class-room disruptions, the stigmatisation, the fights between pupils and the occasional threat of violence which teachers faced. But there were hopeful signs, like the spring flowers poking out of tubs by the main entrance, the absence of graffiti and litter, the fresh paintwork – and Adam. He was a big 15 year old with two large gold rings in one ear who rolled up to Dryden as he stood in the school's reception area. Towering over him, Adam clasped his shoulder. 'Sir, come and look at my maths, five pages, it's really good,' he said. The head, slightly startled but smiling delight-edly, promised he would.

As I walked off for a look around the big council estate from which Alderman Derbyshire draws most of its pupils, I spotted a gaggle of youngsters skulking out-side a classroom block during lesson time. They passed below a window, crouching to keep out of view of any teacher, before swaggering off. They had either sneaked out of class or been thrown out for bad behaviour. What would they learn that day?

Alderman Derbyshire looked similar to the Bromley comprehensive school which our oldest child attended. Both are made up of the same bland, shoe box, post-war buildings found at so many state schools. Perhaps the starkest difference between them exists on paper. It is a gap between percentages which can be found in the millions of newspaper supplements of school examination results printed every autumn. At Alderman Derbyshire 5 per cent of 15 and 16 year olds got five or more GCSE passes with grades from C to A in 1998, putting it very close to the bottom of the achievement range for the nation's secondary schools.[*] The corre-sponding percentage for Hayes School in 1998 was 61 – fairly high for a comprehensive school, but still way below the highest achievers. The average for English state schools that year was 46 per cent – nine times higher than Alderman Derbyshire's score. This is an astonishingly wide range of outcomes for a service which is provided by the state on the grounds that education is a basic right and a basic necessity. That Nottingham school was the nadir of a system which fails millions of children.

Two kinds of explanation are given for this great gulf in educational achieve-ment. One involves the very different social circumstances schools find themselves in. Nearly all of Hayes' students come from owner-occupied suburbia in the nation's wealthiest region. Alderman Derbyshire's come from edge-of-city social housing estates in an area hard hit by the closures of every local coal mine and several firms

*In 2000, having had the fresh start with a new head, mostly new staff and a new name it scored 7 per cent. The chairman of the governors told me he expected three years to pass before a big increase but the school had greatly improved.

that depended on the pits. The gaps between the two areas' levels of male unemployment, car ownership, reliance on state benefits and proportions of single parent
families are roughly as large as the ones separating their GCSE results. Peter
Mortimore and Geoff Whitty of the University of London's Institute of Education[2]
wrote:

> It would be odd if having more spacious accommodation, more nutritious food,
> better health, greater access to books, educational toys and stimulating experiences,
> and more informed knowledge about how the system works, did not confer con-
> siderable advantage in any tests or examination,

The other type of explanation invokes wide variations in the quality of leadership
given by heads (and the local councils that still oversee most state schools) and in the
quality of teaching. The 1997 Labour government's first education White Paper[3]
said:

> We know what it takes to create a good school: a strong, skilled head who under-
> stands the importance of clear leadership, committed staff and parents, high
> expectations of every child and above all good teaching,

Poverty or poor leadership and poor teaching – which is to blame for abysmally low
educational standards and achievements? This is the British education system's
equivalent of the nature versus nurture debate and it is just as stale and frustrating.
The two things surely go together. Teaching classes in which a proportion, sometimes a substantial proportion, of children and teenagers are damaged, unmotivated,
withdrawn or disruptive is hard and frequently unrewarding work. Schools serving
deprived areas are harder to run. The task of finding and retaining good teachers
and leaders is more difficult. The great majority of schools which the government's
Office of Standards in Education (Ofsted) inspection force judge to be failing, or
unsatisfactory, are found in such areas.[4]

Repeatedly, sternly, Ofsted points to the wide variation in exam achievements
between schools whose pupils suffer similar levels of deprivation. It demands
improvements of the worst. But even if they made great advances, schools serving
poor areas would still generally lag behind those with more affluent intakes.
Deprivation makes a rotten foundation on which to build a learning society. Poor
parents and their children usually have little reason to believe in the life-transform-
ing powers of education. Children's intellects are not stretched and stimulated in
homes where parents are struggling to cope, often without a partner. Teachers find
a scattering of their charges on the edges of prostitution, drug dealing and lives of
crime. They, and the school inspectors, speak of troubled urban schools lacking a
critical mass of motivated, biddable pupils. They are battling against anti-learning
peer group pressures. If we want a finer social grain within our cities, and for them

to become prosperous and attractive to people with choices, then most urban schools must raise their game and the gulf between the highest and lowest achievers has to be narrowed.

Some might argue that schools are of secondary importance in sustaining and regenerating cities. After all, three out of every four households are now childless; at any one time only a minority have a direct interest in the quality of the local schools. But many single households will become couples, many of those couples will have children and it is bad news for the city if parents with jobs and aspirations feel they need to move out because they cannot trust the local schools. In large areas of our conurbations almost all of the resident children belong to the poor. That isn't right. An urban renaissance that is based on luring prosperous but invariably childless households into the city isn't worthy of the name; it has to bring in toddlers and teenagers too.

This should increase equality of opportunity, because peer group effects can be benign as well as malign. The social class of a child's parents and the education they received are powerful predictors of its own educational achievements, but so too are the social class and education of its classmates' parents. Large surveys have suggested that the presence of children with affluent, educated parents can raise the achievements of others in the class.[5]

Let's look at how the hierarchy between the highest and lowest achieving schools develops and is maintained. Under a long-established system in which most children receiving a state education went to their local school, those institutions serving the more affluent areas gained better examination results, won the highest reputation and became most popular. Once estate agents start using the nearby presence of a good state school in their marketing of houses, you can be fairly sure it is pushing up house prices within its catchment area. The school is effectively selecting pupils on the basis of how large a mortgage their parents can afford. As its intake becomes more affluent, the lead it has over the rest of the pack will tend to widen. State education is ostensibly free at the point of delivery but a shadow market in house prices has developed, reflecting the widely varying examination results of schools.

Everyone knows this, but it wasn't until 2000 that it was neatly and conclusively demonstrated by two Warwick University economists. They studied the impact on house prices of location in the catchment area of the two most desirable comprehensive schools – the most over-subscribed, and achieving the highest CGSE and A level results – in the western half of Coventry. For one school the effect was a 19 per cent price boost, for the other a 10 per cent increase. Being in the right catchment area was typically adding between £10,000 and £20,000 to property values.[6]

The great school reforms of late Thatcherism, in particular the 1988 Education Reform Act, made state education much more like a marketable commodity. Parents were given a statutory right to choose which school their children should go to. The powers of local councils to control and finance schools and to allocate pupils to them were curtailed and schools were funded chiefly on the basis of how many pupils attended them; 'the money followed the pupil'.

Now if, under this quasi-market system, high-achieving schools with the best teaching and leadership were simply able to carry on expanding in order to meet demand, then more children would have the chance of a good education. But enlargement is no easy matter. A popular school may lack land for growth and, even if it has the necessary space, it will need funding from central government or the local council to pay for extra classrooms and other facilities; not enough money follows the pupil to meet the major capital expenditures of expansion. Besides, the teaching profession is understandably uncomfortable with schools having more than about 1,500 pupils. So successful, high-achieving schools are invariably over-subscribed and have to turn many parents and children away.

The fate of low-achieving schools in this quasi-market system is to become under-subscribed. As their rolls shrink, so does their income. Troubled, failing schools usually do have spare places; Alderman Derbyshire, for instance, had room for nearly double its 470 pupils. The theory was that such schools would then come under intense pressure to raise standards, for if they failed to do so they would eventually close. Improve or die; either way the problem school ceases to be a problem.

The business of getting a child into a good state school can be a more complicated matter than simply living close to it, as many a rueful parent will tell you. Different types of school have different admissions criteria, some set by the local education authority, some set by the school itself. A minority select partly or entirely on the basis of academic ability. And then there are the thousands of church schools which expect the parents of their pupils to profess – to some degree – a faith. Many parents, including Mr and Mrs Blair, succeed in sending their children to popular, high-achieving state schools which are several miles from their home.

Even so, over-subscribed schools mostly continue to select pupils on the proximity principle or some version of it.* The link between affluent neighbourhoods and high-achieving schools persists, and may even have strengthened, despite the introduction of parental choice, which has turned out to be pretty meaningless for most poorer households. They can state a preference that their child attend the area's best state school, several miles away on the other side of town, and then be turned down. Where proximity is not the main factor influencing admission chances, the more affluent and aspirational parents usually still have an advantage over poorer ones when it comes to getting their children into a good state school. They are choosier and more clued up on the hierarchy of local schools and their differing examination performance. They are better equipped to

*Selection based on pupils attending a feeder primary school or on their siblings having attended the target school is closely related to proximity.

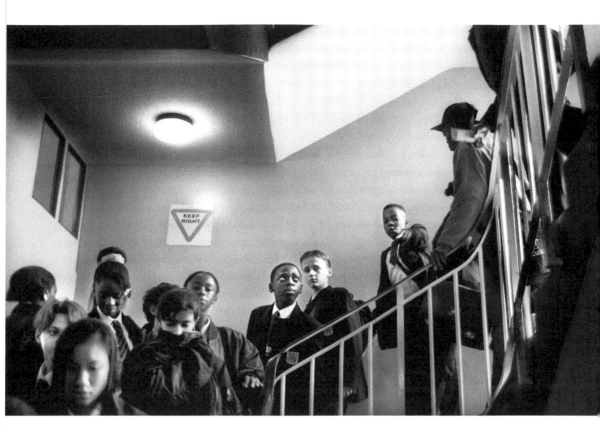

Secondary school, north London

help their children travel miles from home to school and back if that is what it takes. And if selection on academic ability is involved, then their children have a huge head start.

Desirable schools generally maintain their high achievements or improve on them. Poor teaching, criticisms from school inspectors and any deterioration in examination results evoke anxious parental inquiries. But a low-achieving school in a poorer neighbourhood can easily find itself sliding further downhill. Parents usually expect less of their children and less of education. Those who demand more withdraw their children or never send them there in the first place. Once the school has spare places the local council starts to send the waifs and strays that cannot be fitted into the area's popular, over-subscribed schools. In come pupils excluded from other schools for bad behaviour, children from refugee families who cannot speak English and from the mobile poor newly arrived in the borough. They tend to add to a problem school's problems.

in the outer boroughs. So, too, had those other people we knew who were poring over school league tables, getting Ofsted school reports off the Internet, moving out – often clean out of London – as their children neared secondary school age or choosing private education. It could be that our children would have learnt just as quickly and done just as well if we had stayed in Greenwich. The higher GCSE and SATs (the national tests children take at the ages of 7, 11 and 14) attainments of Bromley primary and secondary schools could be entirely due to their pupils coming from more affluent, education-minded households. Teaching in Greenwich might even have been superior to that in Bromley.

But I doubted it. I suppose I had two things in mind when weighing up whether to leave Greenwich. One was that we ought to send our three children to schools where the average achievement was high, because chances were that they would be around average. I had vague notions that peer group pressure and school expectations would serve them well. The second was that I did not want them to go to schools with too many rough kids from poor estates. To be with a few was fine, indeed valuable and, as Ofsted's former chief inspector Chris Woodhead might put it, 'experiential', but not too many. Don't many parents think along the same lines?

School choice is a complex, murky affair in which a school's academic achievement is only one factor, albeit a fairly important one given that league tables of SATs tests and examination results gain enormous publicity. Parents are trying to pick the social class and, sometimes, the race of the children their children go to school with. They fret about drugs and teenage pregnancies. If you ask them what exactly it is they are selecting for, they may not be completely frank. Nowhere is the process of school choice so fraught as in London, where a huge middle class shares a state school system with hundreds of thousands of deprived households. The solution for many is to opt out altogether and use private education.* In the inner boroughs, where deprivation is more abundant and social polarisation at its starkest, 13 per cent of children go to independent schools – almost double the 6.7 per cent for England as a whole and much higher than the 8.3 per cent of the outer London boroughs.[15] Others fake their address and attempt to cheat their children's way into state schools, or move out to the higher-achieving shires.

Here is Trevor Phillips, distinguished broadcaster, would-be Labour Mayor of London and chairman of the Greater London Assembly, explaining his decision to have his daughters educated privately rather than sending them to the local north London comprehensive where only 14 per cent of pupils got five or more Grade A to C GCSEs:

*Independent education fosters enormous inequality of opportunity and is one of the great social fault lines in the UK. As the former higher education minister George Walden pointed out in his book *We Should Know Better*, London, Fourth Estate, 1996, they exempt most of the nation's elite from any personal

This wasn't somewhere where working-class children might suffer but middle-class ones would do okay. Something was so drastically wrong you are effectively saying to your children I don't care if you fail to fulfil your potential.[16]

Across Britain, the school selection stakes are being upped. The government's emphasis on the supreme importance of education is decent and desirable, but it makes the many parents who get the message even more aware of the need to get their children into a good school. Thanks to the decades of expansion in higher education there is also a rising proportion of graduate parents who value education and have high expectations of their children and of the schools they send them to.

The fact that schools tend to drive people with choices out of large towns and cities is particularly tragic because these should be the best place to combine satisfying work with raising a family. They offer children and parents the widest range of things to do and see. By providing a huge concentration of jobs and homes, they maximise the chances for both parents to work close to home. They also provide the largest availability and choice of paid childcare.

The New Labour government's response to this dismal hierarchy of very good to very bad schools was to put increasing effort into making bad schools better while raising overall standards. Much of the blizzard of initiatives following its 1997 election victory has been concerned with improving weak, low achieving schools serving deprived areas.

It started with 'Fresh Start', mentioned earlier in this chapter, and Education Action Zones – which covered parts of every large conurbation in England by 2001. It moved on into the Excellence in Cities programme, launched in 1999, which was due to embrace a third of all secondary schools by 2002. As the election neared a new plan for private-sector backed City Academies to replace some underachieving state schools began to be implemented; these rather resembled the City Technology Colleges that had been the Tories' flagship inner city education initiative a decade earlier. Along the way we have had superheads and superteachers, beacon schools, summer schools, after-hours schools and specialist schools, the latter marked for major expansion in Labour's 2001 election manifesto. There have been all manner of pilot schemes in poor urban areas; £30 a week payments to encourage teenagers to stay on at school after the age of 16, pagers for the parents of truants, second-hand computers to connect families to the Internet. In an education Green Paper issued early in 2001 the government announced that it would run a trial of 'pupil learning credits' – extra funds of up to £360 a year per pupil – in secondary schools in deprived areas, enabling children to benefit from extras such

interest in state education. But as far as pushes and pulls from the city are concerned, their influence seems pretty neutral. A former Conservative cabinet minister suggested to me that urban regeneration should include the foundation of new private schools in inner city areas. I disagree.

as music lessons, museum and theatre visits and additional tuition that more affluent parents pay for out of their own pockets.[17]

Summarising all of this frantic activity is not easy but a few clear themes emerged. There were, increasingly, extra funds and other help for schools with difficult intakes, but no consensus that this was enough to put them on a level playing field. The government was desperate for results and ready to use a variety of sticks – the ultimate being school closure and job losses – as well as carrots. It was willing to fund any number of trials and experiments on a small scale, in and out of schools, not only in the hope of finding measures that worked but also in order to be seen to be doing things.

Ministers hoped that improved schooling could help break cycles of poverty which have revolved for generations. But they were particularly anxious about gifted children failing to fulfil their potential in low achieving state schools. One reason is that as economic growth becomes increasingly dependent on handling knowledge and information, the country needs all the talent it can get. But this special concern to ensure potential high achievers are not let down in school should also play well with earning, ambitious parents who worry about their children learning and passing exams in inner city schools.

Two other core tenets of New Labour's faith were that improved leadership was the key to improving low achieving schools and that diversity was a good thing in itself. As in other fields, Ministers had grave doubts about the competence of local councils to run state schools. A growing number were found wanting by Ofsted and compelled to contract out key parts of their education service. While stopping short of abolishing local education authorities outright – at least for the time being – government encouraged councils to hire outside organisations, or the schools themselves, to take over their education service tasks. Ministers also encouraged Christian Churches and other faiths to relaunch and run failing schools. The six new City Academies it had proposed by the end of 2000 – with more to follow – were independent of local councils. Their running costs were funded by the state with major companies, wealthy businessmen and churches contributing to their management and their foundation costs.

If we needed a modern state-sanctioned religion then we could, for a couple of years in the late 1990s, have found it in the cult of the hero and heroine head rather than the Church of England. Hundreds of press articles were written and items broadcast about these supermen and women turning around sink schools. The typical specimen hit the ground running, persuading several failing teachers to leave and expelling a dozen or more of the most disruptive pupils. The policy on uniforms was tightened, the school had a fresh lick of paint, a wildlife garden appeared in the grounds and graffiti and litter vanished. A new culture quickly spread through the school, infecting the teachers, more and more of the pupils and their parents too. Bad behaviour was no longer tolerated, teaching improved and lessons became more challenging and interesting. The students' work was monitored much more carefully and fewer of them were content to do nothing except mess around in class. The head, an omnipresent, all-knowing figure, 'talked up' the school and its reputation rose off the floor.

Alas, improving schools could raise their exam passes drastically and yet still be performing poorly by national standards.* The successful heroine head has seen the proportion of pupils getting five or more GCSE A to C grades in her school more than double in a few years – from, say, 10 per cent to 25 per cent. That puts it among the usual run of big city comprehensives. That improvement is a real achievement which improves the prospects of dozens of local young people each year, but its results remain far below the national average. The school is applauded for doing extremely well, considering the level of deprivation among its intake. But would you send your child there? Would a parent with choices choose it?

In any event, the cult of the superhead has collapsed following troubles and failings at several 'fresh started' schools and some well publicised resignations. As for the overall policy of combining pressure and support in order to improve low achieving schools in urban areas, the government could claim some success. GCSE results in schools covered by the Excellence in Cities programme improved more than the national average in 2000. So had SATs test results for numeracy and literacy among 11 year olds in Education Action Zones. By the end of 2000, more than 600 failing schools placed under special measures in England had had them removed following improvements in teaching, while 122 had been closed and twenty-five given fresh starts.[18]

But why not tackle the inequality-magnifying fundamentals of the state education system? While eventually committing itself to very large increases in education spending, with a focus on low achieving schools serving deprived areas, New Labour in power made only small adjustments to the polarised, quasi-market system laid out at the acme of Thatcherism.

If, instead, we could realise the old ideal of socially balanced neighbourhoods throughout the land along with another old ideal of universal comprehensive schools, then children would simply go to their nearest primary and secondary schools, which would all be of pretty much the same standard. They would all have the same quality of education and an equal chance to fulfil their potential.

But neighbourhoods are not socially mixed. So delivering real equality of opportunity in education would require either much more redistribution of resources to the schools in poor areas or an evening out of the intakes, ensuring every school had roughly the same mix of social classes and abilities among their pupils. Children would have to be assigned to schools on the basis of a combination of aptitude tests and their parents' earnings, or type of employment.

183

*In 1998 the Ridings School, a failing Halifax comprehensive that had received a great deal of press coverage, was judged by Ofsted to have greatly improved and was taken off special measures. But the proportion of 15 to 16 year olds obtaining five or more GCSE grade C to A passes fell that year from 6 per cent to 3 per cent. By 2000 it was still only at 9 per cent.

Which is quite out of the question. Any government which tried to run such a dirigiste system of pupil allocations with all the assessing and testing involved would face enormous opposition from the massed ranks of middle England. To work fairly and overcome the effects of urban deprivation, such a system would have to operate across entire regions, with children being bussed for miles. It might have a slightly better chance of succeeding if it operated only within single council areas, although this would fail to deliver a fair mix nationally. But here, too, there are big problems. Parents have the legal right to send their children to school in a neighbouring council's area. And many state schools have opted out of local council control and run their own admissions procedure. Such a local allocation scheme would encourage people with choices to move out of the area altogether, or switch to private education.

It did the former in the case of our family. We had been considering sending our eldest child to a comprehensive school which was not the closest to our home in Greenwich but which got better exam results. All Greenwich primary schoolchildren took a test in their final year which sorted them into three ability bands, in order for the council to try and even out the intake of its secondary schools. If your child came in the top band, that reduced your chances of getting her or him into a higher-achieving school. It was a well meaning scheme that aimed to increase equality of opportunity within the borough's school, but it increased our incentive to leave.

The New Labour government wanted to give extra help to poor schoolchildren, but this could not be allowed to cause significant disruption and any dismay for the majority of parents. It wanted to go with the *zeitgeist* of increasing choice and growing diversity. That is why it planned for almost half of all English secondary schools to become specialist schools – specialising in arts, sports, technology, languages, science, engineering and business – by 2006.[19] They can select a small proportion of their pupils on the basis of their aptitude for the specialism. That is why it did little to threaten the remaining grammar schools, despite the then Education Secretary David Blunkett having vowed, in opposition, to end selection.

Many teachers and parents sigh and wish Britain could be like other Western nations where there is said to be far less fuss about variations in school quality and where parents are more content to settle for the local state school. International comparisons show that Britain's state school are among the more polarised. But, as an OECD report has shown, the move towards greater choice in schools has happened in many developed nations, driven by a belief in the improving effects of consumer choice and competition.[20]

An across-the-board policy of fair allocations and evening out of school intakes seems politically impossible. We have to start from where we are now – strong hierarchies, polarisation, and a cloud of anxiety and dissatisfaction hanging over education in the big cities. What more could be done to make choosy parents feel confident about urban state schools? Here is one bad idea for making education work for urban regeneration, followed by six good ones.

NEW GRAMMAR SCHOOLS

We could bring back selective state schools to the more deprived areas of large cities, insisting that parents had to live close to these new grammars in order to send their children there. Children from higher social classes have a better chance of passing the entrance exam for grammars, which cream off a minority of pupils at the transition from primary to secondary schools. Once they are selected, the benign effects of peer groups help them to excel. Such schools could be a very effective way of attacking counter-urbanisation. They may be part of the explanation for Birmingham retaining a larger middle-class population than, say, Manchester with its all-comprehensive schools.

But it is hard to see how a special, urban case can be made for a system which has been rejected nationally by both the voters and most educators and educationalists.* There is no consensus that a selective system leads to higher overall educational achievements than a comprehensive one. Such a system is extremely hard, and often unfair, for the majority of 11 year olds consigned to schools regarded as second class. The best reason for rejecting grammar schools as a tool for urban regeneration is that they offer nothing for the remainder of schools in a city. Indeed, they are likely to drag them down by creaming off the brightest pupils with the most education-minded parents. This was the main criticism made of the few City Technology Colleges established by the last Conservative government, mostly in inner city areas. The CTCs did not actually set an 11 plus entrance exam but they did interview prospective parents and pupils and were seen as being selective.

It may be that selection has a part to play in a secondary school undergoing a fresh start and trying to shake off its previous reputation as a failure. Imagine if, say, it was able to select a third of its students by ability from across a wide area of its city. The chosen third would boost the school's achievements and create a critical mass of bright, motivated pupils. It would still serve its deprived local community, but it would be doing a better job. Selection might, then, be justified if it benefited the local children without doing any significant harm to the remainder of the city's schools.

---

*The Conservatives stood at the 1997 election on a platform of extending the grammar school system. The Labour government that took office has given local parents the right to organise local ballots to abolish grammar schools in their area. The signs are that very few ballots will be organised because so many parents in the area involved – one fifth – have to sign a petition calling for one. In the first of them, held in Ripon in 2000, parents voted two to one to keep the town's grammar school. Nonetheless, the grammar school system does not apply across most of Britain and the ability of those schools which had, under the Conservatives, 'opted out' of local authority control, to select bright pupils for entrance has been curtailed under Labour.

FAIR LEAGUE TABLES

We need a fairer, more objective way than examination and SATs league tables for assessing how good or bad schools are. One method, favoured by the government but not yet introduced, is to compare how effective they are at raising the achievements of the average pupil over an interval of several years – the value added. For example, you might compare each pupil's SATs tests results at the age of 11 or 14 with her achievement at GCSE five or two years later on. The best schools are those which enable their pupils to make the biggest gain in grades compared to national averages.

Another method is to take the levels of deprivation in a school's intake into account when its GCSE grades and SATs scores are publicised. If that was done, it could be shown that pupils in some inner city schools were being better taught, more stretched and stimulated, than 'coasting', under-achieving schools with feeble teaching in prosperous suburbs – even though the latter scored more highly in tests and exams. This would make some parents reconsider their school choice. (But not those who are as concerned about the social class or race of their children's classmates as about the quality of teaching. For them, a school whose raw GCSE results were average but whose deprivation-adjusted results were brilliant would be one to avoid, on the grounds that there were many deprived, albeit well taught, children there.)

Between 1997 and 2001 the government tinkered with making the performance tables fairer, carrying out trials and writing reports.[*] It should consult on and then introduce such a weighting system and then apply it to every primary and secondary school. Most of the data needed to do the job reasonably well are already available. The league tables should cover both the levels of deprivation in school intakes and 'value added' per pupil according to exam and test results. They could either feature both raw and adjusted results, side by side, or give each school a grade from A to F for how much its pupils achieved given their home circumstances.

HIGHER SALARIES FOR TEACHERS

There is an strong case for substantial bonuses to be paid to the teachers working in deprived urban areas, both for taking a job there in the first place and for each year completed. Don't take my word for it:

---

[*]Measures of 'value added' between SATs at age 14 and GCSE exams two years later had been estimated for some schools, and for value added between 16 and 18 (GCSE to A levels) in others. But no nationwide assessment had been published at the time of completing this book.

Any government committed to improving education in disadvantaged areas . . . would need to fund education institutions in those areas much better (*sic*) and allow them to pay higher salaries to attract the best teachers to the places where they are needed the most.[21]

Thus wrote the leading educationalist Professor Michael Barber, then a senior adviser to the Education Secretary David Blunkett, shortly before Labour won power in 1997 (he went on to head the government's new Schools Standards and Effectiveness Unit). Teaching in tough schools is demanding and debilitating and they struggle to find enough high quality applicants. The lack of good teachers, the frequent staff changes and a heavy reliance on supply teachers add to the disadvantages of the children who attend.

Salaries of £80,000 or more are now offered to heads who take on the challenge of turning around sink schools, and some bonus money was also made available for chalk face teachers in the Excellence in Cities and Education Action Zone areas. After great struggles the government succeeded in introducing a modest level of performance-related pay for classroom teachers everywhere. With mounting evidence of teacher shortages not only in difficult schools but across the state system, Ministers were compelled to introduce new incentives for entering the profession everywhere – salaries for postgraduate trainee teachers, 'golden hello' bonuses for teachers in shortage subjects and a proposal that the government would pay off their student loans over ten years if they remained teachers in the state sector.[22]

But it was not until the beginning of 2001, with an election coming up, that Ministers announced further funds earmarked for teacher bonuses in the most low achieving secondary schools. (Its Green Paper of that year also announced that schools would have discretion to offer individual teachers recruitment and retention payments of up to £5,000 a year, although this was intended more for high cost areas – where teachers find it difficult to afford housing – rather than for especially challenging schools.) With at most £70,000 per school, the extra funds worked out at around an extra £1,500 a year per teacher. Rather more than that will probably be needed to bring these schools the good teachers they need, and to keep them there. An extra £5,000 a year could make a significant difference, but it seemed unlikely that most teachers in most deprived areas would be offered such an incentive.

BE OPEN-MINDED

I was talking to a former neighbour in Greenwich who is a schools inspector about this chapter. He had decided to send his eldest son, who is a couple of years older than ours, to the local comprehensive – the one we shunned – and the boy passed

10 GCSEs with good grades. My friend said he had been fairly confident his son could do well there, given some extra private tuition in maths and his own intimate knowledge of the education system. Then he asked me a question: what could that school have done to make us at least consider sending our children there?

I remembered that once I had looked up its GCSE scores in a newspaper supplement I never gave that school a chance. I did not step inside, ask for information or speak to any teacher. That sort of attitude loads the dice against any school whose examination achievements are below or around average. I could and should at least have taken a look at the place.

DON'T DISCOURAGE DEMANDING PARENTS

Urban schools need to encourage parents who have high hopes and expectations for their children. The principle of 'the money follows the pupil' has put schools under pressure to market themselves with glossy brochures, advertising and open evenings. Those in which a large proportion of the pupils come from deprived homes need to think hard, and carefully, about what they can do to reassure aspirational parents that their child's potential really can be fulfilled. Merely declaring it – and they all do declare it in their mission statements and so forth – is not enough. They need to find ways of convincing parents that even though the school's overall attainments are average, or below average, on paper, that is no barrier to individual success. Examples could help.

The last thing any school should do is make such a parent, or groups of such parents, feel they are interfering or resented or a pain. Schools and teachers whose attitude is: 'Our job is to try to educate poor kids as best we can, it's very hard work, society is against us, don't expect too much and just let us get on with it,' are making a vigorous contribution towards urban decay.

NEW URBAN STATE SCHOOLS

If urban regeneration is tied to the construction of new housing for sale (and I argued in Chapter 8 that it almost always should be), then new state schools could be a strong lure for families to buy in the locality. They would have the reassurance of knowing that their children could be educated alongside their peers, the children of other owner-occupiers, in a spanking new institution with up-to-date equipment. The school would help to create ties within the new community and because most of its pupils came from owner-occupier homes, it could expect and maintain high standards. The costs of building the new school would be met as part of a regeneration deal negotiated between housebuilding firms, the local council and grant-awarding bodies such as the Regional Development Agency.

But a new school would only be an option if a substantial number of new homes were being built in an urban neighbourhood. A primary schools with fewer than 150 pupils aged 4 to 11 is struggling to be viable in an urban area. Below that size it becomes difficult to have the usual class sizes for each age group. The costs per pupil escalate and it is difficult to obtain enough teachers to cover the curriculum. If we assume that most of the new housing consisted of family homes with three or more bedrooms, then about 500 new homes could provide enough pupils to justify an entire new primary school.

About 20 hectares of land would provide enough space for 500 homes, most with gardens, along with some public green space and a new primary school provided the housing was built at fairly compact urban densities. Single development sites of that substantial size do not often become available within cities. It would, however, be quite usual to have several smaller derelict or under-used plots amounting to 20 hectares within the catchment area of a typical urban primary school – a circle with a diameter of one and a half kilometres.*

It's possible, then, to conceive of an attractive package in which the local council and housebuilders make a joint pitch to potential house purchasers along these lines. 'We are building 500 new homes for sale in this neighbourhood over the next three years. After two years we guarantee a new infants school will open its doors for children aged 4 to 7. Two years after that a new junior school will open on the same site, taking children aged 7 to 11. We intend this new school to have high standards, and the children attending it will come primarily from the new housing. They are guaranteed admission to it.'

The glaring omission from this scheme of things is, of course, the children who already live in the area. They would continue to go to the existing schools nearby, whose intakes might well be dominated by children from poor households. The new school would create a kind of educational apartheid in the area between the neighbourhood's old and new families. This would be unjust and intolerable.

There are ways of making a new urban school acceptable to all. It could, say, have an admissions policy which took one-third of its pupils from families which were long-term residents of the neighbourhood. The school could set and achieve high standards and be a powerful magnet for incoming families while giving those children whose parents lived in the area before the new housing was built a better educational start than they would otherwise have had. It may be possible to base admissions to the new school simply on the proximity principle and still get this kind of balance between the children of existing and new residents.

But a policy of new schools for urban regeneration has one further major obstacle to overcome. In declining areas the local schools are often under-subscribed

---

*The area of such a circle is 180 hectares, so about a tenth of the land within it would have to be available for development in order to provide sufficient homes and sufficient pupils for an entire new primary school.

and facing closure because of years of population loss. In Hulme, for instance, just south of Manchester city centre, more than 1,500 new owner-occupier and housing association homes had been built by 1999 on land cleared of failed 1960s' council homes, with a further 1,500 to follow. Yet far from this huge investment paving the way for new schools, two of the six primaries serving the area closed that year because of falling rolls across the wider area. It would be difficult to justify the building of new schools for residents if their arrival could instead give a new lease of life to under-subscribed schools facing the threat of closure.

The answer, in this case, is to give schools a fresh start linked to neighbourhood regeneration and the building of new, for-sale housing close by. This is something that hardly ever happens with the current 'Fresh Start' regime. Housebuilders and the local council could put this sort of proposition to potential purchasers: 'The local school is making a complete break with its past and it will achieve high standards. Parts of it are being rebuilt, new equipment is being purchased and all the existing staff – from the head downwards – will have to re-apply for their jobs. We have plans for the pupil numbers to grow and most of the children will come from the new housing being built here.'

Investing in a new primary school or relaunching an existing one could provide benefits for other local schools. They could share some teachers and facilities. The local secondary schools would also benefit, because the high standards set by the new primary school would raise the levels of attainment among their intake of 11 year olds.

It might occasionally be possible to found new secondary schools as part of the process of regenerating wider urban areas. The minimum viable size for an urban secondary is about 500 pupils, which means at least 2,000 new homes would have to be built within its prospective catchment area (of roughly 500 hectares) in order to provide sufficient children. The problems with this policy are just the same as those associated with founding new primary schools in urban areas. But the same solutions apply; if it is not feasible, rational or just to found an entirely new secondary school, then an existing one needs to be transformed.

THE COUNCIL TAX SCHOLARSHIP

One way of attracting the motivated, encouraged pupils, which struggling urban schools will need in abundance if they are to raise their game, is to pay their parents to send them there. Government Ministers or educational think tanks foolish enough to propose such a scheme would have abuse heaped upon them. But here is how the idea might work, in the guise of a council tax scholarship for secondary schools.

A participating school would have to meet three criteria. The proportion of its pupils coming from disadvantaged households (usually registered by the number of pupils receiving free school meals) would have to be higher than the regional average. Second, its pupils' achievements, as measured by SATs tests and GCSE results, would have to be below the regional average. And, third, the school would have to be an active partner in an urban regeneration scheme which aimed to attract new families into a deprived area and to halt the exodus of those with choices.

The parents of any pupil whose abilities were assessed as well above average – say, in the top eighth of the range – would be offered a 100 per cent rebate on their council tax provided they met certain conditions. The child would have to remain at the school, the parents would have to live within a specified distance of it, say, 1 kilometre, and the pupil would have to match his or her strong academic ability and intelligence with continuing hard effort, high achievement and reasonably good behaviour. Each scholar would be appraised at least once a year to make sure they were making satisfactory progress and were remaining in the top ability and achievement band. Scholarships could be withdrawn if the pupil – or their parents – failed to show adequate commitment. This would be harsh but essential if the scheme is to have any credibility.

The average council tax in Britain's big cities was around £700 per annum in the year 2001.[23] Avoiding it through gaining a scholarship would give parents an incentive to stay in the area and use the local secondary school. There are good reasons for advocating a rebate on the council tax (which varies from home to home and almost invariably rises from year to year) rather than simply giving parents a fixed annual payment. The council tax pays towards local community services, so exempting a family from paying it tells them that their presence in the community and their use of the local school are making a valued local contribution. It also holds down the costs of the scheme, since one family could have two or more children who were scholars but it would only get its council tax paid once.

The levels of deprivation around many of the candidate schools imply that a substantial proportion of those parents with scholarship-potential children would already be relying on benefits and would have their council tax paid for them by the state in any case. And, scholarship or not, their child would probably have gone to that school because their parents lack choices in where they live.

In those circumstances, welfare-dependent parents with talented, motivated children should be given half of their scholarship money in cash while continuing to have their council tax rebate paid (and while escaping any reduction in other state benefits). An extra £400 or so a year could mean a lot to a couple or a single parent living on welfare.

Would urban schools offering the scholarships start to attract a higher proportion of children from homes where education was cherished, books read and university entrance contemplated? Would parents with choices in life whose children could win council tax scholarships send them to what, after all, would still be a low-achieving school in a poor neighbourhood?

Some would, for a combination of reasons. There would be the incentive of having to pay no council tax. And the pride and confidence which would flow from having their children earmarked as talented and of high potential. But the biggest lure would be the security of knowing their child's progress was closely monitored, that he or she was expected to excel at school and was being helped to do so. For the school itself would have a strong interest in its council tax scholars making good progress and not forfeiting their scholarships. The more of them it has, the better its exam results and the easier it is to attract parents and good teachers.

Imagine that the introduction of this scheme completely failed to change a school's intake. In that case, the only change would be that a small proportion of pupils were granted scholarships when they would have gone to the school in any case. But even that would be a win. They and their parents would be shown that effort and achievement in school are worth something, not only to themselves but also to their school and society as a whole. This is a message that too few of the children in struggling schools are being given at the moment. Council tax scholars at such a school would be more encouraged, more likely to become sixth formers and to go on to higher education. So even if the introduction of scholarships did not change the intake, the school's achievements could be expected to rise – which might in itself change the intake.

Now imagine that the scheme does succeed in winning more and more pupils from homes where the parents have middling or higher incomes and aspirations. There would come a time when the school no longer met the criteria for awarding council tax scholarships; it would have become too successful. There would be too few pupils from deprived homes and its SATs and GCSE results would be too high. In that case, it would have to stop awarding new scholarships while allowing pupils already in receipt of them to retain them until they left. It would no longer need to offer them because it would have changed from being an under-achieving school to one which was at least average. And average, in national terms, is a standard many urban secondary schools would dearly love to reach.

But how could it ever be right to pay substantial sums of money (in foregone council tax) to comfortably off people in order to give their children what is already a 'free' state education? We know that children from wealthier, more middle-class families would have a better chance of winning the scholarships because of the advantages their home backgrounds give them.

The ends justify the means. If we are serious about turning around failing neighbourhoods and their failing schools, then we do need to think the unthinkable and ask politicians to be courageous. There is, however, a way of making the council tax scholarship slightly less outrageous. You could impose a voluntary means test. The school would ask any scholarship household with a total income of £40,000 a year or more to give half of the money back to the school, where it would be spent on extra teaching and equipment. Any household earning £20,000 a year or less would be expected to keep all of its scholarship money. There would be a sliding scale of requested school donations for those with incomes between £20,000 and £40,000.

Most of the costs of the scheme would be paid for by central government, which already funds the great bulk of state education. These costs would not be huge. If the scholarships were awarded to an eighth of the pupils at 600 under-achieving secondary schools (a fifth of the total of non-selective schools in England), the scheme would cost under £50m a year. And £50m is about a third of the cost of the last Conservative government's Assisted Places Scheme, which used taxpayers' money to send children from low and middle-income households to private schools.

The participating school, or a feeder primary school, would put forward scholarship candidates on the basis of teachers' assessments and written work including classroom tests, homework and SATs tests. Many, perhaps most, would be chosen during their final year in primary school. The secondary school could set scholarship tests. Whatever procedure was used to make the award would have to be robust and proofed against criticism from both educational experts and parents.

Those parents whose children made the grade would then be asked if they wanted their child to be a scholar. To clinch the award they would have to sign a written commitment to support their child's education and the school. Some new scholarships would be awarded to older pupils in each secondary school year in order to replace those scholars dropping out, to encourage high achievement among non-scholars and to lure in some talented older entrants to the school.

The entire programme, including scholars' records, would have to be open for scrutiny and assessment by independent auditors to ensure that it was being run fairly and was an effective use of taxpayers' money. The risk of scholarship children forming a privileged school within a school, along with all the resentment that would cause, also needs to be tackled. But if only one in eight pupils was a scholar they would be too few to fill any classroom on their own, even in subjects where the year is split into sets by ability. They should spend very little of their time being taught in scholar-only groups and the school should not give them any special privileges.

Imagine how the scheme could change a school and a neighbourhood. Bright children burdened with the disadvantage of growing up in a deprived home would get a better education and more encouragement. The school's achievements would rise along with its reputation and the head would find it easier to hire and keep good teachers. As the school's intake changed, more children from middle income and prosperous families would be schooled alongside the urban poor instead of in homogeneous suburbia. The school's catchment area would become less poor, more socially mixed.

Imagine how it could change schools across the nation. The gap between the best and worst would shrink, the arguments and anguish about the variation in school quality would diminish and education could deliver more equality of opportunity. More children would go to their nearest school, making it easier for them to socialise with school friends out of school hours. There would be more walking to class, less traffic congestion caused by school runs. Council tax scholarships would cost no more than £50m a year – about a thousandth of the total UK state education budget. They are surely worth a try.

Chapter 11

THE FRIGHTENED CITY

It was not true that 'nothing works'. The aim, in putting together this report, was to sum-marise the evidence that 'some things work under some conditions'.

Reducing Offending, Home Office Research Study 187[1]

It is time for me to hurry to work too, and I exchange my ritual farewell with Mr Lofaro, the short, thick-bodied, white-aproned fruit man who stands outside his doorway a little up the street, his arms folded, his feet planted, looking solid as earth itself. We nod; we each glance quickly up and down the street, then look back to each other and smile. We have done this many a morning for more than ten years and we both know what it means: all is well.

Jane Jacobs, *The Death and Life of Great American Cities*[2]

Fulham is what they call socially crunchy, which means that it has very rich people with very poor people living nearby to clean their houses in the daytime and burgle them at night.

John O'Farrell, *Things Can Only Get Better*[3]

**Fortified corner shop in England's most deprived electoral ward,
Benchill, Wythenshawe, Manchester**,
overleaf

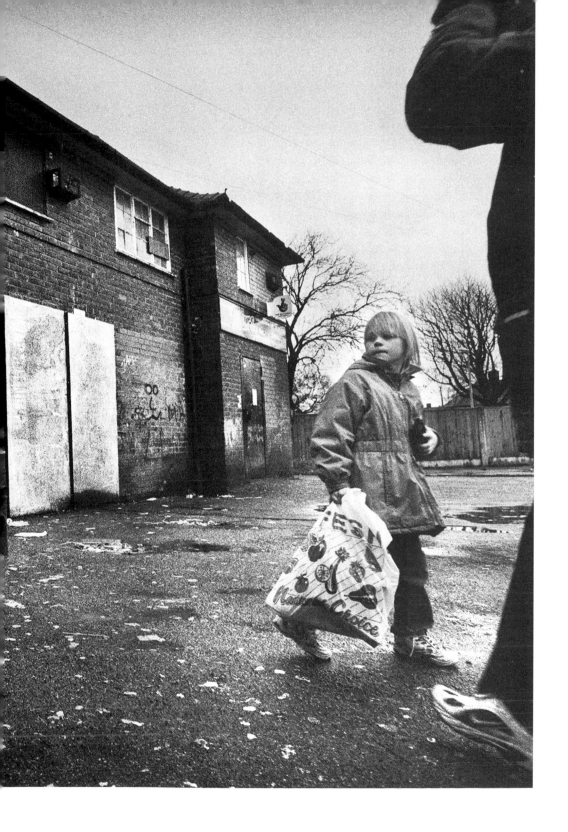

CRIME HAS ALWAYS BEEN the darkest urban nightmare. The fear of crime is, for most of us, a fear that a stranger is going to steal our property, violate our home or – worst of all – hurt us. 'It was the third burglary which really did it for us,' a pleasant man in public relations told me as he explained the family's move from Lewisham, south London, to a Home Counties town. 'The one just before Christmas, when all the children's presents were stolen.'

Beyond anecdote – and most city dwellers have unpleasant crime anecdotes – there is a wealth of survey evidence showing that the level and personal experience of crime is decisive in determining how people feel about their neighbourhood and what kind of place they would wish to live in.[4] It is not just crime, in the sense of reported or recorded breaches of the law, but on-street incivilities such as drunkenness and shouted abuse, drug dealing, vandalised bus shelters and phone booths, which cause this urban unease. Crime turns people in on themselves, erodes trust and co-operation and unravels the social fabric. It wipes out shops and other small businesses and damages crucial state services like education and health, only serving to deepen deprivation. The Home Office estimates that it costs about £60 billion a year in England and Wales; more than £1,000 per person, and that's without taking the unquantifiable costs of fear and reduced quality of life into account.[5]

In this chapter I want to describe a few sure-fire ways of promptly and drastically curbing urban crime such as 'zero tolerance policing', very fashionable with the media for a while. Next I want to show how the successful implementation of these measures would make more people choose to live in cities. And how that, in turn, would make the conurbations more prosperous. Then the numbers of poor and socially excluded would fall, and since many criminals are drawn from among them this would, in turn, cut crime still further.

Unfortunately, I can't do any of these things. If there were guaranteed methods of reducing crime, they would surely have been adopted by now. And if more people with earned incomes and choices were to live in the big cities, there is only the hope – but absolutely no certainty – that this in itself would reduce crime. It might even raise social tensions and the opportunity for crime, thereby increasing it.[6] So perhaps I ought to end the chapter here, declare crime a no go area, and move on to something a little easier such as planning or public transport. After all, this is only one of several pushes influencing people's decisions about where to live. But that would hardly do the topic justice.

The total number of crimes recorded annually – those which are logged by the police – has been rising briskly in Britain and other developed nations through most of the twentieth century. Since the 1920s the rate of increase in England and Wales has averaged 5 per cent, doubling the annual total of crime every 15 years. As recession gripped in 1989, the growth rate quickened and then collapsed; the annual number of recorded crimes peaked around 1993 and then fell markedly over the remainder of the twentieth century.[7]

Recorded crime is only part of the story. Most crimes are never reported to the police and a large proportion of those crimes that are reported are never officially logged. To find out what is really going on requires surveys that ask a large sample of the population about the crimes and criminals they have recently fallen victim to. Britain's Home Office has been doing this for almost two decades, questioning thousands of people in England and Wales every couple of years. Since the British Crime Survey began in 1981, it has shown pretty much the same mostly rising trend as the police records *and* the more recent fall. But it paints a more optimistic picture for violent crime, reckoning that the number of offences fell by a fifth in the second half of the 1990s.[8]

According to the year 2000 British Crime Survey, about one in three people said they had been the victim of one or more crimes during the previous year. In all there were approaching 15,000,000 crimes in England and Wales in 1999 (about four and a half times the number recorded by police) and of these three and a quarter million involved violence or its threat. Robberies — where violence or its threat were used — amounted to 13 per cent of these crimes of violence. Assaults in the home, mostly by men on their wives or partners, and assaults by acquaintances and strangers outside were all far more common. So most of the violent crime in Britain involves people who know each other, casually or intimately; a third of the victims of robbery say their assailant was known to them by name or sight.[9]

The Conservative government responded to the rapid rise in recorded offences at the close of the 1980s with tougher sentencing for criminals and by building new prisons, all under the slogan 'prison works'. The subsequent fall in crime was gratifying for Tory ministers and the policy was not fundamentally changed when Labour entered office. But the fact is that crime also fell through the 1990s in the USA, Canada and several European nations. No one knew for sure why this was so and very few felt confident it would continue.

Understanding the drivers of broad trends in crime is difficult, not least because of the wide and changing gaps between experienced, reported and police recorded crime. Over the decades the growth in insured possessions, the number of telephones at home and then mobile phones, have tended to raise the number of offences reported to the police. As for the underlying level of actual crime, population change, economic growth, the rising volume and variety of possessions, levels of drug abuse and technical improvements in security (cars becoming better protected against theft, more homes with burglar alarms) all appear to exert some influence on property crime figures, rather more so than the old press favourites of sentencing policy and the number of bobbies on the beat.

Since the 1950s the best long-run predictors of the numbers of thefts and burglaries in any year has been the size of male population aged 15 to 20 – the age band
in which criminality peaks – and the total of consumers' spending over the previous four years.[10] The more potential thieves there are, and the more possessions for
them to steal, the more theft occurs. A 21 per cent rise in recorded robberies in
England and Wales between 1999 and 2000 (which caused a predictable crime panic
as a general election neared) was blamed on teenagers mugging other teenagers for
their mobile phones. But the onset of recession and high levels of unemployment
also appear to have a short-term effect, pushing ambient theft up as hardship and
inequality increase.

There is slightly more certainty about what makes criminals. High levels of
unemployment, family conflict and breakdown, poor performance in school, tru-
ancy, suspension, expulsion and knowing other young criminals – all bedfellows of
poverty – are the measurable, objective factors linked with teenagers and young men
taking up crime.[11] We can put more resources into catching young criminals, lock
them up for longer, deal with them more rapidly. We can acknowledge that only a
small minority of the poor are criminals and that plenty of crime – much of it unac-
knowledged – is done by the affluent. But we are left with the fact that children,
especially boys, who brush against criminality from an early age and who grow up
lacking the skills and connections needed to prosper in the mainstream economy are
highly likely to commit crimes.

Just as deprivation is concentrated in large towns and conurbations, so is
crime. But there is something else besides poverty that has long been recognised as
tending to boost urban crime. It arises from cities' seas of strangers and the churning of their populations. In neighbourhoods where households frequently change,
or where residents come from different regions and nations with varying values and
norms, or where many lives are stressed by relative or absolute poverty, trust and
social glue are lacking. When all three of these factors are in play, there is almost
certain to be a palpable sense of insecurity with people unsure of what the rules are
and who will abide by them. Not believing in their own power to change things
for the better, and sensing there are few allies willing to work with them, they
mind their own businesses and keep their heads down. Surveys have shown that
people in neighbourhoods like these shrink back; they don't attempt to mediate in
disagreements, or tell off misbehaving children who aren't their own, or get to
know their neighbours. They are fearful of the consequences of doing these
things.[12]

The police forces that cover the big English conurbations record about twice as
much total crime per head of population as their counterparts in the more rural
shires.[13] For some types of crime the contrast is much starker. Robberies per inhab-
itant are eight times higher in Greater Manchester than in neighbouring Cheshire,
seventeen times higher in Greater London than in next door Surrey. We find similar contrasts in recorded crime within cities. In the inner London boroughs the

number of robberies per 1,000 head of population is four to six times higher than in the most prosperous outer ones, while the burglary rates are about twice as high.

Examining people's experience of crime rather than police records, the British Crime Survey shows levels of crime to be highest in the poorest urban areas – the more deprived council estates and multi-ethnic inner city areas where there is a mix of owner occupation and private renting in run-down pre-First World War terraces mingled with council estates. In some of these places the majority of flats possess steel-barred windows and doors with armour or security grills. Here, local criminals are targeting their neighbours. The poor and unemployed, single parents and people from racial minorities are at a considerably higher risk of falling victim to crime than the average Briton. On the most deprived council estates about one in eight flats are burgled at least once each year and in low-income, multi-ethnic neighbourhoods the proportion is one in twelve (compared to one in thirty in the more affluent types of suburbia).[14] Occasionally crime epidemics, sometimes linked to the arrival of one family, precipitate mass abandonment. In the worst of these places people fear to leave their homes at any time of day or night because of the likelihood of a burglary. Gangs of teenagers and sometimes of smaller children roam streets and walkways deep into the night, smashing windows and carrying out sundry hooligan acts. Local schools are in danger of break-ins and frenzied vandalism. Few dare to call the police for fear of being punished with a beating or a burglary for grassing. Those that do sometimes find that, far from putting extra effort into investigating an offence in such a notoriously high crime area, the police seem lackadaisical. The unspoken attitude seems to be: 'What do you expect, living in a place like that?'

But the British Crime Survey and the recorded crime figures show that crime is also particularly high in inner city areas where large numbers of wealthy and deprived people live side by side, such as Islington, Westminster and Camden in central London. Affluent households and individuals provide a more rewarding target for burglars and muggers. Criminals living nearby will come by. These heterogeneous neighbourhoods often have fairly transient populations – they may, for instance, have large numbers of student residents, or visitors – and there are many non-residents passing through from hour to hour. They may lack the deterrence of more stable communities where people keep an eye on each other's homes and cars and will report suspicious outsiders to the police.

Further crime patterns emerge when we move to higher scales of magnification. There is the vicious brawling that takes place late at night around city centre concentrations of pubs and clubs. And the transient drug and sex marketplaces. And the corners and greens where teenagers gather, around which you will find a little vandalism and much litter. Criminologists urge police and other crime fighters to improve their mapping of all this crime, in time and space, as a first step to reducing it.

Most of us also carry some sort of crime map inside our heads. As we move through the city we're often making subconscious security assessments, picking up clues from the appearance of buildings, pavements and people. How safe we feel depends on the time of day, how well we know the place, what we've heard about it and – most importantly – our sex, age, class, race and what crimes we have ourselves been the victims of. Being middle-aged, middle-incomed, male, white and only ever having been the victim of two or three trivial crimes, my own mental map of danger areas has few warning markings on it. There are many whose maps are spattered with red, telling them to keep out of particular areas, or pass through as quickly and unobtrusively as possible.

It has often been argued that the fear of crime is far more prevalent and damaging than crime itself, that the press and opposition politicians sow panic in order to sell newspapers and win votes. But the British Crime Survey has shown that people, by and large, actually have a rather good idea of their vulnerability to various types of crime, depending on where they live and who they are. Those who are more likely to be victims than the average appraise their risks as being higher and are more fearful.[15] This applies, for instance, to Asians and black people, to those on the lowest incomes and inner city residents. And while, in the second half of the 1990s, the British tended to the pessimistic view that crime was increasing locally and nationally – at a time when it had generally been falling – the survey found that their overall fear of crime has fallen slightly during this period. This underlying public realism and common sense about crime mean that telling people their fears of urban crime are exaggerated is highly unlikely to tempt them into urban living. City dwellers have to be shown that there are ways of controlling crime that will work for them and that the threat is being contained or, better still, reduced.

The easy place to start is with changing things – technology, buildings and streets – to reduce crime, rather than the much harder business of changing people. In the 1990s most of Britain's major and medium-sized town centres came to have CCTV – closed circuit television systems – covering pedestrian precincts, the busiest roads and pavements, car parks, bus and train stations. These systems, run by the police and local councils, have been financed by a mixture of central and local government money and contributions by local shops. Half a dozen operators in a control room linked to a few dozen cameras can cover most of a large town's central shopping and office district, calling on the police to attend any incident in an instant. There are many more private systems in shops and covered shopping centres, in offices, stations and other types of building. Recordings of people's movements can be combed for clues by police in the aftermath of a crime. The use of these images in solving two notorious crimes, the murder of two-year-old Jamie Bulger in 1993 and a solitary, maniacal nail bombing campaign in London in 1999, have shown just how commonplace the CCTV surveillance everyone now undergoes had become.

To date, CCTV is most concentrated in those central parts of towns and cities where fairly few people live. It rarely covers the residential streets and housing estates which make up of the bulk of the urban fabric. The cameras have to be fairly visible in order to deter criminals and to comply with the common sense moral principle that the public ought to be told when it is being observed and video-recorded.* But in residential neighbourhoods their visibility may send out a signal – 'this is a high crime area' – that could discourage local investment and confidence. Much of Newcastle's West End looks eerie not just on account of the streets of empty houses, the lack of people and cars, but because of the prominent cameras stationed on tall black posts at road junctions, fortified with spikes, wire netting and toughened glass to prevent climbers or stone throwers from blinding them.

Does CCTV actually reduce crime and make people feel safer? Since many hundreds of millions of pounds worth of private and public money have been and will continue to be invested in spreading it around the country you might assume the answer is a confident 'yes'. But CCTV's phenomenal growth has been driven by the authorities' need to be seen to be fighting high crime levels and by the marketing efforts of companies with expensive electronic systems to sell. A better one-word answer would be 'probably'. The evidence from Britain is that the introduction of camera systems in public places can bring about marked reductions in burglaries of commercial premises and car break-ins.[16] It can also be effective in cutting robberies. It does not seem to have much effect on late night, alcohol–fuelled violence; drunks who want to attack people don't much care who is watching them, although the cameras can at least help police to get to the scene more quickly. Some studies have suggested that some crime, especially robbery, is simply displaced into nearby areas where there is no CCTV monitoring while others have suggested it can extend its protection to unwatched places nearby.

What seems to be crucial in determining the effectiveness of CCTV is whether the local police can employ it intelligently and make better crime–fighting use of their officers. Cameras on their own will only have a short-term deterrent effect if criminals find out that they make no difference to the risks of arrest and conviction. But if the police can use them to respond to crimes more rapidly, to pursue criminals more effectively and to convict offenders by showing video recordings to the courts, the areas covered will become riskier places for criminals to operate in. There will be a sustained reduction in local crime and people will feel safer. Some offenders will move off and ply their trade elsewhere; a few will give up and go straight.

Advances in information technology now promise to transform CCTV's crime-controlling powers. When it comes to spotting suspicious behaviour and the faces of criminals on video screens, computer software is cheaper than human watchers and

---

*Sometimes, however, mobile and disguised CCTV cameras are set up temporarily to view particular crime 'hot spots', such as drug market places, as a prelude to a police crackdown.

may be more reliable. One thrust of technology is to use software to detect potentially suspicious movements in public places. Scientists at Leeds and Reading universities have developed a prototype programme that tracks the movement of people through car parks.[17] Those arriving to take their own car follow what the software regards as a normal trajectory, even if they cannot locate their vehicle right away. But car thieves follow an abnormal trajectory; they spend longer in the area and the types of pathways they follow are different. The software considers the pixels sliding across the screen, carries out a complicated statistical analysis and decides whether it is being presented with routine or suspicious circumstances. If it is the latter, the control room operator is alerted and told which screen to watch. This type of technology might also be used to warn of lurking muggers and drug dealers.

There is another, even more astonishing way of hitching information technology to CCTV systems. Software can automatically recognise the faces of known criminals. In a windowless room with one wall covered with video screens, a system called FaceIt scans the features of thousands of people walking, loitering and chatting in the surrounding streets, seeing if any of them match with the faces police want. The system's eyes are six cameras that gaze on some of the busiest parts of the main shopping centre in the East London borough of Newham. It continuously 'acquires' as many faces as it can from the throng and compares them, one by one, with several dozen digitalised villains' photographs supplied by the Metropolitan Police. On one screen in the borough's CCTV control room a small red ring springs into view, surrounding one of the faces out of the dozens in view on the street. It stays with the face as the person moves across the screen, signifying that the computer programme has the measure of those stranger's features and is now furiously checking them against its database of mug shots.

If there is no match, the red ring disappears and the stranger picked from the crowd is promptly forgotten. FaceIt swiftly moves on to acquire and check other faces. But if the software thinks it has a match with a mugshot the red ring surrounding the face turns to green and an alarm sounds. With a couple of key strokes a human operator in the control room can then have an enlarged image of the face from the street placed next to the face of the criminal pulled from the database. If she or he decides the software has got it right and they are the same person, a call is made to the police.

The technology was developed by an American company, Visionics, and this was its first, experimental use in crime prevention on city streets anywhere in the world. I watched the system watching the flow of people emerging from the local Underground station. The red rings popped into view, one after the other, checking a substantial proportion of all the people streaming out of the station exit. FaceIt copes with large differences in ambient light levels, the movement of its targets, their varying distances from its cameras. It does not need a head-on view and can acquire

Fortified CCTV camera, Newcastle's West End, facing

faces turned some way to the left or right. Trials with council volunteers showed that it was not confused by hats, spectacles, beards and moustaches or skin colour. The software takes numerous measurements of key distances which make each face almost unique – from outer eyebrow edges to base of nose and so on. Then, having turned the passer by in the street into millions of digits, it rushes through its comparisons with the criminal faces it has on file.

Ideally the system would be linked to several dozen strategically placed cameras instead of just six, including those which cover all the main entrances to a town's busiest public places and its transport interchanges. FaceIt could then scrutinise most of the people moving in, out and through the centre. To be sure of not being seen and checked by it you would have to stay away, or wear a mask or a bag over your head. And that's just the point, according to Bob Lack, the former policeman who runs the system for Newham Borough Council. It is a deterrent to criminals, making them stay away from his patch altogether. During the first six months of the trial shop break-ins fell by 70 per cent in the area covered by the system, compared to the previous half year. He attributes this to the extensive media coverage the system received when it was launched at the end of 1998.

The local police do not tell the council staff operating the system the names or the criminal records of the individual faces that they supply to FaceIt. They may be bail jumpers, people wanted for particular crimes, and criminals with extensive records for shoplifting, burglary and robbery. They probably include some drug dealers and paedophiles. Every few months some of the faces are removed from the database and some new ones are added.

In the nine months after its launch FaceIt had only made two matches which were then confirmed by the human operators in the control room. The low level of identifications may be because villains had heard about the system's introduction and decided to stay away. But with FaceIt being linked to only six CCTV cameras out of 154 covering the borough's public places, there was also a reasonable chance of someone on the police mugshot database visiting Newham without being detected. So, with money from the Home Office, the borough extended the trial and beefed up the software enabling FaceIt to be connected to more than twice as many cameras, acquire more faces and make matches more rapidly.[*]

The system provides little deterrence against new criminals lacking police records and mugshots. Nor can it be guaranteed to acquire and check every passing face. And while it should have a powerful deterrent effect in the place where it operates, it is as likely to make criminals ply their trade elsewhere as it is to make them give up crime.

Imagine, however, what would happen if – as seems fairly likely – this kind of technology spreads to cover the busier areas of entire cities and acquires a nationwide

---

[*]At the time of completing this book, the improved system had also identified twelve suspected football hooligans out of 4,500 people arriving for a West Ham match.

database of thousands of people with serious criminal records. Cameras linked to facial recognition software will not scan every single residential street and corner but they will gaze over all the main thoroughfares, the public transport interchanges and shopping centres. And the system might start to take feeds from privately run CCTV cameras inside shops, pubs and other commercial premises. People with substantial criminal records will begin to feel as if they are wearing an electronic tag, one that is constantly signalling their whereabouts to the authorities. The risk of arrest and conviction for any crime they commit will rise and they will find it much harder to provide false alibis when the system logs all the places and times where it has spotted them.

Some people will find this sinister and Orwellian and talk about Big Brother watching everyone. Civil libertarians, disturbed by the system's potential for misuse, protested at FaceIt's introduction but the Labour-run Newham Borough Council was having none of it. Its leader, Councillor Robin Wales, told me: 'There is a civil liberties issue at stake here. It's the right of our citizens to live in Newham without fear of crime.'

The use the police, other state authorities and private companies make of this technology needs to be kept under scrutiny and there should be open debate before it develops and spreads. How serious a crime do you have to commit before your face goes into a database? How long do you have to stay clear of crime before your face is removed? Do you have the right to know that you are on a face recognition database?

If this technology does become commonplace, it will tend to even out the imbalances in the levels of crime between the big cities and the smaller towns and rural areas, and between poorer or socially mixed areas and more prosperous, socially monotonous suburbia. It will do this because it will be deployed in high crime areas first of all. And if it does succeed in reducing crime in urban areas, it will also reduce the fear of crime.

We move on to the layout of streets and buildings in our search for what might make city dwellers feel secure in and around their homes. Jane Jacob's master-piece on urban non-planning, *The Death and Life of Great American Cities*, opens with an analysis of why her neighbourhood in New York's Manhattan felt safe back in the late 1950s.[18] Jacobs, a journalist covering architecture and town planning, argued that inner city streets, unlike those of suburbs and small towns, were bound to be full of strangers. How could everyone know everyone else when so many thousands were toing and froing along them? But while strangers provided the potential for insecurity, there were always dozens of local apartment residents and shopkeepers on sub-conscious look-out duty on her block in Greenwich Village. Children could play out on the sidewalks. It was safe enough to leave a toddler on the stoop. Even if the gaze of the street's 'natural proprietors', its residents and business people, should drop for the second, there would always be enough ordinary, decent outsiders passing through on the sidewalks to deter criminals.

From her day-to-day observations of the 'intricate sidewalk ballet' of Hudson Street, Jacobs drew up a set of principles for a safe city street. It had to have a clear demarcation between public and private spaces. It needed to be overlooked and patrolled by city dwellers, as many as possible and for as long as possible. So as well as having plenty of people living on it (which meant high densities and apartments), the safe street required restaurants, bars and a variety of shops to keep the pedestrians flowing day and night. It needed people's eyes on it – as many and as often as possible.

Things don't appear to work that way in modern Britain. The nearest we have to the kind of city neighbourhood she wrote about – very high densities of residents, a mixture of incomes and of uses, lots of visitors – are parts of inner London that tend to be rather high crime neighbourhoods. And clearly the urban living Jacobs described failed to enthuse tens of millions of Americans with choices, who continued to pour out of the great cities into suburbs and smaller towns. Even so, she was dealing with ideas that have remained at the centre of thinking about what makes neighbourhoods feel safe. In a city full of strangers, we need space in and around our homes where only we are allowed. It is possible to share some of this space with known, immediate neighbours but any stranger entering this private space uninvited is a potential threat; that needs to be clear to both its owners and to outsiders. And as for public urban places that are open to strangers, these generally need to be busy, or overlooked, or both in order to feel safe.

Oscar Newman studied New York's public (i.e., low-income, state-subsidised) housing estates and concluded that the levels of crime and vandalism correlated with three quantifiable factors.[19] One was anonymity; how often near neighbours were likely to encounter each other as they entered and left their homes and therefore how well they knew each other. This depended, to some extent, on the layout and scale of the housing. Then there was surveillance; how much opportunity there was for residents to see people moving around their homes. Newman's third factor was the number of alternative escape routes; how many different ways criminals and vandals could get out of an estate (or hide in it) and avoid arrest or confrontation with residents. Building on this influential work, Alice Coleman, a geography professor at King's College, London, and fellow researchers compared the amount of incivility and disorder they found in various residential neighbourhoods – from big blocks of flats to ordinary terraced streets – with the design of the housing.[20] Her measures of disorder were graffiti, litter, vandalism, excrement (dog and, alas, human), recorded crime and the proportion of children taken into care. She concluded that there were fifteen design parameters that influenced the level of social breakdown and crime, including such things as the number of stories, the presence of overhead walkways and the positioning of entrances. Bad designs were not only making it easier for criminals from outside to prey on the estates but also helping to criminalise their own residents.

Today that radical notion has little support. Few believe that enormous dents can be made in something as complex and difficult as crime by architecture, town planning and technology. In any one neighbourhood, other factors, notably the number of actual or potential criminals nearby, will have equal or greater weight. Nonetheless, the design of homes and their relationship to each other, to streets, footpaths and other public space have an important role to play in making residents feel and be safer from crime. Big improvements are possible even in housing which has long had a poor reputation, such as low income flats (see text below).

SAFER TOWER BLOCKS

Electronic security systems can help to make low-income, high density housing feel safer. They can also be used to crack down on that small proportion of criminally-inclined residents which makes life hell for the other tenants. CCTV cameras cover the communal entrances to the block, the lobby and the interior of the lifts. Each flat can be put on an internal communications system that links it to all the others, plus a concierge and an entryphone at the communal entrance. When a visitor rings the bell for any one flat, its resident can view the caller on their television, talk to him or her and decide whether to open the door. A computer logs the time of all these door calls. It also takes note of each occasion when a resident uses her or his own electronic key to enter the building. If residents or the landlord fear a particular flat is being used for drug dealing or prostitution, the computer record can show whether there have been numerous visits by outsiders.

Each flat can also be fitted with infrared motion detectors linked to a centralised burglar alarm system. When a resident is out and movement is detected inside the flat, security staff are alerted to the potential break-in. This system can also be set to sound an alarm if the resident *is* in and no motion is sensed after many hours. The absence of movement may be a sign that she is seriously ill or dead. (These security systems also allow the council or housing association landlord to manage the buildings more efficiently. If a tenant does a moonlight flit, never to return, then the computer can notify the concierge that a flat has been empty for some time and it can be re-let.)

Brian McGrail, an Open University academic who has studied the effects of installing these electronic security systems in high rise estates in Edinburgh and Glasgow, says they promise to make it easier to manage high-turnover, low-income housing while making the residents feel safer. He found that the great majority of tower dwellers support their installation but the cameras, alarms and computer records may prompt a few suspect tenants to leave. 'Once people start to feel more secure about their surroundings, the residents begin to chat,' he says. 'This technology can remove fear.'

For some years, British police forces and the Home Office have been encouraging developers to build more crime-resistant homes to a 'Secured by Design' standard; those that meet it can use this as a marketing tool.[21] The standard covers principles of layout as well as the number and quality of locks on windows and doors. The guidance it contains accords strongly with the established ideas of private, or 'defensible' space around homes where non-residents are seen as intruders, and the need for good surveillance. Entrances to homes and groups of homes should be well overlooked so that thieves have little opportunity to break in unobserved. Footpaths should be as direct, as overlooked and as busy as possible, to make people using them feel safe. 'Secured by Design' also encourages a variety of homes in every new development, in the hope that this will result in more residents being at home to look out for burglars through the day (rather than everyone leaving at the same hour to go to work).

Unfortunately, 'Secured by Design' has been interpreted as demanding a minimum of pedestrian routes through any new housing development, or none at all, on the grounds that public paths give strangers an excuse to enter the development legitimately – 'just passing through' – and that they allow burglars more chance of entering and exiting unobserved. The anti-crime drive to eliminate through routes has been one of the justifications for cul-de-sac street plans, the roads to nowhere much favoured in housing developments of recent decades.

This has been a mistake. As we shall see in later chapters, the fewer pedestrian routes there are, the more difficult it is to move through neighbourhoods on foot, inconveniencing both residents and visitors and making them more inclined to use cars. Nor are such impermeable developments necessarily safer; indeed, they may be less secure than ordinary streets. Researchers from University College, London, have studied house and car break-ins on individual estates in great detail relating them to lay-out, the number of passers-by and the chances of a thief being observed.[22] They have demonstrated that maximising surveillance by residents *and* passers-by is what deters crime. As Jane Jacobs observed, strangers can enhance as well as threaten security. The inner parts of no-through-route, modern housing developments lack passers-by and can suffer excess crime as a result. The classic Victorian and Edwardian streets of terrace houses with front bay windows, shared back garden fences and no rear access have proved to be an intrinsically safe design. They provide ample surveillance of entrances by residents and passers-by and a clear demarcation between public and private space.

In making homes and neighbourhoods feel safer for their residents, we should not compromise the convenience and attractiveness of the public realm for citizens. Some developers and house buyers have seized on gated estates as the obvious solution to urban crime, ensuring that new homes and owners' cars are protected behind high, blank walls and gates that no uninvited, unscrutinised outsider can penetrate. Some council planners have found certain developers will only build in the urban heartlands if they are allowed to put up fortifications.* But high walls and gates will not make the neighbourhood as a whole look or feel any safer. They will do nothing to reduce local crime levels. They may fuel fear and resentment among existing residents.

Technology, architecture and urban design have only so much to offer. What can be done to change people, to make fewer city dwellers commit fewer crimes? Several years before the 1997 general election, Tony Blair promised that Labour in government would be 'tough on crime, tough on the causes of crime'. That struck a chord with an electorate that feared that British society was becoming more barbaric and less communal. People wanted the hard law and order policies that the Conservatives had offered but they also sensed these could not work on their own. Polarisation and poverty were part of the problem. After the change of government the Home Office conducted a large review of research into the most cost-effective methods of reducing crime. Its wise women and men examined hundreds of research papers from around the world. Experts from universities and other institutions in Britain and overseas were called in to help the civil servants. The report they produced concluded that while some of the approaches and trials they had considered looked promising, none were likely to have a major impact on crime on their own.[23] There were no quick fixes; measures that appeared to cut crime in the short term would probably not sustain the reduction. Furthermore, some of the most potentially powerful measures were unlikely to have any impact on crime levels for several years after implementation – bad news for politicians. There was a need for more trials, more research, more careful evaluation. The quotation at the head of this chapter captures the cautious mood of their document.

The government's response to that review was to fund a £250 million crime-reduction programme, starting in 1999 and spread over three years, while continuing to expand the number of prison places. The money was funding thousands of local and national projects judged to have potential for curbing crime. Most of the effort would go into the more deprived urban areas which suffer most from crime. There are three kinds of approach. First, there are initiatives concerned with the ways in which criminals are handled by the police, the courts, the probation and prison services and by society after their first and subsequent encounters with the law. Then there are those based on trying to change childhood and teenage years, to make young people less likely to enter into crime in the first place. And third, there are projects that try to make any particular crime in a particular place more difficult, more risky and less rewarding. The spending was planned to increase with each passing year. As evidence of success or failure accrued from the

---

*In 1998, Birmingham's then Director of Planning told me the city council had initially resisted the gating of Symphony Court, a large development of expensive townhouses on the edge of the city centre's big Brindley Place offices, leisure and culture complex. He had to warn the planning committee's councillors that if they did not permit the fortifications, the developer would not proceed. They backed down and the walls went up.

early projects, the intention was to invest more in duplicating and disseminating those measures which appear to work best while dropping the ineffectual ones. Even if total crime was neither capped nor reduced, Britain should become much better informed about how it might be (or whether it could be).

There was more. After a couple of years of economic growth and deficit-reducing budget surpluses, the New Labour government began large increases in expenditure on law and order. It wanted to hire more policemen and women but – as was the case with teachers – it struggled to recruit them in the face of tough working conditions and tough competition from private sector employees. Several major new Acts of Parliament gave police and local authorities more powers against adults and youths who were drunk, disorderly, anti-social and criminal. All councils and police forces were put under a legal duty to form local crime-reduction partnerships embracing the health authority, the probation service and others they chose to include. Every three years they had to carry out an audit of crime and disorder, publish it and then consult local people and organisations in drawing up a strategy which made the priorities for curbing crime and disorder clear and established targets for improvement. The hope was that these partnerships and strategies would enable new, more effective ideas and working methods to emerge while making the authorities more accountable for the services they delivered and the taxpayers' money they spent. Hopefully at least some of this will work. In 2001 it could, at least, be said that crime reduction has more political, intellectual, and financial support than ever before.

How might all of this crime reduction effort be tied to local attempts to regenerate urban quarters by attracting and retaining people with choices in where they live? If crime and criminality in such neighbourhoods remain at high levels, then any middle- and high-income newcomers may find themselves targeted.* Incomers would be inclined to live in gated housing estates or fortified apartment blocks and use cars rather than their feet to move around the area. If some of these new residents decided to raise children in the area, then the likelihood is that those children would travel by car to schools outside it. All of which would be a mockery of sustainable, inclusive urban regeneration.

Part of the answer is to prevent very local places from becoming crime hot spots where anti-social behaviour and incivilities are tolerated, even encouraged. This is the 'no broken windows' approach to curbing crime which originated in the United States.[24] The theory is that dilapidation, graffiti and accumulation of litter all emit signals that a corner of the public realm has been abandoned by the forces of law and order. Soon al fresco alcoholics start to hang around the place, followed perhaps by small-time drug dealers and prostitutes. Residents and passers-by feel intimidated and begin to leave or shun the area. Property values fall, there is further

*On the other hand, the presence of newcomers might tend to reduce the total number of offences by providing criminals with richer pickings, thereby encouraging them to lay off the local poor a little. It would surely be preferable for a heroin habit to be financed by a couple of burglaries a week of wealthy,

neglect of the buildings, the place becomes more attractive to the lawless and a spiral of decline sets in with crime becoming a prime mover in urban decay. The British Crime Survey has shown that crime and the fear of crime are elevated in areas of 'high physical disorder' where there is litter, graffiti, vandalism and dilapidation.[25] The theory holds that councils and police forces need to be vigilant and nip this kind of social decay in the bud.

It is easy to make this approach sound oppressive and intolerant. And it is not easy to eliminate all such places; once tidied up, they can quickly degenerate again. The litter-dropping, gobbing loiterers who populate them will often be bored teenagers with nowhere better to go for chat, fun and fornication. That said, the 'no broken windows' approach is essential to urban regeneration. If a council, or any other landlord, allows public places to become squalid and litter-strewn, then it has fallen at the first regeneration hurdle. Litter signifies disrespect for people and place and begets more litter. There are laws which allow councils to issue fixed penalty notices to litter droppers and laws which enable people to take action against councils and landlords that allow litter to accumulate.[26] Both are used far too little. As for the knots of loiterers, their presence should never be allowed to intimidate anyone walking through a neighbourhood, be they visitor or resident, black or white, elderly or young. Gaggles of bored teenagers are less likely to form and less likely to constitute a threat in busy, high density urban neighbourhoods.

Deliberate attempts at urban regeneration (as opposed to unplanned gentrification) usually involve a strategy covering entire neighbourhoods of at least several thousand residents. Crime reduction needs to be part of that strategy. This could, for instance, include a CCTV system, or offering grants and advice to local homes and businesses to improve their security and their resistance to burglary. An evaluation of 300 such schemes in Britain found that in high crime areas the cost of preventing each burglary – in improved locks and so forth – was about £300 compared to the £1,100 costs of the average break-in to the state and the victim (the latter figure ignores the considerable but hard-to-measure costs of distress and increased fear of crime).[27] An urban regeneration strategy could involve allocating two or three police officers to work full-time in the regeneration area and perhaps to live there, getting to know its people, places and priorities. With an average of one officer for every 420 residents across the nation and government plans to increase the total strength, there are enough police to make this possible in hundreds of urban neighbourhoods.

Alternatively, or simultaneously, uniformed security wardens could patrol the area and complement the police. They would be recruited from among local residents and be employees of the local council rather than private firms. Their job would be to deter vandals, robbers, burglars and drug dealers in high crime areas by

contents-insured homes instead of a dozen less lucrative break ins at the uninsured flats of hard-up pensioners and single parents.

their presence, and by being in radio contact with a controller who was, in turn, in contact with the local police. The wardens would make law-abiding residents feel safer, talk to them, report damage and dilapidation to the local council and, possibly, collect some useful information for the police. In the late 1990s there were several such warden schemes operating in Britain. By early 2001, the government had promised sufficient funding to pay for about 1,000 wardens in dozens of schemes around the country.

Wardens complement, rather than substitute for, police officers, and they need no special legal powers. To be effective and trusted they must be well trained and remain in the job for long enough to be known to locals. They, and their bosses, need to have an excellent working relationship with the local police and a clear understanding about their respective responsibilities. The hope is that a patrolling warden can be almost as effective in deterring crime and increasing public confidence as a police officer on the beat, but at nearly half the cost.

Most attempts at crime reduction that could complement urban regeneration have already been tried somewhere, at some time, in Britain in recent years. What appears to be needed is for the police and local councils to work together *at the neighbourhood level* across tracts of urban Britain, despite their very different responsibilities and command structures. The crime and disorder reduction partnerships that police and councils were compelled to form at the end of the 1990s need to devolve strategies and targets down to this level, with the objectives of making crime more difficult and risky within the neighbourhood and of preventing residents taking up crime in the first place.

This needs to be done in consultation with local people, giving them every chance to influence the strategy. It would, for instance, be foolish to introduce neighbourhood wardens without seeking residents' views. Most people will probably not want to be involved in the strategy and there is some danger that efforts to interest them could stoke up the fear of crime. But there will be some residents and local business people who care passionately about crime, who have useful ideas and who could influence the strategy without compromising its accountability to the neighbourhood as a whole.

In many deprived areas police effectiveness is hampered, sometimes crippled, by the fear and hostility many residents feel towards officers. Covert and overt racism in forces have to be challenged and more policemen and women from ethnic minority backgrounds recruited, things the government pledged to do in the wake of the Macpherson inquiry's report into the police handling of 18-year-old Stephen Lawrence's racist murder in south-east London in 1993.[28] None of this is easy, and even if it was, the problem of poor relations between police and residents in deprived urban areas extends beyond racism. Police officers dispatched into these places usually live outside them and find themselves dealing with the worst of their residents. This leads to the worst of them disrespecting, even despising, the entire community. They visit poor neighbourhoods as little as possible and are offhand or rude when they do. That should not be tolerated; it undermines any attempt at regeneration.

The first issue in linking crime reduction to local regeneration is the one I come to last. Young people need to be prevented from being inclined, or motivated, to commit crime. The kind of measures judged most promising are to offer counselling and practical help to struggling, low-income parents, to increase opportunities for toddlers to go to pre-school, to mount intensive anti-truanting campaigns and to prevent badly behaved youngsters excluded from school from spending months on the streets. Expanding leisure and out-of-school learning opportunities for teenagers and ensuring that all young people, including those with criminal records, are offered work training if they fail to gain employment on leaving school are also means of fighting crime.

Local people, especially school-leavers and the long-term unemployed, need the jobs created by local regeneration. Construction firms putting up new homes and other buildings, supermarkets opening up, and any other enterprise employing people in a regeneration area should reserve as many of their jobs as possible for local residents. But employers operating in, or coming to, deprived areas are likely to find local people lacking the skills needed to take jobs. Some of them will lack the necessary habits, too. A youngster who has a drug problem, or whose parents and grandparents spent most of their lives unemployed, may find it difficult to get to work on time each morning. So as well as training in work skills, people may need pre-training training to help them with self-discipline and presentation.

Why should employers locate or remain in decayed neighbourhoods in the first place, let alone hire less than ideal local labour? It is a question of balance. Business will attempt almost anything provided there is a reasonable prospect of profit. There are various inducements for locating in down-at-heel urban areas – relatively cheap land, a shortage of alternative development sites, nearness to large pools of customers. A few business people actually want to be involved because they see regeneration as a kind of moral duty. But there are also opposite forces pushing businesses away from deprived and troubled areas; high risks of premises being burgled or vandalised, skill shortages, the fear that clients and customers will be discouraged by the area's poor appearance and reputation. The task for central and local government is to tip this balance in favour of locating in deprived urban areas, and devise a mixture of incentives and pressures which inclines employers to hire local labour.

One way of doing this is to make the granting of planning permission for new developments conditional on the training and hiring of local residents. Council planners are able to do this using the planning gain agreements mentioned in Chapter 9. But only a small minority do so. Research carried out for the Joseph Rowntree Foundation has shown that many councils are unaware that they can use such agreements to create employment for local people.[29] Richard Macfarlane, the researcher, concluded that government ought to make a change in its written guidance to planners, clarifying that such agreements were legitimate.

Regenerating areas need firms that are committed to training and recruiting local people. The benefits are multiple. Reduced unemployment boosts the local economy, increases the local labour pool (which will ultimately benefit other employers), raises social capital and reduces criminality.

Crime reduction programmes will often fail or provide only temporary relief in neighbourhoods that remain dominated by poverty. But imagine that we can succeed in creating more socially mixed, stable urban neighbourhoods with families from all income groups sharing schools and other local facilities, walking around more, working nearer to – or in – their homes, using their cars less. How safe would they feel?

If many people knew their near neighbours – not intimately, but well enough to smile and exchange greetings, to pass the time of day and to know which child belonged to which parent – we might expect them to display some of the civilising, crime-resistant properties that Jane Jacobs celebrated. Long-term residents, natural busybodies, and children whose play and schooling bring parents into contact would help to weave a social web. Residents of such urban neighbourhoods could feel a sense of mutual obligation. They would be willing to report suspicious-looking strangers to each other and the police, to intervene when children who weren't necessarily their own did blatant wrong, and to keep an eye on each other's homes, belongings and progeny. Criminals would find such an area risky and uncomfortable. They would either desist or divert to other places.

Even if we can create more such neighbourhoods, crime and cities will continue to stick together in people's heads and in reality. Ours is a fiercely competitive society with stark inequalities of opportunity and wealth. It places an all-consuming importance on the consumption of material goods. Aggression seems to be part of many, if not all, of us. Perhaps, given the inherent conflicts, we should be surprised at how little crime there is. Racism, family breakdown, drug dependency and the long-standing, increasingly questionable policy of criminalising all drug use (save nicotine and alcohol) are added ingredients in the stew of criminality. Big towns and cities are the cauldrons in which it is most vigorously stirred.

Crime is one more urban difficulty, large but not insurmountable. We can work at the margins, trying to reduce it, fixing those other things which are wrong in cities but are easier to put right, and making it a little harder for people to opt out of urban living. We cannot defeat crime but it cannot be allowed to destroy cities.

Chapter 12

MIXED USES AND HIGHER DENSITIES, OR MUHD

The scenes that illustrate this book are all about us. For illustrations, please look closely at real cities. While you are looking, you might as well also listen, linger and think about what you see.

Jane Jacobs, *The Death and Life of Great American Cities*[1]

WE COULD DO WITH a vision of the good city we are aiming for, as it might appear a couple of decades from now. A vision is more inspiring than incremental improvements that start from the sorry present. It is much more exciting than the drab matter of policy options and debate about which instruments – regulations, taxes, subsidies, and so forth – might do more good than harm. A vision cuts through all of that argument and reasoning and reaches our need to hope and to imagine.

But the twentieth century gave us every reason to be sceptical about urban visions, for the two which were most admired and influential seem to have led us astray. Writing just before the century's dawn Ebenezer Howard believed his garden cities would represent the perfect combination of town and country.[2] Perhaps they would have turned out better if they had had the fairly high densities he envisaged, as well as the self-governing and largely self-financing status he intended. In Britain, as state-sponsored new towns, the worst of them have turned out to be fairly dire and the best, while being prosperous and even popular, are inclined to be insipid and placeless. They fail to inspire. But the real failure of the garden cities and the new towns is that they have neither saved, nor transformed nor taken over from the great towns and cities that had already been built. And today they have little to teach them.

In the 1920s and 1930s the young French Swiss architect Le Corbusier drew his cities of 60-storey cruciform skyscrapers amid parkland, of expressways and gigantic slab blocks of housing with every urban function rigidly separated into zones. Here was the garden city on stilts and steroids, pumped up into the sky. There were plenty of followers who tried to construct little portions and scaled-down versions of this awesome vision, both within and outside existing settlements. Le Corbusier is universally blamed for their failures in high-rise housing set amid useless lawns, for the brutality and lack of human scale found in so many post-war developments.

Jane Jacobs neatly bound these two quite different yet related visions of a fresh start into one source of all the ills of post-war urban planning. Howard, she wrote in *The Death and Life of Great American Cities*, had wanted to 'do in' the great cities which – whatever their ills – remained places of enormous opportunity and promise, the engines of civilisation and economic growth.[3] 'Virtually all modern city planning has been adapted from, and embroidered on, this silly substance' of garden cities, she wrote. Le Corbusier's vision of *La Ville Radieuse* was 'so orderly, so visible, so easy to understand . . . it said everything in a flash, like a good advertisement'. But, she concluded, 'as to how the city works, it tells, like the Garden City, nothing but lies'.

Death and Life is among the most important and hopeful books ever written about cities. Instead of a grand vision of a better future, it offered Jacobs' own observations of what worked well and what was going wrong in her Greenwich Village neighbourhood and other quarters of American conurbations she had visited. She urged change but her proposals were rather modest. The main thing was for planners and builders in both city and state governments and the private sector to desist from grandiose schemes like expressways, huge housing estates and shopping centres, and from zoning more and more of the jumbled, multi-purpose urban fabric into mono-functional areas. The planners, politicians and developers were acting like dynamite fishers on a coral reef, blasting great holes in an immensely complicated and poorly understood ecosystem.

A flourishing urban economy was, for her, the highest priority. To prosper, an urban area required lots of people passing through from dawn to dusk and deep into the night. The constant toings and froings and mingling of people found in a thriving city neighbourhood multiplied opportunities to trade, to exchange and refine ideas for making things and providing services. These were the crucial fertilisers of economic growth. The creation of this essential mixture of large numbers of people in pursuit of business and pleasure required four factors:

- First, every urban place had to have two or more functions, to give citizens a reason to visit at different times of day – unlike, say, a downtown devoted only to office buildings which becomes deserted at night and weekends.
- Second, the city needed a dense network of streets to keep journeys on foot as short as possible, to give people many alternative routes for moving between their usual destinations and many chances to turn corners. Today this is known as permeability.
- Third, neighbourhoods needed a mixture of buildings of different ages in order to stimulate a diversity of uses. The older buildings, whose purchase costs had been paid off, could be occupied by businesses that could only afford fairly low rents; this raised the local diversity of employment and enterprises.
- Finally, the city needed high population and housing densities. She suggested *at least* 250 homes per residential hectare; the density for a typical British post-war suburb is around twenty to thirty. The very high densities she advocated gave neighbourhoods the numbers of residents needed to sustain a wide variety of businesses and it equipped those neighbourhoods with sufficient 'eyes on the street', enough informal policing by local residents, to handle the large number of strangers passing through.

Together, these four conditions could give a district the custom and income, the exuberant diversity and the public safety it needed to flourish. If any one of the four was missing, it was at risk of decay and eventual slumdom.

Jacobs' approach was essentially conservative and anti-visionary. Her focus was on the opportunities the metropolis offered to the individual, the small businessman, the newly arrived immigrant. Just let cities get on with what they did best, she wrote, which was bringing together lots of different people to trade and seek their fortunes. She combined a cool, detached style with quirky personal anecdotes and observations to create a highly readable and convincing polemic.*

Although her main concern was to reveal what maximised a city's potential for creating wealth for all its citizens, the book's enduring success derives from its promise and its celebration of how rich, exciting and plain good life in a great metropolis can be. The city is – or should be – a place of endless surprises, of enormous social and business and pleasure possibilities when you had such huge numbers to draw your friends and contacts from. You could have friendly, co-operative relationships with your neighbours but keep as much distance from them as you wanted. There was always somewhere new to go, something new to see. And you only had to find your way to the best-functioning parts of any large city to see this anti-visionary's vision made flesh.

For at least a decade after publication Jane Jacobs' ideas were almost completely ignored by planners, developers and most architects on both sides of the Atlantic. They carried on blowing up the reef, lobbing in roads and enormous housing estates, gigantic office and shopping developments. But the book gradually acquired followers, many of them left of centre, as the harm this was doing became more and more obvious. *Death and Life* has been translated into a dozen languages and become indispensable reading for every student of urban design. She founded a school of thought about how cities should be planned and rebuilt which has become a new orthodoxy. Her status as founder of what Americans call 'the new urbanism' is akin to Charles Darwin's as founder of the theory of evolution. There were others who had glimpsed parts of the new truth before her, or were seeing it at the same time, and since she published a great deal of fine work has been added. But even though her receipt of the founder's accolade is a simplification, a matter of convenience and a little luck, it is nonetheless richly deserved.

Forty years on, Jacob's followers comprise a broad church and sing from a variety of hymn sheets. The urban virtues she celebrated are still relevant in an age of majority car ownership and mass access to the Internet. In academic circles geographers and other social scientists praise the compact city. A growing band of architects and developers practise the new urbanism in the USA. Britain has its urban villages movement; here the Prince of Wales' patronage of Luxembourg-born architect and town planner Leon Krier and others has played an important role.

*Like Ebenezer Howard, she was essentially an amateur in the field in which she won fame.
**Jane Jacobs saw high permeability as a way of helping small businesses, which rely on passing trade to survive.

The denser and more interconnected the street network, the more even the distribution of pedestrians around it will be. Which means more passing customers for businesses not sited on corners or main thoroughfares, but

Jacobs' anti-visionary work remains the best vision on offer for the future of large towns and cities, even if it is a little confused and contradictory in places. And things seem to be moving in the direction of the new urbanism. In Britain, UK government policy has swung behind it.[4] The report by Lord Rogers' Urban Task Force gave the compact city movement a new manifesto and put this notion as the heart of the urban renaissance.

What is it building? What does it want? There are ten key ideas that apply equally to redevelopment and refurbishment of existing towns and cities and to construction on greenfields outside of them. (And there is one principle that binds them all together; an urban area is only any good if walking through it is pleasant for residents and visitors.)

1. Workplaces, leisure facilities, schools, shops and other communal facilities should be as close as possible to housing, giving people options to walk or bicycle instead of using their cars. This will reduce traffic congestion, noise and pollution and the time wasted in commuting. Neighbourhoods ought, therefore, to have mixed uses and never consist only of large tracts of housing. People who begin new, small enterprises ought to be able to locate them near to or even in their homes. Many individual buildings should also have mixed uses, with flats located above shops and offices and houses containing studio, office and workshop space.

2. Nearness to communal facilities implies a shift towards higher residential densities as well as mixed-use neighbourhoods. Higher densities bring more people within walking range of these local facilities, such as small shops, increasing their economic viability. It also increases the viability of public transport because more people will be within reach of bus, train and tram halts. By increasing the accessibility of public transport, higher density encourages people to use it instead of their cars for longer journeys as well as short ones.

3. If people are to take up alternatives to the car, they need a high standard of pedestrian and cycle routes. These can be along pavements and roads, so long as the traffic is not so fast and heavy as to make walking or cycling unpleasant and unsafe. These routes must also be convenient and direct, enabling people to get from anywhere to anywhere in the neighbourhood without long detours and diversions. That requires a fairly dense network of public streets, giving high permeability.** Outside of urban parks and woods, pedestrian and cycle routes need to be overlooked from buildings in order to give both passers-by and residents a feeling of security.

located midway along quieter streets. This kind of high permeability network also makes it quicker and easier for people who want to visit two different shops on separate streets to travel on foot between them. In the context of a gridiron layout (like New York's) high permeability meant not allowing the long sides of the blocks to be too long. She advocated cutting new streets through long blocks to open up new routes and connections.

4   If people can work and meet more of their needs in the immediate area where they live, they will be more likely to walk around their neighbourhood. They will see more of their neighbours and their sense of belonging in the place will increase. They will need to spend less time commuting and making other journeys, and they may therefore spend more time in their own neighbourhood. These shifts could strengthen community ties, reduce crime and improve the upkeep of the area.

5   Neighbourhoods should have a mix of housing types, from one-bedroom flats through to four- and five-bedroom houses. Some of the homes need to be affordable to people whose incomes are insufficient for them to rent or buy in the open market. This can take the form of subsidised renting, subsidised ownership or some combination of the two.

6   Buildings, and groups of buildings, must have architectural good manners. Back garden fences and windowless gable ends should not front the street or be placed opposite junctions. Houses and other buildings should be close to the street rather than set far back. This allows residents more private space in back gardens while enabling the public realm of the street to be better framed and enclosed. Larger buildings should never offer pavements long, blank frontages lacking windows and doors.

7   More people are spending more time working from home using computers, telephones and the Internet. This should be encouraged because it enables people to waste less time and cause less pollution and congestion in commuting. Housing must be flexible and spacious enough to accommodate this telecommuting demand and neighbourhoods ought to be sufficiently attractive and stimulating to make people want to spend their working time in them.

8   The diverse ecology, wildlife and valued landscape features found within cities (or on greenfield sites chosen for development) should be respected, protected and enhanced whenever possible, particularly in the case of species which are rare or vulnerable. The best of them should be celebrated. This means conserving watercourses, mature trees and hedgerows and other features of the landscape. Generally, these should be made accessible to all residents of a new housing development rather than being enclosed in private spaces. In some cases it may be appropriate to create new habitat elsewhere to substitute for habitats destroyed by new housing development.

9   Neighbourhoods must be legible and have some sense of place. They need landmarks large and small, a variety of buildings, an avoidance of standardisation and a network of streets that enable people passing through to know where they are and where they are headed. So retaining old buildings of character and history is worth the extra effort and expense. Neighbourhoods must have a high quality, well-managed public realm of streets and street furniture, squares, communal buildings and green spaces.

10 Residents and businesses should be encouraged to be involved in the management and appearance of the communal areas and the public realm. That may require very local, democratic structures such as a community trust in addition to the local council.

Most of these qualities were to be found in Jacobs' mid-town Manhattan neighbourhood in the late 1950s. In Britain, urban village protagonists such as the Prince of Wales have pointed to much older, European urban quarters such as Sienna and Rye as their source of inspiration. The differences in continents and centuries do not matter; what does is that such places all seem to offer possibilities for a good urban life.

But there is something slightly uncomfortable about the phrase 'urban village', even if estate agents have begun to seize upon it as a way of marketing inner city developments. The two words jar alongside each other. Combining the idea of the city (big and bustling) with a small place on its own in the country (cosy and community-minded) in one phrase might not work. Mixed use, high-density urban neighbourhoods do need their own sense of identity and community but, as Jacobs recognised, they must also be completely open to non-residents who come in to work or play or are just passing through. Otherwise they won't deliver, economically or socially. After all, the residents of an urban village would want access to the entire city with all of its opportunities for business and pleasure; that is the point of urban living. If you want to reside in a village, head for the countryside, or live in a suburb and convince yourself that it is just like a village.

I sought my heroine's views on this question. Jacobs lives in Toronto, widowed and well into her eighties, having left her beloved New York long ago to avoid a son being drafted for the Vietnam War. She was not keen on 'urban village' either. 'If they think of them as neighbourhoods, woven into the fabric of the city, then that's fine,' she told me. But I'm not sure neighbourhood will do either. Any residential urban area can be called a neighbourhood. It does not bring out the key attributes of Mixed Uses and Higher Densities, or MUHD, which conform to the ten principles set out above. Someone is going to have to come up with a good, short name for this old and new kind of urbanism that everyone can instantly recognise and understand. Until then, I'll stick with MUHD.

MUHD's time may have come, for several reasons. Let's start with the one that is the most obvious and politically pertinent, thanks to NIMBYism in the English shires. It saves countryside from being consumed by new homes because each new home takes up less land than it would do if it was part of conventional suburban sprawl.

But while sparing countryside from being built on is a hugely popular cause, it offers the weakest justification for a shift to higher densities. To start with, there is plenty of undeveloped land left. Government-commissioned research has shown that if we continue with the most recent pattern of housebuilding (at densities well

Architectural bad manners: new estate in Milton Keynes

So there are powerful arguments for this new, yet essentially old, way of building cities. And mixed use, higher-density developments are starting to happen. The Prince of Wales's Duchy of Cornwall estate was the pioneer in Britain with construction of its Poundbury extension of Dorchester, masterplanned by Leon Krier, starting in 1993. This development, on Duchy farmland on the edge of a handsome market town, will eventually grow to more than 2,000 homes over a 30-year period, along with factories and workshops, shops, schools and other facilities. At the time of writing only one corner of some 200 houses (mostly for sale, but with forty for rent to low-income tenants) had been built and praise has been heaped upon it. Most visitors love Poundbury. Housing and town planning professionals are impressed, even inspired, and the houses fetch premium prices. A few scorn this painstaking effort to create an instantly ancient Dorset town, with neither telegraph poles, nor TV aerials nor twentieth-century architectural styles permitted.* But Poundbury's density is considerably higher than that of conventional housing development (front doors open straight onto the pavement), visitors cannot tell the social and owner-occupied housing apart and it's a pleasant enough place to stroll through where the car does not dominate. Several other MUHDs – usually called urban villages – have also begun in the 1990s, notably the Crown Street project in Glasgow's inner city Gorbals and the massive redevelopment of Hulme, an area of failed 1960s' council housing just south of Manchester city centre.

Ministers have made speeches damning 'soul-less, little box' private sector estates, praising the high-density middle-class housing developments of the nineteenth century and calling for a return to terraced housing with communal squares. The government has also invested some of its own land and credibility in building mixed use, high-density developments. Soon after Labour came to power in 1997 the Deputy Prime Minister, John Prescott, said he was planning up to five 'millennium villages' on large, derelict sites controlled by government bodies within or besides existing towns and cities. The intention is for these developments to set new standards in construction efficiency and in environmental and social sustainability, and to demonstrate how MUHD and mixed tenures can be made to work and to sell on a large scale. The first of these is Greenwich Millennium Village, which is being built on an enormous, abandoned gasworks site in East London enfolded by a sharp meander of the Thames. Unfortunately it got off to a rather uncertain start (see text below).

---

*Lord Rogers is not a fan. In a letter to *The Observer* (21 May 2000) he dismissed Poundbury as 'a questionable exercise in Hardyesque nostalgia' which should not be held up as a paragon of modern urban planning. At the time of completing this book he had never visited the place.

Urban villages taking shape – Hulme, Manchester (above) and Poundbury, Dorset, facing

THE FIRST MILLENNIUM VILLAGE

A few weeks after the 1997 General Election, the new government announced plans for the first in a series of pioneering 'millennium villages' to be built on derelict sites around England. A little more than three years later, the Deputy Prime Minister John Prescott handed front door keys to the first families moving into the 1,400 new homes being built beside the Thames in Greenwich, London. In the intervening period the project had begun to disappoint. As was the case with the Millennium Dome half a mile to the north, it seemed as if initial hopes had been raised recklessly high.

The idea was a brave and good one: to set the private sector the challenge of building a high-density residential neighbourhood on an enormous gasworks site as a showcase for sustainable development in return for getting land at an attractive price. It would benefit the environment by making large savings on energy, water and other natural resources compared to ordinary new housing – both during construction and over the lifetimes of the buildings. Excellent public transport links, including a nearby Tube station, and a local school, health centre and shops would make residents less dependent on their cars. Innovative (for Britain, at least) construction techniques would be used, reducing build times and costs. The village would benefit society by being a mixed community with a range of sizes and types of housing including subsidised homes for families with low incomes.

A competition was held to choose a development consortium to build on the site, which had been purchased and prepared by the government regeneration agency English Partnerships. The winner, announced early in 1998, included two leading developers, Countryside Properties and Taylor Woodrow and two housing associations that would take responsibility for the social housing. The village's master planner was Ralph Erskine, an octogenerian born in England, long resident in Sweden and celebrated architect of the Ark office building in London's Hammersmith and the Byker Wall council flats in Newcastle. The project's main architects were London-based Hunt Thompson Associates, whose work in social housing is highly regarded.

Their plans and sustainability targets looked radical; a mixture of apartment blocks rising up to ten storeys high and terraced houses, built around squares and courts which in turn flanked a village green and watery ecology park. The housing would be at around 100 units per hectare, four times the density of conventional greenfield housing development. The homes would be flexible, allowing owners to change the number of rooms and bolt on balcony conservatories as their space requirements changed. They would all be linked by the village's own computer Intranet. A fifth of the units would be reserved for low-income families. Energy consumption would be only a fifth that of conventional housing.

But the site itself is awkward and isolated. To the east is the Thames, to the west, a motorway to the Blackwall Tunnel, to the south, industry, and to the north, land awaiting development. Sandwiched between the Millennium Village and this motorway is classic car-dependent, edge-of-town style development; a superstore, three big retail sheds and a 14-screen cinema fronted by an enormous car park. The only pedestrian link to the rest of Greenwich is a footbridge across the whooshing, droning motorway.

Hunt Thompson resigned in 1999 after rows with the developers, claiming that the project was failing to live up to the original high ideals and was sliding towards conventionality. The construction programme slipped. The first phase was completed over a year later than was planned, missing the millennium. It fell well short of the environmental, social and innovation aspirations for the overall development although the aim was to make up for this during later phases of the 5-year construction programme. To take one example, the very first 100 owner-occupied homes to be completed, in two apartment blocks, were well separated from the first affordable units – terraced family houses. The developers did not want to challenge purchasers with a mixture at the outset.

In 2000 these first for-sale homes sold like hot cakes, off-plan, with the largest penthouse apartments fetching over £400,000. The Millennium Village seemed to be guaranteed commercial success. It will be some time before we know how near it comes to realising the high ideals of 1997.

All of the larger MUHD-type developments that have been built in the UK since the compact city made its comeback have depended on heavy subsidy from central and local government (saving Poundbury, a greenfield development which would not have been built without royal patronage). The property industry has shown, however, that higher-density and mixed use developments that include dozens and even hundreds of homes can be built profitably *given the proximity of a thriving town or city centre*. It has converted thousands of old warehouses, factories and office blocks into loft apartments and built striking new apartment blocks on cleared urban sites. The housing high rise is making a come back in the private sector. More and more developers are finding ways of mixing shops, hotels, apartments, offices and restaurants together within old and brand new buildings.

Developers are making fortunes out of Jane Jacobs' vision of the good city teeming with people, with an intricate, fine-grained mixture of different activities and a variety of buildings old and new. It is these older, still useful, buildings that often give them their opportunity. One that I found particularly interesting was a nondescript, ten-storey shoebox just north of the City of London.

Here, in a speculative office block chucked up in the 1960s, *The Independent* newspaper was born in 1986. I was its cheapest and most junior reporter. Its first 8 years passed by on the City Road, with the paper occasionally seeming to flourish but mainly struggling, ever in need of injections of fresh finance – but all the while keeping plenty of journalists amused and busy. The newspaper moved to Canary Wharf on the Isle of Dogs one dark day in 1994 and, when I visited its old home

4 years later, the last of the luxury penthouses and apartments that now filled it were being sold. I had to pretend to a bored young salesman that I was the sort of person who could afford £445,000 for a two-bedroom pad on the top floor, along with £20,000 for a secure car parking space (both would cost a great deal more whenever you are reading this). The building's original steel and concrete frame still remained somewhere behind copious cherrywood and slate flooring, etched glass, mosaic tiling and acres of white walls. Inside and out, not one bit of the old office block was visible behind this 'Mondrianesque minimalism' that, a brochure from Metropolis Developments assured me, was 'edging out the cluttered chic of the hectic eighties'.

The Lexington penthouse apartments were splendid. But they could at least have had a little plaque in the lobby to say that Britain's first quality daily to be launched in the twentieth century had once filled the place with energy and hope.

This new market in loft and apartment housing is well established in London, growing fast in every large city and has started to spread to smaller ones. It is a market that will probably expand so long as house prices are stable or rising. But without deep shifts in culture and attitude it will remain in a niche, confined close to city centres and purchased largely by the childless, the young and those with fairly high incomes.

MUHD's time may have come. Then again, it may not. For the concept is either unknown to or rejected by most of the people who really matter – the consumers of new housing. If you speak of mixed uses, high density, reduced car dependency and the new urbanism, most buyers will be baffled, suspicious or both.[10] The average housebuilding company shows every sign of wanting to carry on doing what it knows best: putting up estates on greenfield sites on urban fringes, preferably in more prosperous areas and certainly not in declining inner cities. Many housebuilders complain that the high price of land caused by planning restrictions forces them to build at densities that are already too high. Gardens and floor plans are squeezed, detached homes pushed too close together, placing their products at a disadvantage compared to the more spacious housing built earlier in the twentieth century.

The typical new home going up in Britain in 2000 is decidedly, determinedly anti-urban. Drop into the show home on your local greenfield building site to see what I mean. Usually all the houses for sale on a brand new estate will be aimed at one sector of the market – cheap, middling or expensive. Housebuilders believe that purchasers like all of their neighbours to be in roughly the same income bracket. The homes will probably be detached or, failing that, semis because that is what the punters prefer; only if they pitched at the bottom end of the market will they be joined into terraces. There will be nothing on the estate but housing – no shops nor workplaces mingled in. It will, then, be not a MUHD but an OULD – One Use, Low Density.

It gets worse. Walking or cycling right through the estate, entering at one end and leaving by the other, will probably be impossible. Its internal road system will consist of a spine road and cul-de-sacs with just one link to the public highway. Developers lay out estates in this impermeable way because they assume, usually correctly, that most residents will come and go from their homes by car and because the police tell them it makes the houses more crime-resistant (intruders and burglars allegedly prefer a multitude of entry and exit points). Each house will be set several yards back from the street, allowing owners to feel they command a little more private space and giving them room to keep a car or three off the street and on their own land. There will either be an integral garage in the body of the house, or one proudly jutting out of its front. Here the cars can be tucked away in their own bedrooms, really making them part of the family.

It is easy to be scornful of Britain's housebuilders and the dreary stuff they churn out. There is their use of flimsy, mass-produced external details, known as 'gob ons' in the trade, to fake up a little of the traditional and the vernacular on standardised, mass-produced brick boxes. Back gardens still manage to be minute despite the rather low housing densities. The roof spaces are so crowded with structural timbers that you can't put in a loft extension. And they tend to have a fairly high level of defects for something new.

But, reply most builders, they know their market. They put up what the customers want and they have no trouble in selling it; scorn us, they say, and you scorn the people. The critics and knockers are upper middle-class snobs trying to impose their own minority tastes and frustrate the public's decent and commendable home-owning aspirations. 'These people despise the majority of their fellow citizens for wanting their own suburban houses with gardens and for preferring a suburban, lawn mower and barbecue culture instead of living in flats and cities and enjoying a metropolitan pavement culture,' says the industry's trade body, the House Builders Federation.[11]

In fact, opinion research shows that the public have a rather poor opinion of new, private sector homes.[12] They think they are cramped, boxy, and lacking in individuality and they generally associate them with the bottom end of the housing market. But this research has also shown that most people who would consider buying a new home would strongly prefer to have off-street car parking and garages – preferably a double garage – and at least 5 metres of private space between their front door and the pavement. They want a back garden, and they strongly dislike detached homes being too close together – they ought to be separated by at least 2 metres. As for appearance, most potential purchasers prefer traditional-looking homes. They want them to be in keeping with surrounding buildings yet have a measure of individuality.

They prefer, in short, the larger kind of new homes at the top end of the market. If, as seems likely, economic growth continues and living standards carry on rising, people will probably continue to want even larger houses and gardens. These things will continue to be important status symbols. But the need to signal one's wealth is not the only reason why large homes will remain in demand. Gardening's

Spot the gob-ons: new private housing, Essex

enormous popularity seems to be growing, if the number of green-fingered television programmes and garden centre sales are anything to judge by. People gain great pleasure out of relaxing, eating, playing and growing things in their own fragment of the great outdoors; the passion runs deep and, as prosperity increases, more people will want more garden. They will want extra indoor space, too. We can be fairly certain about an increasing demand for home office space as more and more people spend at least part of their working week working from home. Households also tend to purchase more and more leisure equipment and toys for adults and children as affluence rises; they will want space to put it all in. Already some 60 per cent of households with garages never use them to shelter their cars because there is not enough room.[13]

People may demand more space, but will they be able to purchase it given the planning constraints and high land prices in these densely populated islands? Whatever the government's new, pro-MUHD guidance on planning and housing says, it may be difficult to resist such a deep-seated trend so long as economic growth continues. Why is it that new homes in England's South East have been built at a *lower* density than those elsewhere despite this region having the highest land prices, the greatest degree of planning constraint and among the highest population densities?[14] The answer, I guess, is because it is the region with the highest per capita income.

There is, then, a disjuncture. Superficially, the arguments in favour of high density, mixed uses and a social mix seem to be winning the day. There is a lot of enthusiastic talk and much excitement; things are starting to happen on the ground. Government and a growing number of local authorities make approving noises. The more enlightened housebuilders want to disown the poor layouts, the standardisation and the sheer dullness of much post-war private housing development. Alan Cherry, the chairman of a large housebuilding firm (which is heavily involved in the Greenwich Peninsula development) and a member of the Rogers' urban task force, told me: 'We are not going to be allowed to carry on building anywhere type homes everywhere.'

But the majority of producers and consumers of housing seem perfectly content with suburbia and the OULD. There are thousands of local councillors sitting on planning committees who are willing to support them. And while the government may have changed its planning guidelines, the planning system gives town halls some scope for foot-dragging (which, if you believe in local democracy, is right and proper). Research commissioned by the Department of the Environment, Transport and the Regions has confirmed the expected; that residents are, on average, more likely to be negative than positive about the intensification of development in their neighbourhoods which compact city policies imply.[15] The Department itself has never published this research and one can only suspect that this is because it is off-message. So the new urbanism still has to win hearts and minds. How can we make MUHD stick?

HIGHER DENSITIES AND MIXED-USE neighbourhoods have to be made to fit with aspirations for good-sized homes and gardens. They must accommodate people's wishes for privacy, tranquillity and contact with the natural world. I think they can; in this chapter I want to show how.

Density is merely a number – the number of homes, or people, or habitable rooms per hectare. But high density is a term swarming with impressions and clotted with history. For the pioneers of town planning it meant over-crowding and a dearth of sunshine and sanitation. For today's urbanists, high density is a guarantor of urban vitality and ecological responsibility; cities cannot function properly without it. For much of the public and the politicians, and for planners of the old school, the term continues to be associated with the concrete jungle – too many buildings and people and too little greenery, 'over-development', 'town cramming', a sense of being hemmed in.

While I was researching this book I heard one of Britain's more persuasive urbanists, Martin Crookston of town planning consultancy Llewelyn-Davies, eloquently expounding the case for higher densities to the House of Commons' Select Committee on the Environment. Some of the most desirable and expensive residential areas in Britain, he pointed out, had always had densities far above the suburban. Spacious rooms, large homes and the highest environmental quality were all possible. The committee's chairwoman, a redoubtable Old Labourite called Gwyneth Dunwoody, was sceptical and challenging. Those living in posh, high-density areas had the benefit of second homes in the country, she said. When rats were over-crowded they became stressed and vicious and turned on each other. 'Human beings are not very different from rats, in my experience,' concluded the Honourable Member for Crewe and Nantwich.

As residential density rises, private gardens and public parks must shrink and eventually disappear. At densities of around 10 units per hectare you find big detached houses and very generous gardens.* At about 30 it becomes difficult to have only detached houses (although developers try, with sadly squeezed outcomes); semis and terraces begin to enter into the picture. Jump up to 80 units per hectare and it is no longer possible to have only two-storey terrace houses with gardens; buildings must reach higher and flats and maisonettes make their appearance. Although there can still be some single houses, the neighbourhood has an urban,

---

*Actually, density is not a simple number. There are several ways of measuring it. The figures I use throughout this chapter are one version of *net*

residential density – the number of dwellings per hectare including their gardens and their share of the immediately surrounding streets, pavements and

rather than a suburban, feel. At above 200, everyone in a Western city lives in multi-storey apartment blocks; if they do have any private outdoors space it will consist of a balcony or roof terrace rather than a ground level garden. In Third World nations much, much higher densities are found in shanty towns made up of tiny, single-storey, single-room homes. The table gives an idea of what various density levels look and feel like.

Net density, units per hectare	Typical housing type
0–20	Leafiest, most expensive suburbia; detached houses with large gardens. A small minority of British housing is of this density, but it is common in American suburbs
20–40	Most British housing built for owner occupation in the twentieth century falls into this density band, as does a great deal of council housing. It ranges from smaller and medium-sized detached houses through semis and terraces. In the middle lies the 30 homes per hectare standard set in the Tudor Walters report of 1919 (see Chapter 3). The average net residential density of homes built in England in the 1990s was 25 homes per hectare
40–60	Detached housing drops out in this band. Spacious Victorian and Edwardian houses built in terraces, more compact modern houses built in the last 40 years in terraces and semi detached, both council and owner occupied. Some blocks of council flats also fall into this band, thanks to generous quantities of courts and lawns around them
60–80	Typical inner city residential densities. Includes fairly spacious, high income three and four storey houses built in the eighteenth and nineteenth centuries as well as flats, maisonettes and some modern high density terrace designs. The semi drops out
80–100	Working class by-law housing of the nineteenth century (see Chapter 2), still abundant in inner cities. No front gardens (front doors open onto pavements), very small back gardens or yards. Denser, large post-war estates of council flats, including tower blocks, in inner city areas
100 plus	Unusually high densities for Britain today, although private sector apartment block developments (new build and conversions) in city centres are being built in this band. The earliest council housing and housing association mansion blocks, put up around 100 years ago, were built at these densities

High housing densities and mixed uses are closely linked. At the high end of the density range shops, schools, cafés and all the other things people need from day to

green spaces. Net residential density does not take into account the area of larger parks, shops, schools, principal roads and so forth that serve these homes. If you include the latter you have the gross density.

day must be mingled with the housing. The large numbers of clients and customers within a few minutes walk provide the demand required to make these services viable at a very local level. If they were not local and walkable, the extra car and public transport journeys that would have to take place in order to get people to them would impose huge strains on both the city and the citizens.

A law of diminishing returns applies as residential density rises. Any given number of homes and residents requires some space to be devoted to communal assets such as parks and play grounds, schools and shops, roads and pavements.[2] As density rises and the housing is squeezed into a smaller space, the shared, communal area cannot shrink correspondingly. Indeed, it may actually need to grow, because as the number of private gardens falls, the amount of public green space ought to rise. At very high densities, even the pavements need to be wider because more people will be passing along them.

People's individual wants and needs – for private space indoors and outdoors, a place to park their cars and a share of communal, green space – tend to bring densities down. So, too, to some extent, does the need for permeability, for space to move through and around a neighbourhood. Yet densities are pushed up by our collective need for cities that are vital and varied, that have frequent and dense public transport services, where many services can be obtained a few minutes walk from home. Push, pull; a balance has to be struck.

Architects and planners can argue about where the optimum lies until the cows come home. And they do; a mass of material has been written in this debate, much of it by people with strong views for or against a shift to higher densities.[3] I decided to try to make a contribution by designing a space-saving, primarily residential block. It would meet people's desire to own cars and have larger homes and gardens as well as providing some smaller homes for those who did not need (or could not afford) as much space, or who needed subsidised housing. The point of the exercise was to find out whether this kind of spacious, aspirational housing could achieve a high enough density to favour walking, cycling, public transport and mixed uses. Having absolutely no training or skills in architecture or urban design, I was very fortunate to find Paul Drew, who works for Llewelyn-Davies and has plenty. The initial, flawed concept was mine; the effort and intelligence required to make it into something that could actually be built (and would, we think, be splendid) were his.

We wanted our block design to be instantly familiar to housebuilders and house buyers, something both could feel comfortable with. We chose the three-storey terrace on a gridiron street layout as our starting point. The back gardens of two rows of houses meet down the middle of a rectangular block while the house fronts face onto the street. This is a design that has been flourishing in Britain and other nations for some 200 years. Such homes remain in high demand in prosperous quarters and housebuilders still build them in places where land is more expensive.

But there are problems with this basic block design. If you have the sides of the end-of-terrace houses (the gable ends) and the sides of their back gardens lining the short sides of the rectangular block, this makes the street that runs along it look second rate. It has no front doors and windows looking onto it. It also provides a potential access point for burglars, because with no houses fronting this street there is far less surveillance by residents. So you tend to need high, blank walls (which you sometimes find topped by broken glass or barbed wire). It looks unfriendly and uninteresting. If, however, you solve this problem by lining the short sides of the block with houses, you find there is not enough room for every home to have a back garden. Problems of overlooking also crop up. The corners, where the back walls of houses come closest, are particularly difficult to sort out.

The terraced home also suffers from a lack of direct access from the street to the back garden. Unless it shares a back alley, or there is a space-wasting opening penetrating the terrace, nothing can be got in and out of its garden without it having to pass through the house. People prefer not to have to trek through their homes with garden waste and bicycles.

More and more households own boats, caravans and other bulky, transportable leisure equipment that needs to be kept in the garden, or a garage or some other outhouse with access to the street. This is possible with detached or semi-detached homes but difficult or impossible in terraces. You could park your dinghy or your camper van in the front garden, provided the latter was large enough, but you would probably worry about it being stolen from there. Items as large as caravans stowed in the front garden mar the look of the street, block out sunlight and give cover to burglars.

These inherent terrace problems can be solved by giving the houses their own back alley running between the two rows of back gardens and connecting them to the streets. People with bulky playthings can keep them in their back gardens. But it would have to be fairly wide alley and it would also provide access for anyone seeking to sneak through the shrubbery and break into the backs of houses.

How can these shortcomings of the terrace and the gridiron street layout be overcome? The first step is to line the streets running along the short sides of the block with homes, vastly improving their appearance and surveillance (see plan). In our design the units located here are either flats without gardens or compact, two-storey, two-bedroom homes with small back yards. The larger houses on the long side of the block lose little garden space and the amount of overlooking is minimised.

The second step is to have access from the street to the back gardens. Our design has a 3.5 metre-wide alley running between the two rows of back gardens. This is secure because at either end, where it meets the streets, the alley has a substantial, hard-to-climb gate that would be kept locked (and which no thief would attempt to scramble over because he would be easily spotted from the houses opposite). Each resident would have an electronic key for this gate.* Down this alley would pass bicycles, bulky garden rubbish, boats, caravans and the occasional car, although most motors would be kept on the street. People could keep large, mobile items in their gardens (within reason) but they would need wide gates in their back fences in order to do so. The block's residents would have to agree that the alley must be kept unobstructed with a strict no parking rule.

The homes along the block's long sides have good-sized back gardens, 18.5 metres long and 6 metres wide. Each has room for an outbuilding, which could be a garage, a workshop or an office well separated from the home. Trees will grow up in some of these gardens to reduce the amount of overlooking and to cast summer shade (although *leylandii* and other anti-social large evergreens will be forbidden). The three-storey houses lining the long sides of the block are big, too. The plot width is 6 metres, compared to less than 5 metres offered by today's developers in their typical terraced products. The ceilings are 2.8 metres high, bringing back the tall, airy rooms that vanished from new homes after the First World War. The internal layout is flexible but the usual arrangement is to have the kitchen and a large living and/or dining room on the ground floor, another living room, a bedroom and a bathroom on the first floor and three bedrooms, a small study and another bathroom on the second floor. The loft space above the second floor has been designed for easy conversion into a new top floor, with sufficient space for another flight of stairs and two attic rooms or one large one.

The block's houses have no front gardens but this is a virtue, not a defect. It makes for a handsome, well-proportioned streetscape with excellent surveillance. Instead of a front garden, each house has its own small, paved external space behind a low wall. This prevents strangers on the pavement pressing their noses against front room windows while providing enough space for a couple of dustbins, a bicycle and a potted shrub or two. The block's flats all have balconies or roof terraces large enough to get half a dozen people round a table. None of these balconies are on north-facing walls, so they all catch at least several hours of sunshine a day (should the sun deign to shine). Some look over the streets, some over the back gardens.

Many housebuilders and buyers will find one glaring defect in these houses. They fail to be detached. But the units on the long sides of the block each have 130 square metres of floor area, rather more than most detached new homes being sold today. They provide more generous space for growing families and for working from home. Their heating costs are substantially lower than those of conventional new detached and semi-detached homes because shared walls mean shared warmth. And they have the high levels of sound insulation needed to overcome purchasers' fears of having to listen to their neighbours' noises.

*Note that this does not make our block a gated housing estate, of the kind that was frowned on in Chapter 11. All of the front doors open onto the public street.

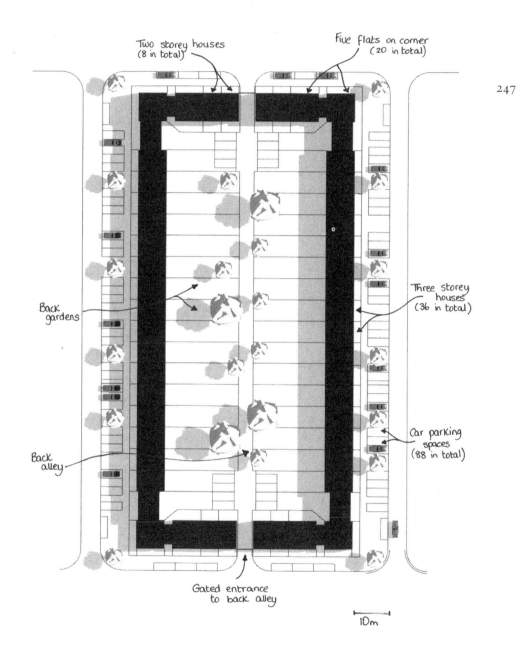

Two storey houses
(8 in total)

Five flats on corner
(20 in total)

Three storey houses
(36 in total)

Back gardens

Car parking spaces
(88 in total)

Back alley

Gated entrance
to back alley

10m

Residential block combining high(ish) densities with generous gardens

On each short side of the block there are ten one- and two-bedroom flats and four of the small, two-storey, two-bedroom houses. On each long side there are eighteen of the large, three-storey houses. But any number of these could, instead of being a single home, become two. The ground and first floor would provide a three-bedroom maisonette with a garden. The top floor would hold a one-bedroom flat with a balcony which, with a loft conversion, could expand into spacious, light-filled two- or three-bedroom duplex apartment. The width of these houses is large enough to accommodate two front doors and there is enough room for both the maisonette and the flat above to have their own staircases.

In total, then, an entire block in which six of the thirty-six large houses on the long sides were split into two homes would contain seventy homes* In Chapter 9 I argued that every new housing development should contain some subsidised housing if, as will usually be the case, there is a local need for it. Up to a third of the homes in the block we have designed could be in this category, scattered through it, giving people on lower incomes (families with children, pensioners and young singles and couples) good-sized, affordable housing mingled among owner-occupied property.

A communal oval could be provided in the heart of the block if twelve of the thirty-six houses on the long sides lost part of their back gardens. This 550-square metre outdoor space would be big enough for a shared garden and playground, or a tennis court, or a swimming pool, solar heated through the summer months and covered over by a platform in winter. This oval could also contain a miniature gas-fired combined heat and power station in a small, sound-proofed shed to generate most of the electricity and all of the heat and hot water required by the seventy dwellings. This would slash their heating and electricity bills and reduce the emissions of fossil-fuel carbon dioxide that are changing the earth's climate.

One of the principal aims of higher density designs is to reduce people's need to use cars – those other great greenhouse gas emitters – by increasing the viability of public transport and by putting shops, schools and other frequently visited places within walking distance of homes. But if most households were to choose to settle or stay in the compact city, they would probably still want their own cars, even if they used them less. The car-free housing that some urbanists advocate will only occupy a niche market in the foreseeable future. In our block design there is enough space for every home, flats and maisonettes included, to have one reserved car parking space. Almost all of these are sited immediately outside people's front doors, where owners prefer them to be. Residents can ensure that no one else uses this space when they are away by using one of those lockable bollards on a hinge that swing up out of the road surface. And those without a car could rent their

---

*Thirty four-bedroom houses with large gardens, six three-bedroom maisonettes with gardens, eight two-bedroom houses with small back yards, and twenty-six one, two or three bed flats with balconies.

parking spaces to neighbours who needed more than one. Almost all of the block's parking spaces are located in bays along the streets. Residents could also build garages at the end of their back gardens and park their cars in those, but few would either need or want to do so.

Our block would need some sort of communal management to ensure the security gates at either end of the back alley were working properly, to keep it clean and unobstructed and to ensure the system of reserved parking slots on the street parking bays was running smoothly. It would benefit from a few rules concerning the upkeep of shared areas and the downkeep of late night and early morning noise. The task could be handed to a management company set up by the developer and run by a committee drawn from among the residents (including tenants as well as owner-occupiers) and elected by them. If there was a communal oval in the centre and a combined heating and power system, then the committee could also take responsibility for these. This management would be financed by an endowment from the developer, or service charges, or both. The block should also have its own electronic notice board, in the shape of a computer intra-net with a terminal in every home. Here residents would find details of babysitters and car pools and neighbours with shared pastimes. The block should be the sort of place that encouraged neighbourliness without making it compulsory.

Consider the streets surrounding this block. Neighbourhoods based on gridiron layouts are highly permeable, with the criss-crossing streets enabling you to take a fairly direct route from A to B *provided* the blocks are not too long. In our design they are not; the long sides measure just under 140 metres.

But council highway engineers tend to baulk at the gridiron; too many cross-road junctions, too many stretches of straight road encouraging drivers to speed. Their fears would be soothed away with traffic calming devices such as raised road surfaces at the junctions, different colours of paving, and street trees planted close to the kerbside to give an impression of narrowness. There would need to be 20 mph speed limits on the streets around our block and priority for pedestrians, cyclists and children at play, making it a 'home zone' of the kind described in the next chapter.

The carriageways on these streets measure 5.5 metres from kerb to kerb, giving enough space to allow a car and a dustcart to pass in opposite directions. But while this narrowness of road would discourage traffic from speeding, the street as a whole is generously proportioned. The distance from house front to house front on the long sides of the block is 23.5 metres, wide enough to allow sunshine into front rooms even in winter when the sun is low in the sky.

The basic design does not have to conform to the exact dimensions shown here. The long sides could be a little shorter (but not longer, for that would erode the permeability of the street network). The short sides could be a little longer, allowing even larger back gardens (but not any shorter, for that would make these gardens too small). Nor need the block have precise right angles; it can be warped a little. It could, in other words, be made to fit into a real urban site which is irregularly shaped and sloping. It could accommodate existing buildings up to three storeys high. Nor need it be entirely residential. There would be some small office spaces in the houses and back gardens and there could also be a few one- or two-person workshops, so long as they did not generate obnoxious noises and smells and vehicle traffic along the back alley. A parade of shops could occupy the ground floor space along one of the short sides of the block with flats above.

This kind of block is not the only way of achieving higher densities while giving families plenty of space, privacy and the private car. Instead of houses in terraces you can have taller apartment buildings with generous balconies and roof terraces for the units above ground floor. They can be built from scratch or be created by converting buildings. The car parking can be situated in courtyards behind the housing or in undercroft basements rather than in bays off the street. And instead of private gardens, all the dwellings in one block can share a large communal garden, accessible only to residents and perhaps with the car parking beneath.

But what I like about our block design is that it dispenses with lifts (which consume lots of electricity and need maintaining) and that it is based on a popular, enduring, tried and tested design; the three-storey, middle-income town house. And while its outward appearance can be as traditional as developers and purchasers want it to be, it can meet the present's changing needs within. The layout of each storey could be flexible with moveable internal walls to change the number and size of rooms. There could be solar panels on the roofs, and the rainwater that falls on them could be collected for the flushing of toilets and the watering of gardens.

The block also offers a variety of dwelling sizes and designs in close proximity. Neighbourhoods that do this enable residents to change their housing to suit their changing needs while staying in the area, close to friends and family. As children grow up, as households expand and then shrink, people have the option of moving to somewhere suitable in the next street rather than having to leave the area altogether in order to find the right size and type of home. This local variety can help to foster the extended family, allowing grandparents to live near enough to help raise their grandchildren and then be helped themselves in advanced old age.

The average net residential density for new housing developments in England in the 1990s was 25 homes per hectare.[4] The design described above achieves 60 per hectare, twice the minimum density set out in the government's latest planning guidance for housebuilders and council planners. It does this without any squeezing of floor areas or room sizes. And despite this increase in density, most of the homes in our residential block get a generously sized garden while all of them get

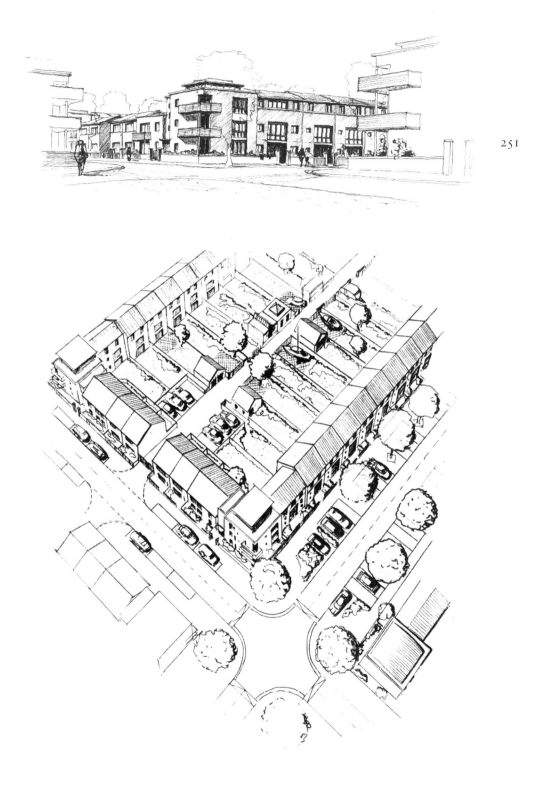

251

some private outdoor space. If we could build the millions of new homes Britain requires over the next quarter century at this higher density, it would save more than 500 square kilometres (double the area of Birmingham) from being built on.[5]

But is 60 per hectare a high enough net residential density to bring about a significant reduction in car use, a flourishing of mixed uses and of urban vitality and diversity? To answer that question, you have to imagine a circle 1 kilometre across enclosing an urban area of 80 hectares. The distance from the centre to the rim of this circle is 500 metres. Provided the street network is reasonably permeable, anyone anywhere inside this circle can reach its centre with a stroll of no more than ten minutes, even at a leisurely pace. The circumference marks out a pedestrian catchment area or 'pedshed' within which residents will find it quick and convenient to walk to centrally located services and public transport halts. As we shall see in the next chapter, people tend to walk rather than use their cars for such short journeys.[6]

Now, if we could fill this pedshed circle completely with homes at the same average density as the block described above, it would contain more than 4,700 units or, given the UK's average of 2.4 people per household at the turn of the century, some 11,000 residents. But of course there are many other things that will also need to be fitted inside the circle besides homes. Shops and schools will take up some space although there could be flats above the shops. So, too, will parks, squares, playing fields and playgrounds. Some land will have to be set aside for the extra road widths needed to handle through traffic, although with gridiron street networks there is no need to waste space on intermediate-sized distributor roads. We also need to set aside some land in the circle for employment uses in order to expand opportunities for people to work near their homes.

TWO LONDON PEDSHEDS

The sketch maps illustrate two London 'pedsheds' surrounding public transport interchanges and shops. The edges of both circles are 500 metres from the centre. Every home inside them is within 700 metres, or ten minutes walking, of the centres.

The inner London pedshed (on the left), straddling Highbury and Islington, is dominated by fairly high density three- and four-storey Victorian and Georgian housing. It contains roughly twice the residential population of the second, outer London pedshed. There are railway and Underground stations and numerous bus routes pass through. There are also three schools, a library, a post office and dozens of shops, pubs and restaurants. This locality offers very little off-street car parking and all on-street parking is tightly controlled; it is not an easy area to visit by car. Despite the high densities there are several hectares of greenery within the pedshed, in Highbury Fields.

The outer London pedshed (on the right), in Hayes, Bromley, consists almost

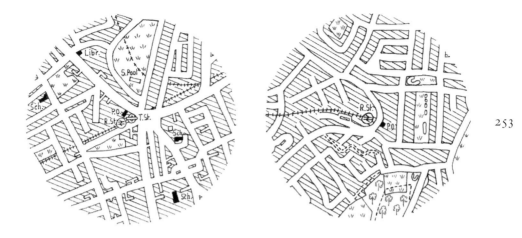

entirely of two-storey semi-detached housing built in the 1930s. It has five bus routes passing through, but the surrounding estates off the edge of the map are not so well served. The pedshed contains no school although there are three a few hundred metres outside its rim. The road running past the station is flanked by some twenty shops, a handful of cafés, takeaways, restaurants and a post office. Not enough people live within ten minutes walk of this local centre to support these facilities, so many of their customers must come by car. There is abundant parking space on and off street.

Which of the two works better? The inner London pedshed has smaller gardens and less public open space. With main roads running through it suffers much worst traffic. It is, however, much closer to central London and as well as offering a much wider range of local facilities it provides many more options to travel by public transport. Its house prices are about twice as high as the outer's.

It seems reasonable to devote roughly half of the circle's area to non-housing uses giving it an overall or gross density of 30 dwellings per hectare which means there are 2,400 homes and some 5,500 residents inside its rim.[*] That is sufficient to support a couple of primary schools and even a small secondary school, a small

*For comparison, the gross residential density of Greater London, with all of its parks, woods, and other spaces and tracts of Green Belt countryside in the outer boroughs, works out at about 20 dwellings per hectare.

group of shops including a post office, a pub and a doctor's surgery.[7] The residents could easily walk to these essential services nearby, but they would also need to travel further afield at least once a week, and probably once a day, to go to work, to the supermarket and to reach a wider range of shops, the cinema, the leisure centre, and so on.

And here public transport can come into its own. The pedshed circle contains enough people to support a few bus services running through its centre and, maybe, a tram line provided it is part of a larger town that supports the wider range of facilities. So some, but not all, of these journeys outside the circle could be made quickly and conveniently by public transport. Substantial reductions in car dependence become possible. Most residents would find it convenient and cost effective to run one car instead of two while a minority would need no car at all.

Net residential densities of around 60 homes per hectare (or 30 per hectare, as a gross density) do appear to be high enough to support essential local services within a few minutes walk, along with good public transport services. They could bring about a big reduction in car dependency. But they are probably not high enough to foster the compact city virtues of intense vitality, variety and thoroughly mixed uses. I am thinking of the qualities Jane Jacobs so admired in her mid-town Manhattan neighbourhood circa 1960: a rich mix of buildings old and new, most of which contain workplaces as well as homes, masses of little shops, bars, cafés and restaurants, lots of hustle and bustle and people on the streets.* You can find this intensity across tracts of central and inner London and scattered, here and there, in parts of other large British cities. It is fairly common in continental conurbations. But it tends to be sadly lacking in smaller and medium-sized British towns, which are usually divided between an almost entirely non-residential centre and an almost entirely residential remainder.

We need more of these urban quarters that mingle living, working and playing at high densities. But we have to recognise that most of our towns and cities could never be like that. Most people want more garden room than can be found in such quarters. Besides, much of the vitality you find in these high density places depends not on their own residents but on the flow of visitors from quieter, duller parts of the city and beyond who come to shop, dine, drink and enjoy the atmosphere.

Neighbourhoods made up of housing at net densities of around 60 units per hectare can accommodate some mixture of uses as well as a social mix. There are more people out and about on the streets than are seen in low-density suburbia and the housing is more broken up by schools, shops and so forth. But, like the great bulk of our urban fabric, such places look primarily residential and feel fairly quiet and unexciting.

---

*Jane Jacobs thought cities could only flourish with residential densities above 240 dwellings per hectare; neighbourhoods, which dropped beneath that level, were in danger of degenerating. They lacked the numbers of residents needed to maintain surveillance of passing strangers and sustain a large number of

Which is fine, because this does not prevent them from being cherished and successful. The more ardent compact city enthusiasts need to be a little more realistic about the limitations Joe and Joanna Public would wish to impose on high densities and mixed uses. How many people, for instance, would be enthusiastic about mixing their housing with a pub or a fast food restaurant? History tells us that people with choices have opted overwhelmingly for unmixed, low-density suburbia because it offered privacy, tranquillity and a garden of their own. Neighbourhoods built at middling densities of around 60 homes per hectare can offer their residents all three, but with a sufficient concentration of people for public transport and local services – reached on foot – to flourish. The h in MUHD stands for higher, not high. Such neighbourhoods would be less dull and less car dependent than twenty-five houses per hectare suburbia.

Block designs that have never been constructed, perfectly circular neighbourhoods that have never been built . . . this chapter has, so far, tended to the theoretical. We need to get off the drawing board to show how MUHD might be applied to real towns and cities.

First, a quick revision of the essential features of Anybigtown, UK. It has a centre consisting largely of shops, leisure uses, offices and various seats of authority – the main police station, the law courts, the town hall. Few people live here because the centre's high land values and its busy, built-up character have pushed housing out. Almost all manufacturing industry has also fled to cheaper sites which are more accessible for lorries. Seas of housing surround this centre with residential densities tending to decline towards the urban rim. In the inner city lie Victorian and Edwardian homes built for the working and middle classes, mostly laid out at fairly high densities. Great chunks of this older housing have been torn out by various twentieth-century slum clearance programmes and Luftwaffe bombs, to be replaced mainly by council homes.

Then, further out, lies generally lower-density stuff built since the First World War. Albert Square and Coronation Street give way to Betjeman's Metroland and Brookside Close. Large, peripheral council estates lie next to swathes of owner occupation. Other land uses intrude into the housing: industrial areas, parades of shops strung out along the busier roads, parks, golf courses, allotments and cemeteries. Here and there lie the remains of small towns and villages engulfed during two centuries of rapid urban growth. These have become microcosms of the entire city with shops, offices and other businesses remaining clustered in their centres.

Throughout this urban patchwork are structures rather like the circular pedshed neighbourhood we conjectured earlier in this chapter: little hubs of services and

small, local businesses. Once you fell below 240 you had to drop right down to 50 to find a viable housing density, but it was one that could only work in

suburbia and 'other, more inert type of settlement', she wrote in *Death and Life*. I believe she was wrong about this.

public transport stops amid primarily residential areas. Being evolved rather than planned they are irregular in size and shape and not all of the communal services and facilities are gathered conveniently at their centres. They exist because this was the obvious way to organise things at a time when most people did not have access to a car and because town planners have, for most of the post-war years, tended to reinforce this kind of neighbourhood layout. But rising levels of car ownership and the spread of the superstores have put these little, local hubs in danger of losing their customers. If they decay, that ratchets car dependency up another notch.

How do we build the benefits of mixed uses and higher densities into this complex, heterogeneous stuff of cities, this shifting mix of general unease, roaring success, solid prosperity and abject decay? First, we look at those crucial, zero-density places within cities where there are no buildings at all.

THE COUNTRY IN THE CITY

As people become more affluent, both *en masse* as the economy grows and as individuals passing through life and seeing their earnings rise, many of them want to spend more time in the countryside. They like to cycle, ramble, picnic and carry on all kinds of leisure pursuits in fields and woods and along the coasts. Many hope to move out of town for good, or to purchase rural second homes. They also want to see an improved countryside. That means one that is more wooded than today's, that has more semi-wilderness and flourishing hedgerows (instead of the neglected and dying specimens so common today), where the variety of flora and fauna is protected and enhanced.

Towns and cities need to compete with the countryside by enhancing, joining up and even enlarging the myriad green places that lie within them. This urban greenery will never feel as rural and remote as the real thing but city countryside can make up for that by being on everyone's doorstep. Local councils should aim to have chains of undevelopment shot through the urban fabric, such that no citizen is more than a few minutes walking or cycling from a quiet, green route which will lead towards the heart of the city in one direction and out towards the open countryside in the other. These green chains can be quite narrow for most of their length, following rivers, canal towpaths and old railway lines. Every now and then they should broaden out into a larger green space such as woodlands and parks.

This is far from being some unrealisable dream. Many big British cities have great stretches of greenery extending close to, even up to, their centres; think of Cardiff's Cathays Park, Bristol's Clifton Downs, Newcastle's Town Moor. Taking the train into central London or coming into Heathrow on an unclouded day, I'm always struck by how green the capital is despite being Britain's most densely populated large city. Greater London is, in fact, less dense than greater Tokyo, Paris and even New York.[8]

One summer Bank Holiday Monday we decided to cycle in to London Bridge from our home on the outermost edge instead of joining the car-borne hordes heading out to country and coast. Our aim was to make as much of this urban journey as possible through greenery, beside water or both. It was one of the best days out that year. We started by cycling through the scruffy farm near our home, a wedge of fields poking into the urban fabric. The customary huge car boot sale was in progress. Our winding route then took us over wooded hills and dales bursting with leaf and birdsong, through Britain's busiest golf course at Beckenham Place Park and along the banks of the Rivers Ravensbourne and Quaggy that flow down to the Thames. We struck that mighty stream at Deptford, just inland from Greenwich. There was a salt smell, lots of seaweed and strong sunshine turning the wide, wave-flecked, tea-brown water blue. From then on we followed this surrogate coastline into the heart of the capital, aside from a brief incursion at Rotherhithe where part of the filled-in Surrey Docks have been turned into an ecological park. At its heart lies an artificial hillock, Stave Hill, which offers one of the finest but least known views over London.

We managed to keep beside water, or amid greenery, for well over half of our meandering, 16-mile urban transect from fringe to core. Where we had to cycle on roads we managed, for the most part, to stick to quiet streets. The only really unpleasant non-green experience was the fording of the South Circular Road. Much of the day's pleasure and interest came from watching the city and its families, out enjoying the sunshine, endlessly changing around us as we neared the teeming core.

Considering nature is one of the most precious and loved of urban resources, council planners ought, in general, to keep buildings off the grass. Now it could be argued that in so doing they will frustrate the re-creation of mixed use, high-density cities by putting lots of useful land off limits for development. The government and Lord Rogers' Urban Task Force place great emphasis on the need to redevelop derelict 'brownfield' sites, many of which have turned themselves from brown to green. Scrub and then trees grow up. Birds, mammals and insects – including rare native species and exotic foreigners – make themselves at home and the derelict, abandoned land becomes cherished by local residents. Should such sites, which were once developed, be off limits to future development?

And what of the other underused or useless greenery we find dotted around cities? Such as bleak communal lawns where no one walks or plays, or deserted and overgrown allotments. Are they not a waste of precious space?

The aim should be to ensure that all existing urban greenery, whatever its origins or its status, performs – as a wildlife habitat that is also accessible to people, as movement corridors for humans and animals, as peaceful spaces and playful places. It can usually carry off several of these roles at the same time. Only when, for what-

Primrose Hill, London, overleaf

ever reason, there is little chance of an open, undeveloped place ever being able to perform any of these functions at all well should it be built over.

Such cases will be rare. The government and local councils ought to adopt minimum standards for urban greenery, which would drive the creation of new open spaces in towns and cities. The government's Countryside Commission (now the Countryside Agency) recently proposed that every city dweller should have an accessible green space within 400 metres of their home and that all towns and cities should have safe routes free of road traffic leading out to the countryside.[9] The Wildlife Trusts, a leading charity in this field, and English Nature (the government's nature conservation agency) have proposed that every urban resident should have a green space of at least 2 hectares within 500 metres of their front door, and that for every 1,000 residents there should be at least 1 hectare of nature reserve.[10] These neighbourhood levels of greenery do not prohibit the compact city, nor the higher housing densities we should be adopting. They are easily reconcilable with an overall (or gross) urban density of 30 homes per hectare.

SUBURBIA

Great tracts of our towns and cities have been built at low densities that encourage car dependency and discourage public transport and local services. In some places the small shops are struggling or have given up the ghost while the through roads just get busier and busier.[11] But Britain's huge belts of owner-occupied, post-First World War suburbia are, for the most part, either holding their own or prospering. They provide millions of households with privacy, space, gardens and trees, along with the freedom to alter their houses to meet their changing needs.

In flourishing suburbs densities tend to creep up of their own accord. Developers try to make as much money as possible from any site that becomes available; if the local planners allow them to build blocks of flats they will generally do so. If a neighbourhood has particularly roomy gardens then some owners are bound to want to build another house or two on their plot, making a small fortune in the process. And if the existing houses are large, then some owners will want to make money by converting them into flats.

Usually the neighbours will lobby council planners to refuse the necessary planning permissions. There are likely to be angry accusations of town cramming, of degrading amenity and – most heinous, this – of lowering property values. The neighbours' resistance is reasonable because as well as eroding space, greenery and privacy, the shift to higher residential densities brings ever more cars and traffic. The wealthier a neighbourhood is, the more powerful and effective this resistance to densification is likely to be.

There is a need for planning policies that allow a careful, controlled increase in suburban densities in order to bring the benefits of the compact city to its outer belts – less social segregation, less car dependency, increased overall prosperity. Politics decrees that this will have to be a sensitive, gradual process if it is to have any chance of actually proceeding. The insertion of extra homes should only be allowed within a few minutes walk of public transport stops, and each one should be allocated only one off-street parking space. On-street parking controls must be introduced once cars start to line the pavements. Near larger public transport nodes (railway stations and clusters of bus stops) higher-density flats should be permitted with only one car parking space per unit at most. There is also a strong case for encouraging higher-density development – low-rise flats and terraced housing – to front onto suburbia's parks and woods. The residents will have green views while adults and children walking and playing in the greenery will be overlooked by the housing and feel safer as a result.

One day housebuilders might make money by buying entire streets of suburban semis and detached homes then replacing them with double the number of homes in brilliant modern versions of the urban terrace. Tired, drab little shopping parades scattered thinly through suburbia will blossom and grow, with new cafés and food stores and other shops. More, and better, bus services will appear and then trams. The urban periphery will become more densely populated, more vibrant, a more interesting place to live without any diminution of its environmental quality.

INNER CITIES

Inner cities usually have sufficiently high population densities to sustain good public transport networks and a variety of local services. Where they are flourishing, the task of local councils is to apply the lightest of touches, conserving environmental quality, understanding what makes them successful and trying to replicate this elsewhere. Planners should use the opportunities provided by development sites to bring in a range of new housing – from family homes with gardens to studio apartments. Houses should always be built in terraces; semis and detached properties with more than one off-street car parking space are inappropriate in the inner city. But nor is it right to cover every available site in multi-storey, gardenless apartment blocks; that is no way to retain or attract families with children.

Most of the inner city in Britain's big conurbations is, however, far from flourishing. It is poor – poor in terms of its environmental quality, the income of its residents and levels of investment. In many places it is rotting despite repeated attempts at regeneration. In the worst parts of the inner city, high housing densities are being thinned out by piecemeal demolition; councils can find no other way of dealing with abandoned homes.

Planning policies that encourage mixed uses and higher densities cannot, on their own, save these places. What they need more than anything is for people with choices to want to live there. And more income. And new development to give them desirable homes along with workplaces, shops and leisure facilities. Achieving these things will involve spending taxpayers' money to improve local schools and public transport, to reduce crime and the fear of crime and to subsidise the price of new, for-sale housing.

If inner city regeneration is to work, local councils, private sector developers, landlords and local communities will have to work together to create, or recreate, neighbourhoods with a critical mass of population and prosperity. Neighbourhoods that can support a range of local services and frequent public transport services within a few minutes walk of every home. They will need to think in terms of sorting the inner city into pedsheds with at least 2,000 homes and 5,000 people while giving their residents access to green spaces and green corridors.

It may actually be best to de-densify parts of the inner cities and, once again, to demolish substantial quantities of housing. Besides blighted post-war estates and tower blocks, there remain thousands and thousands of streets of pokey, by-law Victorian and Edwardian terraces which have survived the last century's slum clearances. These houses have a front door opening straight onto the pavement, a tiny yard out the back and half a dozen little rooms at most.[12] Over the years they have been modernised, with indoor bathrooms, lavatories and central heating installed. They could stand for another century or two if they were properly maintained. But should they? Dark inside, leaking heat through their solid masonry walls, the best thing would be to demolish the worst of them and build more spacious, sunny, better insulated homes in their place – like the ones in the residential block advocated earlier in this chapter. The densities will be a little lower than that found in the cheapest by-law terraces but the housing will be more varied, allowing for a social mix.

What is required is a voluntary purchase scheme, backed up by a tougher, faster system of compulsory purchase, which enables local councils to buy entire streets of privately owned houses in which most of the properties are vacant. They should be purchased at slightly above the rock bottom prices they would fetch in the open market, in order to encourage owners to sell voluntarily. Some of the owner-occupiers remaining in these sad streets will also need help in being rehoused. There will need to be extra for those unfortunate owners with negative equity – whose mortgages are larger than the current value of their property – and rehousing help for tenants of private sector landlords. Of course this could induce speculators to buy empty houses in the hope that the council will seek to purchase then demolish them; I discuss ways of overcoming that in the final chapter.

Buying entire terraces would help councils, housing associations and private developers to build up the land banks they need to recreate mixed use, high-density neighbourhoods in decaying inner cities. It would add to the stock of land

Abandoned terrace of by-law housing, Salford, Greater Manchester, facing

occupied by social housing which has failed and needs redeveloping. At the time of completing this book, this kind of approach was being pioneered by five local housing regeneration companies, set up in 2000 with the backing of the government's Housing Corporation, housing associations, local councils and other agencies. They aimed to improve private and social housing in selected inner areas of five English towns and cities characterised by dilapidation and abandonment across the different types of tenure. Demolition was to be one weapon in their armoury.[13]

The rebuilding of a decayed inner city quarter needs to be based on a flexible masterplan drawn up in consultation with local people, implemented with their support, and which sticks to the ten principles of MUHD outlined in the previous chapter. New and refurbished housing for both incomers and the existing residents and environmental quality will be this masterplan's most important elements. It must provide a vision of what the neighbourhood could look like in ten or more years but also allow for extensive redrawing as circumstances change and new opportunities arise. Older, landmark buildings that are familiar friends to the people who have grown up in the area should be conserved and, if necessary, re-used. So, too, should any high quality housing which, while remaining structurally sound, has come down in the world. It won't be easy to secure the private and public funds needed to secure this scale of change. There is also a risk of repeating the mistakes of the great slum clearances of the 1960s and 1970s, dislocating and alienating local people − who won't necessarily take kindly to the idea of bringing in new, more prosperous, residents.

But demolition and rebuilding on their own can rarely bring about regeneration; throughout this book I've argued that it requires change on many fronts. Once social and economic, rather than merely physical, regeneration begins to happen, with residents, jobs and investment drawn in, the urge and the need to demolish tracts of by-law housing, run-down council estates and tower blocks should greatly reduce. They might gentrify, or cry out for publicly subsidised refurbishment, instead.

TOWN AND CITY CENTRES

Plenty of people live in London's very large central area (the West End, the City and their immediate surroundings) although the density of residents there is much lower than is found in the core of most European capitals. Some big British cities, notably Manchester, have seen spectacular growth in the number of city centre homes. But, by and large, a distinguishing characteristic of the central business districts of UK towns and cities is that very few people actually live in them.

This is unhealthy and potentially dangerous. To understand why you have to consider why centres are centres. Being in the middle, they are the locations with the largest catchments of people. So long as they can be reached reasonably conveniently by everyone in the city, it makes sense to concentrate much of its leisure, shopping and workplaces within them. But their very attractiveness for these uses pushes land values up, forces housing out and leaves centres full of buildings which are empty of people for most of the time.

If no one lives in the centres then the great majority of the people who use them have to travel in by car or public transport. This adds to the environmental stresses and damage caused by transport, the subject of the next chapter. The other reason why centres badly need more residents is that many of them cannot be assured of retaining their status as places where large numbers of people come to shop, to be entertained and to work in offices. All of these activities have already spread into the suburbs and out to the edge of town. The growth in car ownership makes it ever easier for people to use facilities outside of the centres. The rise in the purchasing of goods and services via the Internet – E-tailing – will take business from some types of city centre store. Once leisure places (from restaurants to cinemas) and shops start to abandon centres, they also become much less attractive as office locations.

The average town and city centre is at its worst after darkness falls. Its restaurants and pubs have few potential customers living close by, so these must either rely solely on the lunchtime trade or attract customers in by car or public transport at night (or catch people on their way home from work). The result is two kinds of evening urban scenery, both fairly obnoxious. Either the centre is deserted, and therefore desolate and frightening. Alternatively, a few streets teem with crowds of people pubbing and clubbing. The result is mass drunkenness concentrated in space and time, along with violence, foul behaviour and bodily effusions; few things are more unpleasant for non-participants, as any policeman and casualty nurse will tell you. I don't think one could, or should, stop people spending a pleasant evening getting drunk in company but this would be less lethal and destructive if it was more dispersed in space and time.

Many towns and cities have awakened to the range of threats facing their centres, with councils, shops and other businesses co-operating to increase their attractiveness and improve their management. Raising the number of residents would help them to deal with these threats, giving centrally located businesses more potential customers and making them more humane and civilised places. If people with choices choose to live in our urban cores, they will help to look after them, reducing road traffic and crime.

Because land values are highest there, it makes sense to have only high-density housing – either in new buildings or converted ones. Houses built at medium to low densities would be extremely expensive. They would also look out of place among the larger, more closely packed buildings characteristic of city centres. Designing this new city centre housing at high densities (of, say, 80 homes per hectare (net) or above) does not force the individual apartments to be small and cramped, but it does mean that the buildings they are in will generally have to be at least three or four storeys high.

There may be an overwhelmingly strong case for having more city centre homes; it does not follow that they will actually be built. If developers believe they can make more money out of new shops and offices, then they will never build housing. Flats can be built into new shopping, hotel, office and leisure develop-ments but the property industry has generally been wary of such mixed use buildings. Town and city centre developments are largely owned by financial insti-tutions that want a predictable, fuss-free stream of rental income. They seek to minimise the complexities and risks associated with keeping their buildings ten-anted, so the notion of having lots of individually rented or leased homes within them does not appeal.

But there is no fundamental reason why homes cannot be incorporated into city centre development (see text below). A growing number of developers are building these mixed use schemes, including supermarket chains. Councils should insist, in their development plans and in their own guidance to developers, that almost all new town and city centre developments above a certain size (with plot areas above a quarter hectare, say) include more than a specified minimum number of new homes – either within them or on another central site.* New planning guidelines from central government are needed to back them up. A few councils, such as Camden in London, have been pioneering pro-mixed use policies since the early 1990s.[14]

BIRMINGHAM – BRITAIN'S MIXED USE CAPITAL

In Birmingham, what used to be the largest postal sorting office in Europe has been converted into an extraordinary city centre building containing two hotels, 144 apartments, dozens of shops, restaurants and bars and 200,000 square feet of office space. The development has come to be regarded as one of Britain's most bril-liant mixed use schemes.

The Royal Mail sorting office, completed in the early 1960s, was a hulking, 50-metre tall oblong covering nearly 2 hectares in the city centre. When it became redundant enterprising local developers set up a firm, Birmingham Mailbox, to put it back into use. A partially covered, pedestrianised street 12 metres wide was cut through the middle of the building, giving the city centre a new connection to its his-toric and reviving canal network. The grey concrete which coated the building disappeared behind colourful new cladding. The steel-framed structure was strong enough to take three extra storeys of apartments, the most expensive of which sold for over £300,000. It was completed in 2001.

---

*There may be one or two types of city centre building
inherently unsuited for housing, such as bus stations.

Next door to the Mailbox lie the disused Central TV studios where *Crossroads*, the motel soap, used to be filmed. Here another developer, Hampton Trust, is planning to build Britain's first mixed use skyscraper as part of its £450 million Arena Central scheme. The development was approved in 2000 by the Deputy Prime Minister, John Prescott, after a public planning inquiry, with a 175-metre height limit imposed on the tower.

The plan was to build this gigantic development in phases, beginning with a large block of 350 apartments. Five other substantial new buildings would accompany the rather graceful, rocket-shaped skyscraper, wrapping themselves around an existing luxury hotel, office tower and a Grade II listed bank building. Masses of office space, more than one thousand new city centre homes, a multitude of shops, bars and restaurants were all planned, along with new pedestrianised streets and dramatic open air public spaces in the heart of the new development.

Martin Field, managing director of Hampton Trust, said institutions funding large-scale commercial property developments did not like the extra complexity, construction costs and tenant management problems which arose from a mixture of uses within a single building. The value and the returns from a scheme like Arena Central were lower than that of conventional, single-use development. But there was no alternative; Birmingham's planners rightly insisted that major city centre developments should include housing and uses which brought life to the urban core outside of office hours. He hoped Arena Central would be complete by 2005.

Mixed uses and higher densities are fundamental to an urban renaissance. With good design they are compatible with tranquillity, generous quantities of private space and access to nature. But if we allow the urban environment to be destroyed by road traffic, and if we continue churning out the same old car-dependent suburbia outside our towns and cities, the rebirth will never happen. It is to these issues that we now turn.

Chapter 14

EROSION OF CITIES OR ATTRITION OF CARS

What if we fail to stop the erosion of cities by automobiles . . . In that case we Americans will hardly need to ponder a mystery that has troubled men for millennia: what is the purpose of life? For us, the answer will be clear, established, and for all practical purposes indisputable: the purpose of life is to produce and consume automobiles.

Jane Jacobs, *The Death and Life of Great American Cities*[1]

Cycle tracks will abound in utopia.

H.G. Wells, *A Modern Utopia*[2]

The big thing about living in the country is you spend your whole life in the car. You have the idea of this healthy, open-air type life and it isn't like that. I've never driven so much in my life.

Rural resident[3]

Cars, cars, fast, fast! One is seized, filled with enthusiasm, with joy . . . the joy of power. The simple and naive pleasure of being in the midst of power, of strength. One participates in it. One takes part in this society that is just dawning. One has confidence in this new society: it will find a magnificent expression of its power. One believes in it.

Le Corbusier, *The City of Tomorrow*[4]

THE NIGHT BEFORE I BEGAN writing this chapter I played a game of squash at the local council leisure centre. Then I went on to someone's leaving do at Canary Wharf on London's Isle of Dogs. And, naturally, I used the family car to get to both events from my home on the capital's far-flung edge. A 2-mile dash to the centre of Bromley, a heavy defeat, then another 9 miles of motoring in the gathering dusk, past the Millennium Dome, under the Thames through the Blackwall Tunnel and briefly out into the open again before plunging into the car park caverns beneath the office towers. It took less than thirty minutes to get there.

On the return journey there was quite a snarl up around the tunnel's mouth. It took fifteen minutes to crawl back under the river. And then on through the night, stepping on the gas to make up for the delay, speeding on the long and by now fairly empty stretches of dual carriageway that cut through the slumbering suburbs. There I sat in splendid isolation amid 18 times (18 times!) my own weight of metal, plastic, rubber and fuel, rushing along in my archetypal mid-range saloon.

Mondeo man fancied a face full of greasy mutton and unleavened bread would round off a pleasant evening. My local kebab bar lay but 6 minutes walk from home and in the opposite direction to where I was coming from. Naturally I sped on past my house to get to it, for was I not in my motor already? Alas, in outer Bromley these places close early; midnight had chimed and they were swabbing the joint down. Mildly disappointed, I swung the trusty Ford around and zoomed off to bed with a final blast of climate warping, Nox-laden, ozone-generating fumes.

Public transport? No thank you. Forsaking the car that evening would have involved catching a bus, followed by three train and tube rides to get to Canary Wharf and half a mile of walking at various points along the way. My total travelling time that evening, including waiting and walking, would have been over two and a half hours; more than twice as long as I spent in the car. The fares would have cost more than I had spent on petrol and parking. And, besides, those late night trains out of central London can be insalubrious. They are full of commuters who have stayed in town to drink. Some are loud. Others snore, or devour bagfulls of smelly fast food.

I too commute daily into central London. Take the car? No thank you. I would be as likely to use it for the journey to work as I would for a visit to the moon. With all that traffic the motor would take much longer than the train does to get me there. I have no idea where I would park it. And even if I could find a space it would cost much more to occupy than the price of a return rail ticket, and be a fair trek from my office. The train may have standing room only and be subject to baffling delays, cancellations and halts but it seems the only sensible way of making the journey.

I am a four-wheeled hypocrite, one of those people who think cars have a great deal to answer for but who are happy to use their own when it suits. I am not alone. I want to hate motors, did not own one until my late twenties, have cycled for all or part of the journey to work for most of my life and will never buy a new car. Mine is a slob with faded, scratched paintwork; neighbours who spray and spruce up their prized possessions every weekend baffle us.

Particularly anti-social are those boxy, four-wheel-drive monsters whose sales have been soaring lately and which are taking over my neck of suburbia. What is going on in the vestigial brains of their purchasers? 'I really need this all-terrain capability to get the kids to school and do the weekly shop. And this car is so big, so heavy, so unslippery in shape and so capacious in its cylinders that everyone will see it's a blatant gas-guzzler. They'll know that I'm rich enough to waste petrol and cause extra pollution by driving at less than 20 miles per gallon. Sitting in it makes me two foot taller than other drivers, which is handy because I like to look down on people. And, as a final gesture of contempt for my fellow human beings, I'll attach some extra metal bars to the front so as to improve my chances of maiming or killing any pedestrian I hit.'

I want to hate cars but I know they are a wonderful tool for getting more out of life and making it easier, for visiting places and people, for saving time and having fun. I don't think there is any hope of persuading those with choices to abandon these splendid freedom machines. Cities need people with choices. Cars and traffic are among the things that damage cities so severely that people end up choosing not to live in them. Can cities ever cope with cars?

There are reasons to believe they cannot. Pause to consider your size, the size of a car and the physics which have underpinned our biological evolution, and you will see that the automobile is bound to be a fundamentally threatening and obnoxious object. Anything that large moving faster than what, for the driver, is a crawl (20 mph, say) will seriously harm or kill you if it strikes your body. It makes a lot of noise and even if it had a near silent electric engine there would still be a loud rushing from its slipstream and rolling tyres. For the driver the car is cosy, comfortable, quiet. Thanks to decades of technological progress, he or she is increasingly safe from harm in any crash. For those on the outside it is a wooshing, hurtling menace; almost a third of the people killed or seriously injured on Britain's roads each year are pedestrians or cyclists.[5]

The last place we need these most anti-social objects is the place where we have most of them; the city. In some of the quietest residential areas cars and traffic are still scarcely noticed and children play in the road but the number of such streets has been falling as car ownership and traffic volumes rise. As you move up from this base, through increasingly busy roads, the environment for the people on the streets outside of vehicles becomes less and less tolerable. It becomes harder and harder to sleep, play and enjoy a home life beside them. The busier a street, the cheaper – in general – are the prices of the homes lining it.

The press and the green pressure groups give the impression that pollution from exhaust fumes and crashes are traffic's main threat to people. The government's experts estimate that at least 12,000 lives are shortened each year by air pollution, to which traffic is the main contributor, while ten people die each day in accidents in Britain.[6] This matters a great deal but it is the noise, the ugliness and the inhumanity of busy roads that do far more economic and social harm to us all.

For decades, traffic flows have been rising by around 3 per cent a year – fast enough to more than double each quarter century.[7] The highway engineers and planners responded by providing more space for vehicles – car parks, urban ring roads, roundabouts and dual carriageways, sometimes raised on stilts above homes and streets. People and buildings, the essence of the urban, had nothing to do with these conduits built solely to speed the flow of vehicles around and through the cities. They sterilised and sucked value out of their immediate surroundings and obliterated huge quantities of that most precious urban commodity, space. Instead of solving congestion this huge expansion of road capacity merely encouraged extra traffic.[8] Wide new highways installed to handle through traffic and freight became clogged with cars and vans making short, local journeys.

But many people still wanted and needed to walk around the densest, innermost parts of the cities. The solution here was segregation; banning vehicles from large central areas which were then hemmed in by big roads and lots of car parking space. The inner ring road, the pedestrianised shopping centre and the multi-storey car park have been adopted almost everywhere. It was not only the planners who zoned cities; it was traffic. Urban cores became sealed within road barriers, accelerating their decline into places with no homes and little life after dark. The government now defines a town centre as a place where hardly anyone lives.[9] If you look at aerial photographs of the hearts of many medium-sized British towns you see they are dominated by car parks and roads, with buildings dotted randomly on this background.

Walls of traffic now exclude the homes and businesses of the inner urban areas just outside the centres from the investment and wealth still flowing into busy centres. Pedestrians wishing to move from the inner city to the core have to negotiate ugly, threatening underpasses or ascend to footbridges in order to cross the highways. They are usually forbidden to cross at street level because that would block the flow of traffic. The result is that much of our urban fabric now makes no sense to someone on foot. All is confusion and threat. Swirling, curved roads surround you with no clue as to how to cross them – and even if you could you are not sure where you would get to. The old structure of straightish streets and sharp bends that allows you to retain a sense of direction has vanished. Everything tells you that this is a place for cars, not people, and to attempt to get around by any other means is madness.

But even where traffic does not ruin urban walking, most people still prefer to use cars for journeys which take longer than 6 or 7 minutes on foot. Walking a few miles in the countryside may be fine, but walking further than 500 metres merely to get from A to B in town is too slow, or too tiring, or you might get rained on, or something or someone might threaten you.[10] People like my neighbour vote with their feet and their feet go on the accelerator pedal. He drives his children the half-mile to school every day, rain or shine (and was always kind enough to give mine a lift when they were running late). The same considerations apply to cycling. As a leisure activity its popularity is growing. We buy more bicycles than cars each year, and more and more of the motors heading out of the cities at the weekend are laden with two wheelers that will be ridden on rural tracks and lanes. But most people think cycling around the city will be tedious or inconvenient or downright dangerous and terrifying.

Rising automobile ownership encouraged employers and retailers to move into larger buildings with big car parks attached. When most of the employees or customers of a business own cars there is no need for it to be sited within walking distance or a short bus ride of their homes; instead it can seek economies of scale by moving into a larger place within reach of a 20-minute drive. Major retailers used to have to site themselves near town and city centres in order to be reachable by the old hub-and-spoke public transport services. Once the majority of households owned cars they were freed from this need. Instead they could pitch their tents on any suitably large and available site anywhere inside or outside of town, provided it had the population catchment they needed and the council planners would let them. And until the mid-1990s, they usually did (and if they didn't, then central government would often side with out-of-town developers in the subsequent planning appeals). More and more buildings that could only be travelled to conveniently by car sprang up, increasing the pressure on people to own cars. What incentive does business have to cater for the car-less minority? The very fact that it does not own automobiles demonstrates its lack of spending power.

As car use and car dependence grow, people's views of their cities change. Sitting at the wheel we are no longer close to other faces and bodies; that, for many, is one of the joys of the car. There is no possibility of talking to strangers or bumping into an acquaintance or just popping into a shop or pub or café on our travels. In our metal boxes, off the pavements, we are less likely to notice squalor, litter and destitution on the streets and less likely to care. It is not the public realm but just stuff to be passed through. Being at some remove from people, enclosed in glass and metal, also makes it easier to be uncivil. Would you, as a pedestrian, ever hoot, barge past, shout abuse, make rude gestures and furious faces at your fellow pavement users? This is fairly ordinary behaviour in a car.

Cars on top: Milton Keynes, Buckinghamshire

We come to perceive the city as a collection of disconnected places reachable by car. We choose each of these destinations because it answers some need and feels safe and comfortable. These semi-public, semi-private destinations stand apart from the old public realm of streets, parks and squares. We might feel some sense of ownership of our local high street, but we cannot feel any for an enclosed shopping centre or a cluster of retail warehouses, drive-through fast food restaurants and multiplex cinema. They belong to those who built them in order to sell things. We free-roaming consumers merely choose to use these places; others are responsible for the upkeep and cleanliness and the kind of behaviour which is tolerated there. We can always drive somewhere else if they fail to please. These kind of places, surrounded by car parks, have come to be known as 'out-of-town' or 'edge-of-town' developments, but they are now ubiquitous *within* urban areas. They stand apart from the rest of the urban fabric because you would never want to arrive or depart on foot.

The city is attacked on several fronts as rising car ownership and car dependence drive themselves onwards and upwards. Public transport loses custom. Its services shrink and its passengers become poorer, rendering it an increasingly unattractive alternative. Car parks, roundabouts and dual carriageways have to spread, shrinking the walkable public realm and leaving a growing area unfit for talking, eating, sleeping and any other human activity apart from driving. Destinations disperse, making journeys longer and increasing the likelihood that we will need our cars to reach them. Travel statistics highlight the alarming speed of these changes (see text below).

THE RISE AND RISE OF THE CAR

A quarter of a century ago less than half of adults had a driving licence and Britain was not a car-dependent nation. Today 70 per cent do and it is. Government statistics shows how far and fast travel patterns have changed at the close of the twentieth century.[11]

WE OWN MORE CARS ...

Some 71 per cent of households had access to a car in 1999, compared to 63 per cent in 1986, while more than a quarter have access to at least two. And car ownership looks set to go on rising for years to come. Britain, with 406 cars per thousand households, is still some way behind France (456), the USA (481), Germany (508) and Italy (545).

WE'RE WALKING AND CYCLING LESS ...

The average distance each of us walks has dropped by more than a fifth, down to just over half a mile a day, since the mid-1980s. The average distance cycled also fell, by a tenth. Of all the miles we travel each year (excluding international air travel), less than 3 per cent are on foot.

Our trips are becoming longer as development spreads out and car dependency deepens. The proportion of all our journeys that are short – less than one mile – has fallen from 35 to 27 per cent since the mid-1980s. Four-fifths of these short journeys are made on foot but the share of them made by car is slowly increasing. For journeys any longer than a mile, car travel predominates. The proportion of primary school-aged children walking to their classes declined from 67 to 55 per cent; the numbers being driven there rose from 22 to 36 per cent.

WE'RE TRAVELLING FURTHER, AND USING OUR CARS MORE ...

People travelled an average of 6,806 miles each year in 1998 using all modes of transport, 28 per cent up since 1985/86.[12] This was mainly due to the increase in average journey lengths from 5.2 to 6.5 miles; the actual number of journeys barely changed. Journeys for every kind of purpose grew longer. The lengths of the average shopping and commuting trips both rose by one third, to 3.9 and 8.1 miles respectively.

Car travel accounted for four-fifths of the total distance travelled, up from three-quarters in 1985/86. And cars are gradually getting emptier, too. In the period 1995 to 1997, drivers made 63 per cent of trips unaccompanied by a passenger. A decade ago it was 61 per cent.

WHILE PUBLIC TRANSPORT HAS BEEN DECLINING, BUT MAY HAVE TURNED A CORNER ...

The total distance we travel each year by public transport (bus, rail, underground and tram) has been falling – an average of 742 miles in 1985, 730 miles in 1998. Most of the fall can be attributed to the decline of bus travel. However, rail travel grew strongly in the second half of the 1990s. In London patronage of both bus and Underground rose markedly through the decade and between 1998 and 1999 the total miles travelled by bus in Britain actually increased for the first time in decades. At the end of the 1990s road traffic was also growing less rapidly than it had during previous economic booms, at around 1.5 per cent a year.

There is now a fear that it will be impossible to find our way back; that we have reached the point of no return. So much of the electorate now possesses and relies on cars that the sustained political support needed to change things can never be mustered. No government will dare compel people to forsake their motors, nor will it ever find the huge resources needed to make public transport sufficiently cheap, frequent, safe from crime and comfortable to compete with car travel. Besides, our urban structures have become so adapted to high car ownership that it is becoming difficult or impossible for us to go shopping for food and bulky items, to get to many cinemas, leisure centres or offices, in any other way. Even if we did create denser, walkable, mixed use neighbourhoods, there is no guarantee that people would reduce their car use and choose to walk around in them.[13]

Many children already grow up with hardly any experience of using public transport. Which is a shame, because for someone in their early teens the act of crossing town by bus or train unaccompanied by an adult is an adventure, an achievement, a big step out of childhood. More and more teenagers will be given cars as soon as they are allowed to drive and increasing numbers of sixth formers and college students will get to class by car. In a couple of decades the two or more car household will become the norm (it already is in more affluent neighbourhoods). Walking and bicycling will flourish in the countryside (the swarms descending on fields and woods will grow) but in the towns and cities these outdated modes of movement will be left to a dwindling band of well-meaning eccentrics. Buses will be for the poor, to an even greater extent than they are already.

Think and hope that it will not be so. We are, I think, nearing a turning point, not moving further and further past a point of no return. Car ownership *is* likely to carry on rising for years but people living in towns and cities will either be compelled or choose (in a democracy, the distinction must often be unclear) to start using their cars less.

There are several grounds for cautious optimism. The first is that there is no longer any consensus that building new roads and widening existing ones is the best way to cope with rising volumes of traffic. Public opinion turned against schemes like the M3 Winchester by-pass at Twyford Down, Hampshire and the A34 Newbury by-pass in Berkshire, both of which damaged prized landscapes and wildlife. It sided with the protesters who tried and failed to stop these roads. Any big new highway projects planned *within* towns and cities would now encounter even stronger opposition. People just will not accept hundreds of homes being demolished and urban greenery erased for the sake of a few miles of new dual carriageway.

Reinforcing this opposition to local environmental damage caused by road construction is a widespread acceptance of the argument that we cannot build our way out of congestion; more asphalt makes more traffic. If the government were to propose a massive, nationwide increase in road building as a means of curbing congestion, no one would believe that it would work. The grand strategy of *Roads for Prosperity*, a transport White Paper of the late Thatcher period that promised the biggest road-building programme since the Romans, has itself become ancient history.[14] Most of it remained unbuilt.

That said, it has to be noted that the New Labour government's *Transport 2010: The Ten Year Plan*, published in 2000, promised to more than double annual expenditure on roads (both local ones and the inter-city network of motorways and busiest A roads) by 2010 compared to the spending levels of its first couple of years in office.[15] The plan promised 100 new by-passes, big and small, over the 10 years, and to widen 360 miles of the most congested motorways and A roads. However, the bulk of this increased spending was to improve the condition of the existing road system rather than add extra road space. And, furthermore, public transport –

trains, buses, light railway systems and trams – was due to receive increases in investment and government revenue support that were about as large as those going into roads. The plan envisaged that people would continue to travel more and more with each passing year as the economy grew, but that the growth would be mainly through increased use of expanding, improving public transport services. Over the 10 years there would be 50 per cent increase in rail travel nationally and in bus travel in London, a 10 per cent growth in bus journeys elsewhere, a doubling in tram and light railway use. Road traffic would increase too, but more slowly than in the past. And instead of continuing to worsen, road congestion would actually reduce due to a combination of investment in removing bottlenecks and car drivers switching to trains, buses and trams for some journeys.

Scepticism soon surrounded this heroic plan. As was the case with the National Health Service and state education, the huge, multi-billion increases in programmed expenditure depended on the long economic boom continuing. Later that year a fatal high speed rail crash near Hatfield, caused by a broken rail, precipitated an emergency nationwide maintenance programme that caused colossal disruption for months. Many rail passengers switched to cars, coaches and airlines. It was a great setback.

As for roads, we can expect the intense debates about worsening congestion, safety and economic and environmental damage to continue. These problems have spread far beyond the big towns and cities and the major roads that link them – and herein lies another reason to be optimistic about the future of urban transport. People and businesses can no longer flee the conurbations to escape congested roads. Every weekend suburban highways are snarled with drivers on leisure and shopping journeys. Rural and semi-rural places that were quiet little towns a few decades ago have become beset by congestion. Counter-urbanisation and sprawl have caused traffic flows on rural and inter-city roads to rise more rapidly than the roads within cities.[16] Every driver is now familiar with the experience of sitting in a mysterious, unending queue surrounded by woods and fields, miles from any large settlement. In the most popular tourist regions like the Lake District mile after mile of country lanes sport yellow 'no parking' lines. The Council for the Protection of Rural England complains about how busy and intimidating rural roads are becoming. It fears that, given recent trends, traffic flows on these roads could double in the next 30 years and it is campaigning for 40 mph speed limits for all of them.[17]

This is good news for large towns and cities because they have better prospects than small town, shire Britain of offering their residents attractive alternatives to the car. They can lead the way in developing transport solutions and breaking the impasse. By being big and densely populated they can put a good spread of destinations within walking, cycling and bus-riding distance of people's homes. Public transport has a much higher potential to offer fast, frequent and affordable services in big urban places than it does in small ones and the countryside because cities maximise its number of potential customers.

Buses are the most important and least glamorous form of public transport within cities and between towns and their surrounding countryside. They have been losing passengers since the mid-1950s. During the 1980s an industry that had been highly regulated and mainly owned by local councils was almost completely privatised and largely deregulated. Operators were allowed to start a new service anywhere outside of London and 'bus wars' broke out sporadically as rival firms fought to take each other's passengers. Services stopped and started with very little notice. In some places cut-throat competition produced lower fares and improved services but usually it drove all save one operator off the roads. In so doing it merely re-established a local monopoly, one that was now in the hands of a profit-seeking enterprise rather than the local city council.[18] The public sector subsidies, which were used to keep unprofitable but socially desirable services running, were also pared back. Between 1986, the Year Zero of bus deregulation, and 1999 fares rose by about a sixth in real money terms while passenger journeys outside of London almost halved.[19] Many bus fleets got older and scruffier; operators were not keen to invest in new vehicles when business was so tough and uncertain.

Since 1997, the New Labour government has tried to stop the rot and introduce some stability. Subsidies have been upped and a measure of regulation reintroduced. New legislation was enacted in 2000 enabling councils to re-establish control of services in some areas, giving a monopoly to an operator in return for guaranteed frequencies and standards of service.[20] This means that the bus franchising system introduced by the Conservative government in London (because it knew complete deregulation would not work in the capital) is to be extended to some other cities. The government also encouraged Quality Bus Partnerships – agreements between local councils and bus operators which aim to deliver lasting improvements on individual routes. The council's part is to install new, improved bus stops and shelters, improve the pedestrian routes which lead to them, paint bus-only lanes onto the roads (these are meant to exclude other traffic during peak hours) and alter certain key junctions to give buses a higher priority over other traffic. The operator's part is to introduce new buses, or at least freshly painted ones, promise to meet certain standards of frequency and reliability and go in for some aggressive marketing. Central government does its bit by chipping in a little taxpayers' money.

It can work. In Birmingham local councils, central government and the dominant local bus operator together invested £3 million in improving Line 33 which runs from the centre to the huge Kingstanding Estate six miles to the north. The frequency was doubled to one vehicle every six minutes, bus lanes run almost from end to end, and new, low floor buses (which can accommodate wheelchairs and mothers with baby buggies) purchased. The drivers have honed their Brummy charm with some extra training in customer care. Electronic signboards at the stops tell waiting passengers how long they have to wait until the next bus arrives.

Passenger numbers grew by almost a third in the first six months after the improved service was launched (the government claims 10–20 per cent growth is typical for the dozens of Quality Bus Partnerships introduced across Britain).[21] In Birmingham most of the extra passengers were habitual users of public transport and less than a tenth of them had previously used a car for the journey. Which goes to show that motorists are only going to switch to the bus if it can compete with driving in terms of speed, convenience and cost. In Britain's cities it rarely does, so buses are generally for poorer, car-less people.

If, however, car drivers know that parking at their destinations is either non-existent or expensive and if they can see that the bus suffers less from traffic congestion than they do, then they may well change. Bus lanes can only make public transport faster than the car if their bus-only status is ruthlessly enforced in the peak hours, using such devices as cameras on board the buses to photograph the number plates of offending vehicles and large fixed penalty fines for intruders (again, the new legislation allows this). These lanes can never give buses an entirely clear run because they have to expire as they run up to junctions, to allow the general run of traffic turning left to move into the inside lane. Traffic lights can, however, be set up to be controlled by an approaching bus; it transmits a radio signal which gives a green light to itself and a red to the traffic crossing its path.

The guided bus appears to have more pulling power for motorists. This most promising machine quits the public road and travels along its own concrete guide way for small but crucial sections of its journey. This gives it much of the *gravitas* of the tram or the metro at a fraction of the cost. Once inside the guide way its road wheels run on two parallel strips of concrete separated by grass or gravel and flanked by high kerbs. Small rubber wheels turned sideways on make contact with the kerb faces and guide the vehicle; the driver no longer has to steer but simply controls the speed. The vehicles cost 5 to 10 per cent more than ordinary buses while the concrete track systems come in at around £1.6 million a mile.[22]

Unlike a conventional bus lane, the guide way cannot be invaded by any other kind of vehicle. It is too narrow for large lorries and too wide for a car (the wheels on one side would fall into the grass or gravel of the centre strip). At its end, where the bus must rejoin the highway, is a set of traffic lights. As the bus nears the lights it passes a control that turns them to red, stopping the traffic and allowing it to slip off the guide way and back onto the road without having to wait for a break in the traffic. The best location for a guide way is beside a heavily congested stretch of road leading up to a busy junction. The bus glides past the long tailback and, because it controls the traffic lights, is always first across the junction at the head of the queue. The hope is that a light bulb will illuminate in the heads of the queuing car drivers after they have seen this happen several times, and they will give public transport a try.

In Leeds and Ipswich operators and councils have collaborated to bring guided bus services to Britain from continental Europe. The new systems, both of which started in 1995, have attracted large numbers of extra passengers. This is not entirely due to the guide ways, for these cover only a small fraction of the bus routes – just 300 metres at Ipswich. Frequencies have been boosted, new buses introduced with paint jobs that distinguish them from the road-only fleet and bus stops rebadged and rebuilt.

Leeds' Superbus embraces four busy groups of services linking the city centre with suburbs 4 miles out, on the northern edge of the city. Half a mile of guide way on the inbound route allows the buses to save 5 minutes in travelling time during the morning rush hour; outbound there is less than half as much guide way and shorter time savings. Even so, the number of passengers has grown by 80 per cent since the system opened in 1995, at a time when patronage on other city services have been static or declining. The new services are estimated to have removed about 800 car journeys from the roads each week. Ipswich's Superoute 66 guided bus, which runs on a 6-mile route from the town centre to four outlying housing estates, attracts double the number of passengers of its predecessor service. In both Ipswich and Leeds, about a quarter of the new bus users are estimated to have switched from their cars.[23]

These two pioneering systems depended heavily on central government grants to help cover the costs of guide ways. But, having seen their success, two of Britain's largest bus companies have offered to pay half of the £10 million infra-structure costs of a new system running out to the eastern edge of Leeds along the A64, the city's busiest radial road, and the A63. They have also pledged to spend £10 million on new buses running on those routes, confident that a sustained increase in passenger numbers will make that a good investment. The new guide ways were due to open in 2001, as was another running south from the centre of neighbouring Bradford.

Light railways, metros and trams generate much more interest and enthusiasm than either guided or conventional buses. Through the 1980s and 1990s systems have opened in London Docklands (1987), Greater Manchester (1992), Sheffield (1994), Birmingham to Wolverhampton (1999) and South London (2000), all heavily reliant on government (and sometimes European Union) money to finance the infrastructure. The government's Ten Year Plan for transport envisaged up to twenty-five new light railway lines being constructed by 2010, including three more in Manchester and new systems in Nottingham, Leeds, Bristol and Portsmouth–Southampton. As *the* urban transport solutions these new generation railways and trams offer several attractions. The routes are often already there in the shape of abandoned or under-used tracks which can be brought back to life. Trams spend nearly all of their journeys separated from and unaffected by road traffic while light railways are completely isolated from the highway. Both can therefore run faster and more reliably than any car or bus within the city. They also save space compared to roads; area for area, steel rails can move several times more people in an hour than asphalt.*

What counts most in favour of trams and light railways is their ability to get lots of drivers out of their motors. Manchester's Metrolink, which opened in the early 1990s, claimed to remove 7,000 car journeys each day at the end of the decade.[24] Imagine standing by a fairly busy 'A' road with a car passing you every eight seconds. That is the road traffic Greater Manchester's splendid and expanding tram network has removed. But note that when traffic flows and the demand for road space are both rising, new tram systems will encourage some drivers to get onto the highways by taking other drivers off them.

There are other reasons why trams and metros are no panacea for our urban transport ills. They are expensive; each tram costs about £1 million (five times the cost of the latest and largest articulated 'bendy' buses) and the overall capital costs for a system work out at about £8 million per mile of track.[25] Which means they must attract large numbers of passengers to be financially viable, running along the busiest movement corridors. There can never be a tram or metro stop within a few minutes walk of every front door in the city. And however busy it promised to be, a tram or metro line could probably never be constructed as a freestanding commercial enterprise. The capital costs will always require some subsidy from the public or private sectors, justified by reductions in congestion and pollution and by the urban regeneration and rising land values the new line can stimulate.

The bus and the guided bus will usually be more deserving of public subsidy. Not only are they cheaper but their lanes and guide ways can also be used to remove road space from other traffic – which is what we need. Because just as increasing road space generates extra traffic, removing road space makes traffic disappear. If some barrier or disruption, deliberate or accidental, cuts road capacity between points A and B then the *overall* flow of vehicles between them falls. This holds true even if you measure all the traffic taking the long way round, diverting onto more distant highways and byways in order to make the connection.

We know this because the UK government and London Transport commissioned an international study that looked at the effect on total traffic flows of forty-nine temporary or permanent major urban road closures.[26] These were caused by bridge collapses, big repair programmes, town centre pedestrianisations, earthquakes in San Francisco, Los Angeles and Kobe and anti-terrorism measures (the 'ring of plastic' security cordon thrown around the City of London following IRA bomb attacks in the early 1990s). The shortest of these closures lasted a day; the longest were permanent, so some of the case histories covered periods of several years. In each case the traffic counting covered a wide area that took in all of the diversion routes that drivers took.

Typically about a sixth of the road traffic appeared to disappear after a closure. It didn't divert onto other roads, nor travel at a less busy time of day. As far as the traffic counters could see, it vanished. Now, some of these disappearing drivers would have found a new destination in a completely different direction that met their need; they chose, for instance, to drive to another shopping centre. But others decided to use public transport, or walk or cycle. And some chose to drop their journey altogether because of the extra delay and inconvenience the closure caused.

What these findings suggest is that after decades of providing ever more open-to-all-vehicles road space in our towns and cities, the time has come for a freeze. And then, slowly and carefully, we can start taking space away. More and more urban road should be occupied by bus lanes and cycle ways, by pedestrianisation schemes and piazzas and linear parks, even – here and there – by new homes and other buildings. Traffic congestion would remain at the same high levels if this policy were applied, indeed it would probably worsen in places, but that is tolerable provided there is high quality, decongested public transport on offer along with opportunities for safe walking and cycling. Urban public transport does not have to be particularly cheap (its running costs, in other words, do not require colossal public subsidy) nor luxurious (most journeys are fairly short) but its capacity will have to multiply as more and more drivers quit their cars. It needs to be dense, reaching every part of the city with a stop within 500 metres of every home. It needs to be frequent (no intervals longer than 15 minutes anywhere, anytime between 6 a.m. and midnight), reliable, safe from crime, liberated from the congestion afflicting other road users and not intolerably overcrowded during the rush hours. It also needs to be orbital (circling the city) as well as radial (straight in from edge to middle) because people are no longer simply flowing in and out of the centres. More and more work, shopping and entertainment journeys now miss them entirely.

Of course, drivers would hate such a road squeeze. Their lobbyists would campaign against it and politicians would quiver before their wrath. But no one would be forcing people to give up their cars, merely to use them less frequently and more considerately. Most owners are not obsessed with or passionate about driving their motors. Cars are simply the mode of travel that performs best in terms of some combination of speed, price, comfort, convenience and safety for whatever journey they happen to be on. If you reduce the car's performance in any of these respects, improve the performance of alternative modes and ensure drivers know what is on offer, then most would consider switching. Public transport, walking and cycling would not be the only alternatives. 'Quasi-cars', such as safe, weather-protected motorbikes and car clubs, could emerge and flourish in a world where the overall attractiveness of old-style motoring was gradually worsening (see text below). Those who really had no choice but to make their journey on four wheels, or simply refused to try alternatives, would just have to lump the congestion and delays.* Clever electronic information and navigation systems would help them to plot their way around the jams, squeezing a little more capacity out of the road network. Freight carriers and road-based distributors would get smarter at timing their van and lorry drivers' collection and delivery runs for quieter times of day.

QUASI CARS THAT HELP THE CITY

A motorbike or scooter in motion needs about half the road room of a moving car and much less still when it is parked. It also produces fewer global warming carbon dioxide emissions per mile, per traveller. Cars carry only one person on most journeys so if millions of drivers switched to motorised, two-wheeled transport this would be a formidable decongestant to urban roads and a boon to the global environment. They won't because motor biking is often cold, frequently wet and much, much more dangerous than car driving.

BMW's C1 could change that. The Bavarian car giant that failed to turn Rover around has succeeded in designing a motorbike that offers car-like standards of crash safety. The rider is strapped to the seat by two safety belts and surrounded by an egg-shaped, aluminium safety cage. A mudguard over the front wheel is designed to absorb collision impacts progressively, in the same way that a modern car's front does. A headrest protects the driver from whiplash injuries in the event of being shunted from the rear. A further advantage over conventional motorbikes is that there is no need to wear a crash helmet (although British C1 riders, unlike their French, Italian, German and Spanish counterparts, remain compelled by law to do so). The two-wheeler also boasts anti-lock, electronically controlled braking to stop skids, a catalytic converter to cut pollution and a big windshield with wiper to keep the rain at bay. A water-cooled, 125cc four-stroke engine delivers around 80 miles per gallon.

To create an even more car-like ambience, optional extras include a CD player and radio, heated handgrips and seat, a mobile phone holder and a Global Positioning Satellite navigation system. The C1 went on sale in Britain in 2000 with the most basic model costing just under £4,000.

Another way of gaining most of the benefits of a car without actually owning one is to belong to a city car club. These cater for urbanites who do most of their day-to-day travelling by public transport or some other non-car means but occasionally need the use of an automobile for shopping, visiting and getting out of the city. The proposition is roughly equivalent to four to six households sharing one car instead of each owning one, and saving money as a result.

Members pay an annual fee that entitles them to book a car whenever they need

*An RAC study concluded that only 10 to 30 per cent of most households' trips absolutely *had* to be made by car because there was no viable public transport alternative or heavy shopping has to be carried. ESRC Transport Studies Unit, University of Oxford, *Car Dependence*, London, RAC Foundation for Motoring and the Environment, 1995.

one. Electronic systems can take care of all the security, ensuring only paid-up members get access to the cars. You pick up your motor from a parking place close to your home or work and use it for a few hours or a couple of days, paying by the mile or the hour or some combination of both. The costs are lower than conventional car hire and it is also a little more personal since you belong to a club and have a running account.

A large proportion of the members are people who would otherwise have owned their own car, but because they do not they use cars less and public transport more. One of the advantages of membership is that you don't have to worry about servicing, roadworthiness, tax discs, insurance and so on. All of that car palaver is covered by the membership fee and organised by the club administration.

City car clubs began in Germany. They have flourished there and in Switzerland with many thousands of members in each. Britain's first was launched in 1999 in Edinburgh, funded to the tune of more than £200,000 by the government and the City Council and run by the car hire giant Budget. The new club's slogan was 'don't own it, just drive it'. Members were each given their own electronic car keys and were able to pick up a car from one of four street parking stations with as little as 15 minutes notice by telephone. But only 170 people joined – too few to make the scheme financially viable – and less than two years after it began Budget announced its withdrawal. The club seemed likely to fold.

___

As road space declined, the big towns and cities would gradually become less damaged and dominated by traffic. Within them the proportion of urban journeys made by car would start to fall, although car ownership would remain high. Outside of them traffic flows, noise, pollution and traffic squalor would keep growing. The conurbations would no longer be seen as the worst places to drive but as places where you could live a good life without always having to use a car. Where there were plenty of things worth seeing and doing without wasting hours stuck in traffic.

A daft vision? A hopelessly unworkable programme? Not at all. It aligns with policies that were hesitantly and feebly applied by the last Conservative administration and have been applied with slightly less feebleness and hesitation, but some backtracking, by the New Labour government. Removing road space is on the agenda. The latest government guidance to local council planners and highway engineers says they should consider reallocating road space from cars and other vehicles to pedestrians, cyclists and public transport.[27] Many councils are getting on with it. Bus lanes are spreading. When this book went to the printers Edinburgh was planning to remove cars and lorries from Princes Street, its most important central boulevard, reserving its three quarters of a mile length for buses, cyclists and pedestrians. And Greater London's Mayor Ken Livingstone intended to pedestrianise the extremely busy road that cuts Trafalgar Square off from the National

Gallery by early 2003, with the possibility of further traffic restraint in Whitehall and Parliament Square to follow. 'Home zones', neighbourhoods where traffic must give way to pedestrians on all the streets, were also being introduced at nine urban locations in Britain, with central government funding for more promised early in 2001.[28] The idea is to change the look and feel of the streets in these zones so that they belong to children at play, to adults on foot and to cyclists, rather than to motorists. Cars can park and drive in them, but only slowly and carefully. Extra-low speed limits, bumps, pinch points, shrubberies, textured and coloured road pavings and the removal of kerbs separating pavement from road all help to bring about this change of ownership, turning the streets into one large zebra crossing. The concept has worked fairly well in the Netherlands and Germany.

In the next decade or two neither the Internet, nor electronic guidance systems which take control of car steering wheels and accelerators, nor any other techno-logical breakthrough, will fundamentally change the rules of urban transport and send us off in a completely different direction. The central problems will remain the environmental, economic and social damage done by road traffic and the difficul-ties public transport, walking and cycling face in competing with the car. Solving these will depend, as ever, on the application of sticks and carrots. The sticks are worsening road congestion and new or increased taxes on car use. The carrots are improved options for walking, cycling and using public transport. The question is whether national and local politicians will dare to make a difference or dither around with tiny sticks and minuscule carrots.

One big stick, discussed above, is to reduce the road space available to general traffic. The chief alternative to this is to charge users for it. Late in 2000 a govern-ment bill enabling local councils in England and Wales to introduce urban road tolling and workplace car parking charges (a local tax on businesses which give their employees spaces) was enacted.[29] Charging schemes are aimed at reducing conges-tion in the worst affected (usually central) areas and at raising revenues to support local transport improvements, including better facilities for cyclists and pedestrians, new public transport services and improvements to existing ones.*

Early in 2001, when this book was completed, it looked as if the first full-scale charging schemes would not be in operation until 2003 at the earliest. A few dozen councils covering large and medium-sized towns and cities had expressed a serious interest and were involved in complex, difficult negotiations with central government. Ministers and civil servants accepted the intellectual case for charging schemes, recognising that they seemed essential for the long-term management of road traffic. But at the same time they were anxious to place full responsibility for

---

*The legislation to create the post of an elected mayor and assembly for Greater London, enacted in 1999, gave them power to introduce road pricing and workplace car parking charges. The Scottish Parliament was also expected to complete legislation in 2001 enabling councils north of the border to charge drivers for using urban roads. Edinburgh is expected to be first with a charging scheme covering the city centre.

setting up schemes on local councils, partly to avoid being caught up in the inevitable controversy and unpopularity and partly because such a radical local innovation ought to be led by locally elected and accountable bodies. Yet central government also wanted, indeed needed, most of these new charging schemes to work well. That meant judging which councils were likely to do a good job of pioneering them, then granting town halls the millions of pounds needed to set them up.

Those councils who said they were willing to lead the way with road and workplace parking charging schemes also had difficult judgements to make. What kind of scheme and what type of technology offered the best bets for their particular circumstances? How much government finance might be available? And could their voters and the local business community ever be persuaded of the merits of road tolling or workplace car parking charges?

There was an obvious need for large quantities of investment upfront, to purchase the equipment and administration needed to run charging schemes and to make advance improvements in public transport, and pedestrian and cycling routes. The large growth in local council transport spending projected under the government's Ten Year Plan was meant to cover that. Once charging began, there would also be a need for further sustained investment in expanding the capacity of bus, rail and tram lines to carry the drivers quitting their cars.

So congestion charging is politically risky (the Conservatives have promised to scrap the idea entirely). And nowhere were the stakes higher than in London, where the conurbation's first directly elected Mayor, independent, ex-Labour Ken Livingstone, had a mandate from the voters to introduce road pricing for the city centre. His manifesto declared his top priority to be improving London's ramshackle transport systems with a charging scheme making an important contribution. His unsuccessful Conservative and Labour rivals for the job had both ruled road charging out. Mr Livingstone planned to introduce a £5 a day fee to drive a car, van or lorry within the West End, the City and a sliver of the South Bank – an oval-shaped zone 6 miles across its longest, east–west axis, bounded by London's Inner Ring Road and covering just 2 per cent of the capital's total area – between 7 a.m. and 7 p.m. on weekdays. The scheme was to be enforced by a computer system attached to scattered cameras which could read vehicle number plates and check whether their owners had paid the fee. It would cost around £200m a year to introduce and was projected to raise about the same amount each year in revenues while cutting road traffic in the charging zone by 10 to 15 per cent. It was to be the largest, most ambitious such scheme attempted anywhere in the world. The new mayor appeared to be a man in a hurry, wanting it to be up and running and working by the time he faced the voters again in 2004. Greater London's equally new elected Assembly was wary and sceptical. Central government was keeping its views to itself. Would it want road charging to fail in London, before the eyes of the world? Almost certainly not, whatever its feelings for the Mayor. So would it prefer it not to be introduced, or do its best to help it succeed? Would New Labour have Mr Livingstone back in the fold, as its own mayoral candidate, in 2004? The answers were all linked.

At the national level, government may have made a big mistake in exempting the free car parking places provided by shopping and leisure centres, multiplex cinemas and other car-dependent developments from the new charges local councils can now attach to private, non-residential parking places. The result – the loss of a major stick from the new policy and of a revenue stream that could have made a substantial contribution to funding alternatives to the car. It seems unfair to place the entire burden of the new local car parking taxes on car commuters. 289

Stick and carrot policies for reducing urban congestion will be undermined if we continue to allow car-dependent development to keep spreading outside the towns and cities. The new tolls and taxes would become merely one more reason to live, work and drive, drive, drive in the untaxed domain outside them. (Eventually the congestion and the environmental damage done by traffic in this toll-free domain would rise to the point where the taxes were needed there too; people would begin to demand them. But that lies decades into the future.) The last thing councils want is for major employers to leave town. Their fear of this happening is one more reason why urban road tolls and workplace car parking charges will be introduced gradually and with caution. Even before such charges were contemplated, there had been cases of major companies pressuring local councils into granting them the planning permission they needed to pull out of town centre offices and move to spanking new headquarters built in the countryside on the edge of town. These employers threatened to move from the area altogether if they did not get planning permission, to another part of Britain, perhaps even overseas. Faced with the threat of losing hundreds or even thousands of local jobs, councils gave in and granted permission for the car-dependent, greenfield development. The moves out of town centres by Barclaycard at Northampton and Vodafone at Newbury, Berkshire are recent examples. In both cases, central government chose not to use its powers to intervene and reject the new developments, even though they went against its own guidance to local planners. Indeed, the government's own Private Finance Initiative for the National Health Service also ignores this guidance, since it drives the development of new hospitals on out of town sites in the process of replacing old ones in town centres.[30]

This guidance had changed in the mid-1990s to resist the urban exodus, rising car dependency and the growth in road traffic and congestion.[31] Major new developments that generate many journeys – shopping centres, office blocks, hospitals and leisure centres such as multiplex cinemas – were no longer to be sited in locations where almost all visitors and workers come and go by car. Instead, they should be built where people can get to them by public transport; that will usually mean being near town and city centres. The guidance was

Out of town shopping in the inner city: Greenwich, London

strengthened in 2001 – although not as much as the government had initially proposed – and might further constrain the number of car parking spaces serving new development.[32]

Government, local and central, must resist the pressure from developers still seeking to move out of town. It must be resolute in applying the guidance. In the next but one chapter I'll argue that virtually all development, from the largest new electronics factory down to the smallest starter home, ought to be built either within our existing towns and cities or in compact, masterplanned, mixed use extensions on their edges.

The time has come, too, for cities to think carefully about making it a little less easy for people living outside them to travel in. There is a difficult balance to be struck here, because in thinking this way they will be challenging the link between economic growth and rising mobility – a link that has lasted for hundreds of years. As travel became cheaper and easier through technological advance it increased opportunities to trade and learn. It encouraged competition, economies of scale and longer and longer commuting distances. Most businesses have usually seen ever increasing mobility – which, in the twentieth century, mostly meant bigger and better roads – as a good thing. It brought them wider pools of customers, suppliers and employees and made it easier to distribute their goods and services further afield. The received wisdom was that in order for cities to flourish, goods and people needed to flow in, around and out of them with increasing ease.[*]

'Park and ride' has become one big idea for allowing the huge flows of people in and out of cities to keep growing without them succumbing to congestion. You funnel as many as possible of the car-borne shoppers and office workers into car parks on the urban fringe then bus them the rest of the way. The government's Ten Year Plan for transport is particularly enthusiastic, envisaging 100 more schemes to add to the seventy 'park and rides' already in existence in England and Wales.[33]

But experience tells us that the space freed up on the roads by 'park and riders' will soon be occupied by other car drivers, many from outside the city, in the absence of restraint. And while park and ride may boost the economic vitality of the centre, it does not necessarily help the remainder of the city. Indeed, it may harm it because it makes it easier for people to work in, and use the facilities of the centre, while living well outside the metropolis.

Councils should be encouraging people to both live and work inside their conurbations. We do not need a return to walled cities in order to achieve this, merely a policy hierarchy in central and local government that gives the highest priority to increasing movement around and between neighbourhoods on foot and by bicycle, and around and across the entire city by public transport and bicycles. Next comes movement across the city boundary, into its hinterland and across to other cities by public transport. Travel within and in and out of the city by car gets the least encouragement, to the point where drivers are persuaded to try other options. Hierarchies such as this have been part of the transport debate for years, but central government and most local councils have done little more than pay lip service to them. The Urban Task Force recommended that 65 per cent of all public expenditure on transport should be devoted to projects which prioritise walking, cycling and public transport, increased from the government's 1999 estimate of 55 per cent.[34] The government's response was that it would set no specific target.[35]

We could probably halt the city-killing juggernaut of car dependency within a decade or two using incentives and disincentives and the withdrawal of some open-to-all road space. It will require large increases in public expenditure. It will be fiercely opposed. The press will relish the debate and some newspapers, national and local, will

---

*The received wisdom is crumbling. The same government advisory group that established that new roads generated extra traffic has more recently argued that there are circumstances in which restraining the growth in traffic can boost, rather than reduce, overall economic benefit. Standing Advisory Committee on Trunk Road Assessment/DETR, *Transport and the Economy*, London, TSO, 1999.

campaign for the rights of the motorist to carry on driving in the manner to which he (and it still is much more likely to be he than she) has become accustomed.

But we are used to that. New Labour, which banged on endlessly about the need for an integrated transport policy during its long march through opposition, seemed stuck in its own huge, honking jam as it approached a general election. Deputy Prime Minister John Prescott's 1997 pledge to cut car use and increase bus use within five years was soon dropped.[36] A fuel price protest which paralysed the nation for several days in the autumn of 2000 was followed by post-Hatfield railway chaos. In the meantime the Conservative opposition dumbed down, pitching itself as the party which battles for the motorist. In the 1990s Tory ministers had been compelled to accept that things had to change, that the demand for more and more road space could not be accommodated; by 2000 Conservative shadow ministers pretended the problems no longer existed.

If we want to feel hopeful about the future of urban transport amid the pro-crastination and painfully slow progress in Britain, the usual thing to do is to consider continental cities. There are plenty where public transport is both superior and cheaper to use than anything found in Britain and therefore better patronised, where the rise in urban road traffic has been much slower or non-existent in recent years and where cycling is safe, pleasant and popular.* There are cities in Germany, Switzerland, France and Scandinavia – all nations with higher levels of car owner-ship than Britain's – that are less car dependent and consequently better to live in than their UK equivalents. But I prefer to look at somewhere much closer to home.

London has a rather unreliable, underfunded, expensive, grievously overcrowded public transport system which – when it works – still manages to provide a dense network and frequent services from before dawn to around midnight. Contemplated as a carrot for luring you out of your car, the capital's public transport infrastructure is a large but rather mouldy vegetable. But now consider the sticks that beat Londoners out of their car. During daylight hours they are fierce and heavy. The closer you get to the centre the more difficult and pricey parking becomes (unless you belong to that elite given space at its workplace). The congestion in the capital's heart means drivers spend a third of their journey at a standstill.[37] The result is that Londoners, from the prosperous down to the low waged, are less reliant on their cars than the rest of urban Britain and the nation as a whole (see text below). They either leave them at home, own one instead of two or delay acquiring the car habit, rely-ing instead on a second-rate public transport infrastructure which was once one of the wonders of the modern world. Every day you see builders and decorators on the Tube, carrying the tools of their trade to and from work. Londoners put up with the

*In Britain, urban cycle lanes are rare, sporadic and broken by road junctions. They usually consist merely of white lines painted on roads. In the Netherlands bicycles have had equal priority with cars for decades and there is a superb two-wheel network, usually segregated from the road and with its own system of traffic lights for crossing busy junctions.

stress, expense and inconvenience because they live in a global city with good opportunities for making money and pursuing careers.

LONDON'S MISSING MILLION CARS

Greater London has a per capita income that is higher than any other British city and well above the national average. But the number of cars per household is far beneath the national average (0.80 compared to 1.01) and that of the other English conurbations (0.86). The highest levels of car ownership are found in the affluent South East outside of London (1.20 per household) and in rural areas as a whole (1.30).

This means that London, mercifully, has approaching a million fewer cars that it ought to given its population and prosperity. Why? One reason is the lack of off-street car parking spaces at its citizens' homes. Some 60 per cent of Inner Londoners and 37 per cent of Outer Londoners have to park on the street, higher proportions than are found in the other English conurbations (average 26 per cent) and any other kind of place. Londoners also have a comprehensive, if rather ropey, public transport service and they face the nation's worst road congestion. A fifth of all their journeys are by public transport, substantially higher than for all the other categories of town and city in Britain. Nearly two-thirds of those working in central London commute by public transport and just over a quarter by car. In the other English conurbations the proportions are roughly reversed; only a fifth of people working in the centres arrive by public transport while two-thirds come in by car.

Which is not to say that London ought to feel pleased with itself. Just over 200 miles away lies Paris, a city where people use the railways and metros more and their cars less, where fares are two to three times cheaper, where a regional payroll tax subsidises public transport and where investment in new systems (including an entire new regional express metro, the RER) and line extensions has been much higher since the 1960s. Investment in new tube, tram and rail lines in London in the first decade of the twenty-first century may start to bridge the gap.[38]

What London's transport experience is telling us is that car dependence is not a one-way street. If we can upgrade and expand public transport in towns and cities, make driving within them more difficult and expensive, while at the same time improving urban environments and enabling their economies to grow, people will use their cars less.

And this is what will happen, slowly and patchily. When, eventually, road pricing and workplace parking charges are introduced in a small number of cities and more road space is withdrawn from general traffic, such schemes will probably work well here and there. In other places they will make little difference or fail. But once lessons have been learnt and the new policies are seen to be delivering real improvements in the quality of city living, a growing number of councils will adopt the new measures. The most successful cities will be those where almost everyone uses clean, reliable, and safe public transport for at least some of their routine journeys, where cycling and walking are commonplace and where cars are still abundant but know their place. Environmentalists and public transport campaigners will never forget or forgive Margaret Thatcher's praising of 'our great car economy'. They ought to find comfort in another of her phrases. There is no alternative.

Chapter 15

TOWN AND COUNTRY

It is impolitic and senseless to carry the town to Highgate, Hampstead, and Clapham, when so bad a use is made of its internal parts; where whole districts consist almost of waste ground, or are occupied by beggary and wretchedness.

Sir Richard Phillips, writing in 1811[1]

The richest crop for any field
Is a field of bricks for it to yield
The richest crop that it can grow
Is a crop of houses in a row.

Lines from a Victorian property primer[2]

THE FABRIC OF CITIES is fairly perishable stuff. Tattiness and decrepitude come quickly if buildings and streets are not maintained. The fabric requires continuous reweaving to give it new uses and colours as work, leisure, tastes and technology move on. Big commercial buildings must be refitted and remodelled at least once every couple of decades to remain useful; it's either that or be demolished. Houses, too, must change to satisfy changing needs. A home which is exactly the same as it was 20 or 30 years ago is usually in decline.

We conserve the finest buildings and entire quarters of historic towns and cities because we love old and well-crafted things. But conserving them is expensive, and often it only goes skin deep; behind the mellow façade interiors are cheerfully gutted and reworked. The urban fabric must keep changing if cities are to thrive.

For centuries, we have also needed more and more of this fabric in order to accommodate a growing number of households, an expanding economy and a rising standard of living. Until a few decades ago the pattern of urban growth was simple; as mobility grew, development spread further and further out from the city centres. There was really nowhere else to go.

Now the picture is transformed. There is plenty of room for building within our big towns and cities, enormous scope for reweaving worn, tired fabric that is between 20 and 150 years old. The challenge is to shift more of the money that currently goes into new development outside of the cities to redevelopment within them.

Development is the art of gaining control of land, making things happen on it, and making those things make money. This is a £43 billion a year industry, one of the nation's largest, supplying new homes along with all the other buildings we need to replace or rejuvenate the old stock.[3] Development is urban oxygen.

Tour the abundant, growing badlands of Britain's big cities and the scale of their wretchedness will have you asking why we allow *any* new development to take place outside our towns and cities when it is so desperately needed within them. If, however, we introduced a ban and crammed all new building into the existing urban areas, that would soon do more harm than good. As we shall see, the millions of new homes Britain needs cannot all be built within cities without squeezing private and public space to unacceptably low levels. If greenfield land outside of the urban areas was put off limits to development, the price of all land would soon soar, causing some investment opportunities to be foregone and some jobs to be lost. House prices would eventually be forced upwards and there would be a further shift to smaller and fewer gardens, smaller rooms, more apartments and high-rise. Parks, playing fields and woodlands within the cities would come under threat.

There is, then, a balance to be struck between development on greenfields out-side of the cities and development and redevelopment of new homes, workplaces, leisure places and shops within them. This chapter and the next are about where that balance lies and how it might be struck.

Most development requires planning permission from the local council. So let us start with a quick, dirty and very generalised sketch of the town and country planning system in England; the systems in Scotland and Wales are not fundamen-tally different.[4]

The planning system's end product is decisions on whether to refuse or grant applications for permission to build something new or to convert existing buildings for new uses. If a local council's planning committee refuses permission the frustrated developer can appeal to the government's planning inspectors to over-rule the coun-cil. Sometimes, when the proposed development is particularly large or otherwise important, government ministers will make the decision following an appeal. The government also has the power to 'call in' applications, taking the matter out of the hands of the local council. Thus while councils are ostensibly in charge, both they and the developers are always looking over their shoulder at central government and trying to decide what it wants; developers when they draw up their applications, council planners when they decide whether to grant or refuse permission.

These decisions take place against a backdrop of map-based development plans which every council is legally obliged to draw up, showing what kind of develop-ment it wants throughout its entire area. So the first thing any developer needs to know is what this plan specifies for the land he or she owns, or seeks to own. You might imagine that planning is mostly about changing the way things are, and that these plans consist of maps and diagrams with bright colours and bold arrows and three-dimensional models, all showing the way to a better future. But the maps and the fat policy documents that make up plans are nothing like that. They are mostly about *not* changing things. The great majority of the land area they cover is zoned to stay the way it is – open countryside, residential neighbourhoods, industrial belts, town centres consisting almost entirely of shops and offices.

Even so, the planning system cannot ignore the way society, technology and the economy are changing and the impact this has on land use. So the plans cover a period of about a decade and are updated every 5 years or so. At each revision the new plan has to go through several rounds of public consultation in which every-one is invited to have their say. There is a public hearing before a government-appointed planning inspector or panel whose job is to make sure the council has taken local views into account and that the plan does not clash with national planning policies handed down by central government. The inspector or panel suggest modifications that the council then has to respond to and consult on before the plan is finally adopted. The process of churning out revised plans is a slow and laborious one, not least because the forward planners in council offices who work in this field have been pared down during years of fiscal squeezing by central government. Some development plans are finalised just before the period they cover expires. Thus they are out of date almost at once and work has to begin on the next one immediately.

Starting at the top of the system and working downwards, you find central government exerting a powerful influence on these plans and on individual planning decisions by issuing Planning Policy Guidance documents (PPGs). We discussed PPG 3, covering housing, in Chapter 5 and PPG 13, covering transport, in Chapter 13. These tell councils what lines they should be thinking along and what kinds of development they should be encouraging or refusing. PPGs also tend to be revised every few years amid much consultation. Councils know that if they ignore this guidance they risk having their decisions over-ruled by government planning inspectors and ministers.

One level below the PPGs is Regional Planning Guidance (RPG) that lays down broad-brush land use and transport policies across half a dozen counties. In each of England's nine regions this planning framework is drawn up by the Regional Planning Body, representing all of the local councils, in consultation with the regional office of central government and the Regional Development Agency.* Here, too, there is a process of revision every few years, accompanied by consultation and public hearings. And here, too, central government has the final say on the content of the guidance. Dropping down one more level, you find that in the more rural shire areas where there are two tiers of local government (big county councils, smaller district councils), the counties draw up strategic structure plans covering the entire county. Urban boroughs must also produce their own structure plans.

Down on the bottom tier are the local plans prepared by the district, unitary and metropolitan borough councils who actually take most planning decisions. All of these decisions should be in accord with this bewildering multitude of higher level planning and guidance. It should all mesh together into nothing less than a sequential, hierarchical, plan-led system. If it worked well, it would be a thing of awesome beauty and complexity, attuning itself to the views and needs of local communities and the nation, adapting itself intelligently to an ever-changing world.

But it is not and it does not. That is not surprising, for the people and organisations running and using the planning system are often in conflict and are prone to change their minds. Councils quarrel within and between themselves and with Whitehall. Elections bring new regimes and changed priorities at local and national level. The system muddles through the middle of it all and, because it has little ability to anticipate trends, it is often behind the times. Planning never saw the revolution in out-of-town shopping coming. When it arrived, planning failed to contain it and many high streets were eviscerated. The system has allowed sad and stupid things to happen, such as Newcastle planning to build swathes of

*Regional Development Agencies were set up by the government at the end of the 1990s. They have a range of functions and responsibilities but their primary purpose is to foster economic growth across each region by encouraging regeneration and business investment and by boosting employment.

well-heeled housing in its Green Belt while, a couple of miles away, the city's seemingly doomed West End suffers mass abandonment (see text below).

ENDING THE WEST END'S DECLINE

Newcastle's West End ought to be a magnificent place to live. It lies next to a thriving city centre and its sloping streets command sweeping views across the River Tyne. Yet nowhere in England has depopulated as dramatically. This had long been among the poorer places in Newcastle, but the shrinkage of Tyneside industries in the 1980s appears to have pushed it over an edge into a spiral of social and physical decline. You find street after street of boarded-up Tyneside flats – solid, handsome two-storey terraces built a hundred years ago. Large numbers of houses worth next to nothing are owned by private landlords who let them to tenants on housing benefit. Council estates built in the post-war years have been knocked down and replaced by melancholy urban meadows. Housing association homes constructed as recently as the 1990s have had to be bulldozed; tenants refused to move in.

In 2000 there were approaching 1,000 empty homes in the West End, even though several thousand homes – just over a quarter of its total housing stock – had been demolished in the 1990s. Although it had a population of around 70,000 (and falling) there was only one superstore and a single bank branch. No market house-builder would touch the area, while the Housing Corporation refused to invest any taxpayers' money in building new housing association properties. Newcastle City Council estimated that £500 million (at year 2000 prices) had, in effect, been wasted in the area over the past 20 years in a succession of government-backed regeneration initiatives.

While all this was going on, the city had also been pursuing plans to remove Green Belt status from a couple of square miles of countryside just north of Newcastle but within its borders. This de-designation was required in order for 2,500 homes to be built and new, high technology industries and offices to be lured to a large new greenfield business park. The new housing was intended to consist almost entirely of owner-occupied, family homes. The Labour-dominated council's analysis was that Newcastle badly needed to stop losing people with middling and higher incomes to neighbouring boroughs and beyond. The way to do that was to build a large new middle-class suburb.

It seemed strange indeed to extend the city while large areas of it were imploding. And it was not until the very end of the 1990s that Newcastle began to join things up. The council leader, Councillor Tony Flynn, had warned in 1999: 'There are some fundamental and difficult decisions to be made.' In 2000 the city embarked on an ambitious, long-term regeneration strategy concentrated on the West and East Ends, the latter a smaller, but almost equally afflicted inner city area. Instead of allowing continued decline or settling for stabilisation, the council wanted Newcastle to grow.

'Going for Growth' was aimed at increasing the city's overall population by

15,000 (5 per cent) and creating 30,000 new jobs. While 6,600 existing houses would be pulled down, 20,000 new ones would be built. A key part of the thinking was that private housing developers would only be attracted into the city if large cleared sites, each sufficient for at least several hundred homes, were available. At the same time, the city's target was that for each new home built on greenfield land outside the built-up area, two would be created within it.

The plan was for the Green Belt development to go ahead in three large phases – each requiring separate planning permission and subject to central government scrutiny. But an unusual 'planning gain' agreement was made with the two house-builders who had options to build there. If levels of house construction within Newcastle's inner city fell below target levels, those housebuilders would themselves develop new homes there at less than their usual profit margin (albeit not at an out-right loss).

The city council was clear about its aim of changing the social composition of the inner city, boosting the number of owner-occupied, shared ownership and housing association properties and reducing the numbers and the proportion of council stock. It had separate masterplans drawn for the West and East Ends; the ubiquitous Richard Rogers Partnership helped with the former. These projected up to 13,600 new inner city homes, predominantly for sale, including thousands of riverside apartments. For the West End there would be a new tram link to the conurbation's Metro system, a superstore in a new district centre and an Asian business cluster.

But while the council stressed the three Cs – the need to be competitive, cohesive and cosmopolitan – many inner city residents were most concerned about the big D – demolition. If large areas were earmarked for demolition, reconstruction and social reprofiling, what would happen to them? As the council embarked on extensive consultations on Going for Growth, it encountered protest and accusations. It had already made up its mind. It was not consulting with the most marginalised. Councillor Flynn pointed out that two copies of the masterplan had been produced in Braille versions, and the languages it had been translated into included Serbo Croat. Some specific plans for demolition were withdrawn in 2000 but early in 2001, as the council readied itself to publish revised masterplans, it was saying that the total number of demolitions would still be around 6,600. Since many of the homes are privately owned, this will probably involve extensive use of compulsory purchase orders.

Newcastle's big plan is brave and ambitious. Whatever the agreement reached with the housebuilders, will the construction of thousands of new, for-sale homes in the Green Belt undermine the market for new homes in its inner city? And how much choice did Newcastle have, given that neighbouring councils have allowed large quantities of new greenfield housing to be built and may continue to do so?

One of planning's most important tasks has been to supply land for the millions of new homes required by the growing numbers of households. They are growing not so much because population is rising – it is, but slowly – as because the average number of people per household is falling.

Every 5 years or so central government publishes projections from its demographic experts on the likely number of new households being formed over the next 25 years in each region. At the time of writing the most recent estimate was that there would be 3.8 million more households in England between 1996 and 2021, a thumping 19 per cent growth.* Regional planning bodies have to take account of these projections in their official guidance to local councils, telling each county and large city how many new homes ought to be built in its area over the coming years. County structure plans then divide up the figure allocated to an entire county between its constituent district councils.

At the bottom of this cascade of housing numbers lie the district, borough and city councils; they have to identify the land where most of these new homes are to be built in their development plans. They also have to demonstrate to government that at any one time enough additional housing can be provided in their area to meet the next five years' worth of demand. They need to keep on earmarking new greenfield sites in order to replace those being covered by bricks and mortar and to maintain a bank of land for housebuilding. For decades, this system has been functioning as a machine for turning town-fringing countryside into suburbia. The overall objective is to provide the land for new homes fast enough to keep pace with the rate at which additional households are formed. Government's fear has been that if the land supply falls short, there will be a housing shortage which will escalate prices, force households who would rather be separate to share homes, add to the demands for social housing and push up wages.

Central and local government have long seen greenfield development as a most painful necessity. They are damned by NIMBYs if they do permit it and damned for housing shortages if they don't. The fiercest opposition to greenfield development is often found among the people living in post-war, peripheral housing. Their homes blocked other people's view of the countryside when they were built and now they are determined that the same thing should not happen to them.

The centre and the regions try to spread this political pain around evenly by making every local council provide its fair share of new housing land. Some local councils find ways of slowing the rate of land provision, prevaricating in bringing sites forward for development. It is politically more expedient to let greenfield development take place in lots of small, tactical dollops rather than grasping the nettle of allocating larger, more strategic sites. Thus new housing tends to be built in groups of a few dozen or, at most, a few hundred.

---

*This was based on population estimates for 1996. Fresh Office of National Statistics estimates for 1998 showed England's population rising more rapidly than the previous ones, mainly due to more immigration from overseas. The Joseph Rowntree Foundation argued that this implied an extra 4.3 million households between 1996 and 2021 – *On the Move: The Housing Consequences of Migration,* (eds) Richard Bate, Richard Best and Alan Holmans, York, JRF/York Publishing Services, 2000.

These small, field-by-field estates advancing from the edge of town are not large enough to support mixed uses and their own facilities. Their residents have to travel, usually by car, to reach shops and schools. Attempts to expand these new housing areas to the point where they can be more self-sufficient and environmentally sustainable are invariably fiercely resisted, because it would cause yet more countryside to be lost.

But of course NIMBYs are not the sole influence. Councils can be – and often are – over-ruled by central government when it judges that they are failing to allocate sufficient land for new homes. Housebuilders and other business lobbies argue that new homes are essential for economic growth.

The linked systems of planning and housing development form a strange hybrid: direction and regulation by the state, on the one hand, firms competing to buy land and sell houses, on the other. So many things can happen between the government's initial estimate of the demand for additional homes in any one place and the subsequent supply years later that the scope for mismatches is enormous. People's changing propensity to cohabit, marry, divorce, and live alone have powerful effects on the rate of household formation. So does the current price of purchasing or renting in the private sector and the availability of social housing. And so, therefore, do interest rates, economic growth and the level of unemployment.[5]

Internal migration, mainly in the form of the counter-urbanisation cascade from big cities to smaller towns to the countryside, strongly influences the demand for additional homes in any locality. Each year hundreds of thousands of people are on the move within Britain in pursuit of work, or careers, or a better quality of life – often in retirement. The government's latest household projections assume substantially more North to South migration than the previous set because fresh data arrived in the interim indicating that this flow was quickening.[6] Even so, the numbers moving from the big conurbations into their surrounding regions are much larger. And then there is international migration, with the flow of immigrants exceeding the numbers leaving Britain.* People seeking asylum from oppression or a better life arrive in London and the South East of England, adding to the demand for subsidised housing in the region where it is already in short supply. Efforts are made to curb their numbers and disperse them.

Over the past 20 years the government's projections have been too low, with more households forming than were predicted. But the actual number of homes has grown faster than the number of households, and now exceeds it.[7] We should not be surprised at either outcome. Both are highly likely when a trend-based planning system that is essentially 'predict and provide' in nature is combined with a general

*The latest household projections for England, published in 1999, put net inward migration into England at 66,000 people a year over the period up to 2021 (the number of immigrants exceeds the number of emigrants). Almost all of this is accounted for by arrivals from abroad rather than from Scotland, Wales and Northern Ireland. The great majority of immigrants are projected to go to London and the south of England. The forecast increase in population due to net inward migration is roughly equal to the natural

rise in prosperity and strong desires for privacy and choice about whom you live with. More houses give more people more options; to live alone instead of sharing with friends, relatives or partners, to delay marriage, to terminate it, to split families and time-share children, to own second homes. So long as real earnings rise, and as long as they keep pace with rising housing costs (as, in general, they have in recent decades), a growing proportion of the population will have the income to take these options.

At a local level, the system of planning for household growth seems to work in a circular fashion which fuels further growth outside the conurbations, encourages an urban exodus and acts as a self-fulfilling prophecy. This is partly because it takes past migration trends into account. If council planners in a rural area are compelled to provide land for new homes on the basis of a net inflow of migrants in recent years, then this influx can continue in the future. That, in turn, will influence the next set of projections.[8] On and on, round and round, it goes.

But there is another reason why the existing system aids and abets counter-urbanisation. It allocates land for housing in order to meet projected demand but cannot control the type of housing built there, nor ensure those on low incomes get a fair share of it. Nearly all of the growth in household numbers is accounted for by people living on their own or by couples without children.[9] Yet the bulk of what is built consists of detached and semi-detached three- and four-bedroom family homes; that is the housebuilders' main market and what they can be most confident of selling.[10] And while projections of household growth include a substantial proportion of people who cannot afford market prices, land allocated for new homes often has little or no subsidised housing built on it. The result – a lack of provision for local, low-income families while much of the new housing ends up being bought by incomers.

A large proportion of the homes which actually do get built within cities are unforeseen by the planners. They are constructed on 'windfall' sites that crop up when, say, a warehouse or bus depot closes down, an office building is converted into apartments or a vacant piece of land long reserved for industry is used for residential development instead. The developers have to apply for planning permission, but the local council never knew that that particular site or building was available for housing until the application dropped into the planning department's in-tray. If the planning system underplays the supply of these unanticipated urban homes, especially from larger sites, it has to make more greenfield land available than is necessary to meet the projected demand.**

Now we can understand the paradox of depopulating cities and dwindling

increase due to people living longer and births exceeding deaths.
**A study for the London Planning Advisory Committee by consultants Halcrow Fox found that 50,000 homes were built on 'windfall' sites larger than one hectare in London between 1991 and 1996. Planners had predicted that or 'y 12,000 would be built over this period. Halcrow Fox, *Future Sources of Large Scale Housing Land in London*, London, LPAC, 1998.

countryside. Certain features of the planning system give it a tendency to oversupply land for new homes. And so, even though building on greenfields is fiercely resisted almost everywhere, there is no shortage of greenfield land for building. Lord Rogers' urban task force studied the numbers carefully. It estimated that for the quarter century from 1996 to 2021 the planning system had already allocated enough land for 850,000 new homes *by 1998*. So after less than one-eighth of the 25-year period, sufficient land for nearly a quarter of the 3.8 million new homes needed in England had already been earmarked. And nearly 220,000 of them had already actually been built on greenfield sites between 1996 and 1998.[11] Some councils, mostly in the north of England, have designated enough rural land to meet decades of housebuilding. New estates go up in the countryside while inner city terraces a few miles away are abandoned and urban brownfield sites lie perpetually derelict. There is little incentive to invest in new housing in town when such generous quantities of countryside are available.

So much for planning, planners and the allocation of land for new homes; we now turn to the people who build them. Nothing will be constructed unless the developers who own or control that land are confident they can sell the houses they build there at a profit. To keep going, housebuilding firms need a constantly replenished stock of developable sites and a stream of house sales. Much of their skill has lain in choosing patches of farmland that might be allocated for housing at some future date, then wooing the landowner into signing an option to sell them the land or to develop it in partnership. Huge areas of the countryside around Britain's towns and cities are covered by these agreements. Little money changes hands when the option is first signed because at this early stage no one quite knows what the land is worth. That depends on several factors, including the state of the local and national housing market several years into the future. But there can be initial sweeteners for the landowner. I heard of one case in which a developer looking for land around York gave each farmer whose fields he took an option on a Range Rover car.

Next, the housebuilder will try to persuade the local council to earmark its particular greenfield site for housing development during the next review of its development plan or, if she or he is playing a long game, the one after that. Which is why, if you attend a public hearing into a plan, you will find much of its time taken up by planning barristers and consultants acting for various housebuilders and landowners. They will be busy explaining, at great length, why their site is the best for the new housing needed in the area, how it accords with the latest central government guidance, causes the least environmental damage, provides most benefit for the local economy and so forth.

The housebuilders make it their business to know precisely what prices existing homes are fetching in the local housing market. That gives them a good idea of the price at which they can sell their own products, perhaps allowing for a small premium because they are brand new. They also know the profit margin they want to achieve, the number of houses they can fit on a site, and the likely total building costs including the minimum, basic infrastructure of estate roads, sewers, street lights and so forth. Thus the value of the land to the landowner – the sum he or she eventually pockets – should be the total sum the new houses fetch when sold *minus* the total cost of building them *minus* the developer's profit.

Let us run through all of this with some numbers. Halfway through 2000, a new three-bedroom house would have sold for at least £90,000 in most of Britain. But it would have cost less than £65,000 to build, including the infrastructure costs and the developer's profit.[12] The average density at which estates were built in the mid to late 1990s was 25 homes per hectare.[13] So once the landowner and developer have jumped all of the planning system's hurdles and obtained permission to build, the land is worth a minimum of 25 multiplied by £25,000 – more than half a million pounds a hectare. In high house price areas greenfield values of more than £1 million a hectare are achieved. Good quality crop-growing land (the most expensive sort of rural land, worth more than pasture and woods) is worth less than £8,000 a hectare across Britain.[14] You can now see why many landowners would love to switch from growing cereal to growing houses; they are the ultimate cash crop, inflating land values by a factor of dozens.

There is, however, a large spanner in the works of this money-making machine. It goes by the name of a 'planning gain' or Section 106 agreement.[*] Councils have powers to negotiate these agreements, under which developers give something to the community in return for being granted planning permission. They amount to a very haphazard tax on development.

The gains negotiated often include new roads and road junctions, enabling new development to plug into the existing highway network without causing traffic chaos. Section 106 agreements also cover new parks and green spaces (and endowments for councils to look after them) and, as we saw in Chapter 9, free or subsidised land for housing associations to build homes for low-income families. Council planners can use planning gain to oblige developers to build new classrooms at a local school or to donate land for an entirely new school. It can be exploited to provide community halls and premises for a doctor's surgery, or to guarantee operating subsidies for public transport serving the new development.

The justification for these agreements is that since the new development will impose costs on the community – in terms of lost greenery, extra road traffic, heavier demand on local public services – then there ought to be some compensation. Planning gain also touches on the nineteenth-century notion of harnessing betterment which so concerned Ebenezer Howard; everyone involved in growing a city should benefit from the accompanying rise in land values, not just the landowners who were there in the first place.

*Section 106 of the Town and Country Planning Act 1990, as amended by the Planning and Compensation Act 1991, makes such agreements possible, although they began in their current form in 1968. They are usually an undertaking made by a developer to a council in a civil contract.

These planning gain obligations compel developers to spend extra sums on development, which means they cannot afford to pay landowners so much for their sites. Section 106 agreements curb the jackpots that pour into landowners' bank accounts when councils permit fields to be turned into homes. Clever council planning officers know just how far to go, squeezing out gain to the point where the landowner is close to giving up and deciding to stick to cereals or cows. If anyone made the negotiation of planning gain into a board game it would be much more absorbing and challenging than Monopoly.

The agreements themselves are negotiated in secret between developers, landowners and planners. They are an unpredictable, unsatisfactory way of capturing the community's fair share of the huge increase in land values which planning permissions create. There is no set formula to tell the negotiators on either side what to aim for. Negotiating skills and planning gain aspirations vary widely between councils. Some are savvy, some do not have much of a clue. To know what they can reasonably demand from a Section 106 agreement, the council planners need to understand the finances underpinning the development proposal. But they can only guess at the figures when developers keep them secret.*

Landowners usually have professional advisers who do their talking and whose *raison d'être* is to grab as much for their clients as possible. By the time the talks over a Section 106 agreement reach the nitty gritty phase, planning permission has already been granted in principle; the planning gain deal is the only major remaining obstacle to starting work on site. Council planners know that without a stream of houses to sell, developers are in trouble, so they sometimes try to raise the gain by dragging their feet. Time is on their side.

We should note one further defect of this badly flawed system. Councils minded to extract serious gain from developers and landowners through Section 106 agreements will usually find themselves in a stronger negotiating position on a greenfield site outside their town as opposed to a derelict site within it. There is always a strong political case for leaving greenfields green and undeveloped because of the voters' desire to preserve countryside. Planners know that the granting of planning permission on farmland will not only be controversial but will also create a huge rise in land value; they are therefore under pressure to demand substantial planning gain. But the boots are on the feet of the developer and the landowner when it comes to derelict sites and buildings within the city. Here most kinds of development are badly wanted and welcomed by the local council, so it cannot risk demanding much in the way of planning gain.

---

*The stakes are high. A little less than half of all new homes built in England are constructed on greenfields, covering roughly 3,000 hectares a year. Assume that the granting of planning permission for housing raises the value of farmland by £400,000 per hectare – a fairly conservative guess. The task of council planners negotiating with landowners and developers for planning gain is to claw back as much

Let us summarise the problems with greenfield development. It boosts the wealth of landowners without them having to either work or take risks. It damages the environment by creating low-density, car-dependent sprawl. The process of capturing a fair share of the rise in land values for the community – it should be the lion's share – is haphazard, inefficient and obscure. And the planning system's attempt to keep the pace of greenfield development at an optimum level is ineffectual. Worst of all, it has done great harm to most of our large towns and cities by aiding and encouraging the exodus of people with choices.

Nonetheless, we cannot get away from the fact that *some* greenfield development is necessary. There is simply not enough land and buildings available within the cities to accommodate all of the demand for extra homes (see text below). Even environmental organisations anxious to protect the countryside from development such as the Friends of the Earth accept that a significant proportion of the additional homes Britain needs over the next few decades will have to be built on greenfields. The Friends commissioned the planner and urban designer David Rudlin, a leading advocate for urban revival, to write them a report into just this question. He concluded that it was feasible to fit 75 per cent of the new homes in urban areas ('although it is likely to be very difficult'), which implied that at least 130 square miles of countryside would have to be built over during the next quarter century.[15] Unsurprisingly, the Town and Country Planning Association saw things rather differently. It believed most new homes would have to be constructed outside of our existing towns and cities and that the majority of easy-to-develop urban wastelands would already have been built on by 2000.[16]

HOW MUCH ROOM FOR NEW HOMES IN THE CITIES?

There's plenty of room for debate. Estimating how much unused or under-used urban land, and how many empty buildings, could be recycled into new housing is no easy matter. Many local councils keep tabs on wastelands, but they differ in how carefully and how often they do so, the definitions they apply, and the size of plots which they consider viable for redevelopment. The figures keep changing as some derelict sites are built on while new, freshly abandoned ones appear. At the close of the 1990s the government set up a rudimentary National Land Use Database to keep better track of the situation in England. It registered 38,500 hectares – some 150 square miles – of previously developed land that was either vacant or covered in empty, decaying buildings.[17] About a thirtieth of England's built-up area lies abandoned. The proportion was nearly

as possible of that uplift in value for the community. If they do well and win 90 per cent of it then, across the nation, £120 million a year still ends up in the bank accounts of landowners. If they do badly and win only 50 per cent, then £600 million a year goes to landowners. We can only make guesses; the actual figures are not known.

twice as large in the northern regions and rather lower in London and the South East where the land market is generally much more active.

Assumptions have to be made about what sites and buildings could be redeveloped for housing. Dealing with toxic contamination of old industrial sites adds greatly to development costs and this often makes urban land recycling depend on subsidy. House prices and restraints on greenfield developments are also important factors – if either increase, the incentives are raised for developers to find pockets of empty or under-used urban land, or to convert out-of-date office blocks into apartments instead of refurbishing them into modernised office buildings.

In 2000 the government judged that there was enough urban land that was either abandoned or under-used and ripe for redevelopment to provide room for 750,000 new homes.[18] This is only about half the quantity of additional housing needed over the next 10 years. But while much of this empty land will be covered by homes in the coming years, other sites that are now in use will be deserted by industry, commerce and families and become available for re-use.

Further research commissioned by the government has estimated that between 18,000 and 26,500 homes a year could be provided by the private sector by changing commercial and industrial buildings into housing, by subdividing large, Victorian homes into flats and through other redevelopment of existing properties.[19] That amounts to about 15 per cent of the annual demand for additional homes.

You can envisage much larger quantities of new housing going into towns and cities if you ignore market realities and make bolder assumptions about using unused, or under-used, space. There is room for a million homes on empty floors above ground floor shops in town and city centres. And if we used public transport more and cars less, up to 200,000 homes could be built on town centre car parks.[20] One study estimated that land sufficient for some 50,000 homes lies in London's larger suburban back gardens.[21]

Perhaps the largest single potential source of extra homes is the 760,000 empty homes in England, but no one pretends all or even most of these could be easily re-occupied.[22] Some are unfit for habitation and would best be demolished. Others are in run-down, high crime neighbourhoods where no one would choose to live. In any case, several hundred thousand homes are bound to be empty at any one time because the owners or tenants have died or had to move away at short notice.

One way in which the government tried to get a better balance between development inside and outside the towns and cities was to set a national target for recycling unused and derelict land and redundant buildings for housing. By 2008 60 per cent of the new homes built each year should be provided by recycling land or buildings, as opposed to construction on greenfield sites. Because the scale of urban dereliction differs greatly from region to region, each one will have its own target. Taken together, these regional targets should provide the national six-tenths.

Wasted land less than two miles from Manchester city centre

Is this the right target? It is not a heroic one. Deputy Prime Minister John Prescott picked 60 per cent because it was a nice round figure only a little higher than the 57 per cent or so already being achieved in the mid to late 1990s.[23] The hope was that with some policy adjustments and exhortation the existing planning system could be used to hit the target without too much trouble. It is not, however, an entirely urban target, for it includes homes built on previously used land within the countryside such as disused airfields and old mineral workings. And it may prove to be an insufficiently ambitious one. Like Friends of the Earth, the Round Table on Sustainable Development (a government-appointed advisory group drawn from business, academia, local government and the environmental movement) proposed an aspirational 75 per cent target, saying that it was essential to avoid the 'exceptionally large economic, environmental and social costs' that would flow from continued urban decline and greenfield sprawl.[24]

The optimum balance between development on greenfield land and on previously used land will change as time passes. After several years of intensive exploitation of derelict urban sites there might be a case for allowing more green-field development in order to prevent 'town cramming'. The optimum balance will also vary from place to place. In the South East of England, the richest and second most densely populated British region, more than three-quarters of the countryside is protected from development by national government designations which conserve landscapes or wildlife or, in the case of Green Belts, simply contain and separate built-up areas.[25] A combination of high average incomes and land constraints has pushed the region's house prices higher than anywhere else in Britain. Developers there are already under pressure to search out urban building sites, especially in London. If there is any risk of town cramming, it is greatest here.

The situation in the three northern regions of England, the North East, the North West and Yorkshire and Humberside, is quite different. Here the proportion of urban land which lies in waste, and for which there are no known plans, is three times higher than in the land-hungry South East.[26] There is a case for setting regional and local targets well above 60 per cent, in order that one council's abundant greenfield housing sites do not undermine its neighbour's efforts to focus development within a declining city.

If we are to have targets for recycling land and buildings, then getting the right local and regional ones matters most. These ought to be set first, and thereby determine the all-England figure – instead of the other way round.

Targets are no use in isolation. If there is to be any chance of hitting them, then policies that influence the location of development have to change. And change they have, mainly in the form of revisions to the planning guidance – PPGs – handed down to local government from the centre. The process began under the Conservatives in the early 1990s and has accelerated under Labour. Revisions have covered transport, shopping developments and the system of regional planning guidance. But the most radical and important of the changes concern the location and density of new housing.

PPG 3, published in 2000 and first mentioned back in Chapter 5, told councils to find out how much land and how many buildings within existing towns and cities could meet the demand for new homes, taking account of 'windfall' sites that become unexpectedly available. They should ensure this potential for re-using land and buildings is exploited before they allow greenfields to be built on. And if there is to be housebuilding outside of towns and cities, then the first priority should be to locate it in urban extensions on their edge. Only if there is no suitable land available for such extensions should councils allow new housing developments in the countryside, and these should be on public transport corridors. This is known as the sequential test.

There are other good, pro-urban things in PPG 3. Any proposal to build 150 or more homes on a greenfield site must be notified to the Secretary of State for the Environment, Transport and the Regions, enabling him or her to take the decision on whether to grant planning permission. The guidance also asks councils to consider whether they have allocated too much of the derelict land available for development for industrial and commercial uses; if so, they should reallocate it for housing and mixed use development.

But, wisely, PPG 3 says not all previously developed, abandoned land should be favoured over greenfields, for about 40 per cent is in the countryside, much of it unreachable by public transport and far from communal facilities.[27] It makes no sense – environmentally, economically, socially – to prefer housebuilding on old quarries and spoil heaps in remote places above easily accessible greenfield sites next to towns and cities.

Given strong enforcement and support by local councils, these changes to planning policy and other measures covered by the Urban White Paper should secure more development within towns and cities. They are a big step in the right direction. But setting a target for the re-use of urban land and buildings, bolting some new guidance onto the existing planning system and giving some modest tax incentives for development within towns and cities are not enough for an urban renaissance. In order to strike the right balance between development inside and outside the towns and cities, the location, phasing and quality of greenfield development need to be controlled in ways that are decisively different. It is time for a big idea.

Chapter 16

NEW NEW TOWNS

We believe that the only way that the government's proposals for urban regeneration and for greater use of recycled land can be achieved are by restricting the amount of greenfield land brought forward.

House of Commons Environment, Transport and Regional Affairs Committee[1]

Edinburgh's New Town was for the good of the city as a whole . . . it was conceived not just as a piece of enlightened real estate development but as a means of extending rationally a city which had reached the geographical limits of its natural medieval site.

Harvey Sherlock, *Cities Are Good for Us*[2]

We are going to be forced to revisit the housing question . . . because we face a huge gap between demand and provision, a gap that is soon going to bring housing back into centre stage of the political theatre from which politicians of all complexions thought it had been banished. And, when that happens, planning will come back with it, hand in hand.

Peter Hall and Colin Ward, *Sociable Cities: The Legacy of Ebenezer Howard*[3]

WE NEED MILLIONS of new homes over the coming decades and they cannot all be fitted into existing towns and cities. So we need new towns. I don't mean the post-war, state-planned and financed new towns in which small settlements surrounded by countryside were selected for massive expansion. I'm thinking more of Edinburgh New Town, conceived nearly 200 years earlier. A city decides it needs to grow and then sets out an extension which is thoughtful and beautiful, planned yet adaptable, and which benefits the entire settlement. The great houses, squares, parks and streets of Edinburgh have stood the test of time.

Local councils ought to masterplan town and city extensions on greenfield sites. They must capture for the community the huge rise in land values, the betterment, which these new extensions will bring. In order to do that, they will need to buy the sites involved at near-farmland prices and then sell the land on to developers once planning permission for building has been obtained.

Land ownership allows the public sector to enforce the overall masterplan, ensuring the necessary scale, quality and density of development are achieved as construction unfolds over a decade or two. Public sector ownership is also needed to give private sector developers the confidence that there is a commitment to creating a large, viable town extension lasting through the cycle of boom and bust in property prices.

The masterplan for an extension must show how it is to be bound into the town and city it abuts and how it meets the environmental, economic and social needs of the twenty-first century. So its housing is at higher densities than twentieth-century suburbia. It includes subsidised homes for low-income singles and families; their homes are well mixed through the development. Reliable, frequent public transport passes within 500 metres of every house and flat and there is a dense, safe network of pedestrian and cycle routes. The extension has a good mix of uses including shops and schools, local health centres, work and play places. There is also an intimate mingling of house types, from small, cheap starter apartments through to large, upmarket family homes. If this list of demands sounds impossibly ambitious, remind yourself that there are some thriving inner city neighbourhoods meeting most of these twenty-first-century necessities which were laid out more than a hundred years ago.

All of these requirements will be set out in a flexible masterplan and in design codes that can be modified as circumstances change. The council could draw up the codes and the plan after consulting local people but few planning departments will have all of the necessary skills. So, instead, the council could run a competition to choose private sector urban designers to do the job. Either way, public consultation and endorsement is essential and the process of masterplanning must be based on adequate information and funding.

It will, however, be left to private sector developers to construct most of the extension, buying the land from the public sector (probably on long leases), designing the individual buildings and the fine grain of the neighbourhoods in accordance with the code and masterplan, then marketing the properties. Some developers will buy smallish plots and build only houses. Others will purchase larger areas and take responsibility for bringing in other uses and users, such as a supermarket, a primary school and a doctors surgery. Some parts of an extension will be developed as a joint venture between the local council or councils and a private sector developer, sharing the risks and the profits.

At one end of the spectrum of possibilities the private sector can purchase sites already provided with an infrastructure of sewers, streets and so forth at a high price and merely install new houses for sale on the open market. At the other, the developer can buy the land at near agricultural prices in return for an undertaking to install all of the infrastructure.

The increase in values obtainable by turning fields into town extensions can pay for the public sector's initial investment in purchasing the land at farmland prices *and* the building of all the community facilities the new residents could reasonably want in their neighbourhood (streets, schools, health centres, libraries, and so forth). Outside of the more depressed parts of Britain, there would still be some money left over. This could be used to boost the town or city to which the extension is attached, improving its schools, parks, transport systems or town centre. Some of it might, conceivably, go into a national urban regeneration pot as a kind of development tax raised to help revive the most afflicted towns and cities. It would be a progressive tax, since the greatest increase in land values brought about by town extensions would occur in the wealthiest regions.

For this new policy of town extensions to work, virtually all other greenfield development has to cease. It has to be made clear to the housebuilders and other developers that the bad old ways of sprawl are abolished. The one minor exception will be village extensions, which I explain in more detail below. Planning permissions already granted for housing on rural land can be fulfilled (to avoid having to pay out large sums of compensation to landowners), but those that have not been acted on within their five-year time limit should be cancelled. All existing allocations of greenfield sites for housing in councils' development plans should be withdrawn, although some of these will find their way back into town extensions.

How many homes should be built in these extensions? Enough to cope with the projected rise in the number of households; we do not want to force people to share homes when they would rather be on their own or to create shortages which push up house prices. But we also need to be ambitious about the ability of towns and cities to accommodate the lion's share of these new homes within their existing boundaries. The numbers built in extensions would be based on the assumption that urban infill sites and wasteland were sought out and used, that redundant warehouses and offices would be converted into houses and abandoned housing refurbished or rebuilt. Councils need to improve their skills in assessing the capacity of towns and cities for additional homes.[4]

The extensions themselves should be few and large rather than many and small. To permit the latter would be to repeat the mistakes of the past, creating car-dependent estates which can support only a few, feeble facilities. An urban extension should be a place, not a parasite, large enough to enhance its parent town or city, achieve a measure of self-sufficiency, support good public transport services and absorb at least a decade of local household growth. I envisage them being centred on new high streets of three- and four-storey buildings with shops, offices and cafés on the ground floor and apartments above. A town extension of 10,000 people or 4,000 homes would be large enough to support a supermarket and several other shops, one secondary school and three primaries.

While an extension is growing, its residents will need to use nearby facilities in the existing town. Because the growth is anticipated, this can be planned for; local schools, for instance, can be temporarily or permanently expanded. But once the extension is complete it will provide its residents with most of the basic, local facilities they require while creating some which serve the entire city such as a new shopping centre, office district, country park or cemetery.

Given that the extensions ought to be large – at least 6,000 residents – most towns and cities will not need them between 2000 and 2025. Unextended settlements will have to accommodate growth within their existing built-up envelopes. The urban areas that should qualify for extensions are those with strong endogenous household growth, relatively high gross housing densities and a dearth of vacant and derelict land, parks and other urban greenery. London would probably qualify for several, which would mean breaking taboos and building homes on some of its Green Belt. Conversely, towns and cities with shrinking populations, low densities and ample development sites should have no chance of being granted an extension.

Let us assume that 152,000 new homes are required each year in England,* and that 60 per cent of them are accommodated by recycling previously developed land and buildings in line with the government's target. That leaves 61,000 a year to be built in greenfield extensions to towns and cities. Assume that each extension grows by 400 homes a year (allowing it to acquire a critical mass fairly swiftly, but meaning it won't achieve its full, finished size of 4,000 homes for 10 years). That suggests about 150 town extensions would be under construction at any one time and that something like 400 would be built in total over the next quarter century – or between a dozen and sixty in each of the nine English regions.

*This squares with the government's projection that the number of households in England will grow by 3.8 million between 1996 and 2021.

A MUHD MASTERPLAN FOR NORTHAMPTON

This is a MUHD masterplan for a large western extension to Northampton with 6,400 new homes. It was designed during a five-day collaborative workshop held in the town in 1999. The aim was to demonstrate how the economic, social and environmental performance of new housing development might be radically improved – largely through a shift to mixed uses and higher densities. The extension would have four neighbourhood centres, each with its own cluster of shops, other communal facilities and bus stops. Many of the streets are aligned to give views to the open countryside of the Nene Valley to the south of the extension.

Source: The Prince's Foundation, *Sustainable Urban Extensions: Planned Through Design*, London, The Prince's Foundation, 2000.

The number, size and location of this new generation of new towns need to be decided at regional level, with local councils collaborating to make the choice. Initially this could be done through the regional planning bodies or regional chambers. These are currently small and rather feeble organisations but this new task would give them real influence and power. If, however, there ever were to be directly elected regional assemblies in England, the responsibility should be given to them. In Scotland and Wales the Parliament and Assembly should select the extensions.

The choice of which towns and cities need extensions ought to be a fairly straightforward matter, crunching through the data on projected household growth, housing capacity and densities for each settlement, and then selecting those whose figures combined to make the strongest case for an extension. But of course it won't be straightforward. The process will engender bitter disputes between some towns and cities, fuelled by daft, ancient rivalries and genuine conflicts of interest. Some towns and cities that ought to grow will be run by councils that do not want them to, for the usual NIMBY reasons. Some towns which need and want to grow will find that the only suitable, adjoining greenfield site lies within the boundaries of a neighbouring deeply NIMBYist council. Cambridge has found itself in just such a situation.

In the least prosperous regions, towns and cities might demand extensions in the misguided belief that these will guarantee inward investment and solve their chronic economic and social problems. There will be pressure from growth-fixated politicians for the region's extensions to be too large and too many. Were they to get their way they would shoot themselves in the foot, for if too much greenfield land is released, it obliterates the rise in land prices required to create viable, desirable extensions and to finance regeneration.

Environmental groups will argue that too many extensions are being designated. The Housebuilders Federation and the Confederation of British Industry will doubtless say there are too few. Roughly the right number will emerge, provided there is plenty of argument at national and regional level and the debate goes on in the open. Central government can help the regions to get it right by setting out ground rules and national objectives. Westminster and Whitehall can also arbitrate in disputes and ultimately impose, or threaten to impose, solutions if a region's councils prove incapable of coming up with a rational, defensible scheme of extensions.

Once a list of towns and cities to be extended has emerged, it should be left to their local councils (district or unitary) to take overall charge of each individual extension. In those cases where council borders are crossed by the new growth, it would be best to redraw the boundaries, putting the entire extension under the leadership of one municipality rather than risking all the nonsense which could result from two councils trying to plan one together. There might be places where the most sensible, environmentally friendly way of building on greenfields is to let two towns separated by a narrow, jealously protected belt of countryside merge.

Overseeing the planning and building of one of these extensions will be a challenging responsibility and a great privilege. Unfortunately council planners – be they full-time officers or elected councillors – do not inspire confidence in their ability to undertake this task. Look at the disastrous slum clearance and council flat building projects of the 1950s, 1960s and 1970s. And look at the quality of most of the post-war private sector development permitted on the urban fringe.

Yet who is better qualified to do the job than the elected local authority? If central government, or some central government agency, had the responsibility it would be seen as a dictator and resented, just like the New Town Development Corporations and the Urban Development Corporations were. If the task were handed over to a development firm or private sector consortium it could not command public trust. The capture of betterment for the community would be fatally compromised, because the private sector is in things to make money for itself.

Only local councils can be locally accountable and act for the entire community. Their planners will have to raise their game, becoming creators rather than controllers. They will need to seek outside professional help in drawing up masterplans and design codes and they will have to become better talkers, listeners, persuaders and partners. There can be no repeat of their tragedy of imposing bad housing on people, as happened in the post-war slum clearances, because councils will not actually be building, renting or selling any of the homes. The majority will be sold, the remainder rented out by housing associations. That will impose a powerful and generally benign discipline on the entire process of planning and building extensions for, above all else, they will have to be places where people want to live. Their creation will give local government a chance to regain some of its long lost power and prestige.

A council would, however, be wise to create a semi-independent body – a new town trust – which had the day-to-day job of planning the extension, monitoring its construction, negotiating with developers, buying and selling land and revising the masterplan. This would protect the project from becoming a political football, kicked in wildly different directions whenever political power changed hands after local elections. This body would, however, still be close enough to the council to be under constant pressure; pressure to bring about development that benefited the entire town. The same organisation might also be given the task of masterplanning and securing development in the more run-down areas of a town. Its board would be chosen by the council and consist of councillors along with other representatives of the existing town – its education and health services, community groups, a local MP, industry and commerce. It would hire a small staff, some of whom could be seconded from the council and some from private sector developers who were partners in building the extension.

Once it had been decided at regional level that a particular town needed an extension, the local council's first task would be to designate the land required. This need not consist entirely of fields and woods. It should embrace scarred, degraded land on the fringe such as old mineral workings and capped off landfill sites. There are buildings with extensive grounds that could be used for development, such as Victorian lunatic asylums. The new town trust would not necessarily have to purchase all of the land to be covered by the extension. But any land that it did not own within this area would have to be developed in line with the extension's masterplan and design codes, and with planning gain obligations attached.

Candidate areas for extensions will be as close as possible to town and city centres and other employment areas. They will either be readily plumbed into existing bus and rail networks or capable of supporting new transport infrastructure linking them to the rest of the city. Land alongside a working or disused railway line would be a good bet. If the line is in use, the town extension could justify the construction of a new station and any expansion in track capacity that was required. A disused line might be resurrected as a tramway or as a road devoted to buses.

There will be strong arguments for constructing extensions next to the more depressed areas of cities, close to large sites vacated by industry and peripheral council estates. The new development and the jobs, wealth and new transport links it will bring can be tied into urban regeneration. Extensions can be used to revive run-down district shopping centres and industrial estates by bringing more residents within range and by raising funds for refurbishment and for improving access by road and public transport.

But there will also be a strong case for building extensions around the big, out-of-town and edge-of-town developments which have sprung up in the past two decades, the multiplex cinemas, retail sheds, warehouse boxes and office parks. These are dull and often ugly places surrounded by seas of car parking. Outside of their operating hours they are a lifeless waste of space where no one lives and no one passes through. If, however, these car-dependent places were engulfed by town extensions it would bring more public transport services to their doorstep, enabling people to reach them without using their cars. There would also be more potential customers and employees within walking and cycling range. The travel and pollution which out-of-town centres generate would fall. The space occupied by car parks could shrink if multi-storey parking was constructed and apartment homes were then built on the freed-up land. Eventually the edge-of-town developments would be remoulded into vibrant, attractive district centres used throughout the day and night. They might come to resemble thriving traditional high streets.

Urban extensions should preserve a memory of the countryside they cover leaving mature trees, woods, streams and the best of their hedgerows intact and integrating them into streets, parks and private gardens. Greenfield land of high ecological and landscape importance and river floodplains should never be chosen for urban extensions, although new homes and workplaces may frame such areas and give residents views and ready access to them. A wooded ridge visible from much of the city ought never to be earmarked for development because that would destroy a communal asset. Some towns and cities with a good case for an extension might run up against environmental constraints which make it impossible to find sufficient greenfield land in the right place.

What should happen to the development-free Green Belts, which surround most of Britain's big towns and cities? The extensions will have to intrude on them. After all, if a decision has been made to extend a town, then a land designation intended mainly to prevent towns from expanding makes no sense whatsoever. The policy I advocate would result in the loss of about 5 per cent of England's 16,500 square kilometres of Green Belt over the next quarter century.[5] That is acceptable. Yesterday's symbol should not stand in the way of creating viable, sustainable extensions which minimise the environmental damage caused by household growth while maximising its economic and social benefits. Almost everyone who has thought seriously about these issues believes some building on Green Belts is acceptable, but they are frightened to speak out because these belts are the high altar at which all NIMBYs worship.[6]

Once the sites have been earmarked, the land has to be purchased by local councils or their new town trusts, with central government providing initial funding. Let us assume the extensions are built at gross (overall) densities of 20 homes per hectare, providing for fairly dense housing along with sufficient land for schools, shops, workplaces, transport infrastructure, parks and other greenery. Let us further assume, as we did earlier in this chapter, that 61,000 homes a year need to be built in these extensions. Putting these numbers together, we find that 3,000 hectares are required for extensions each year. The bulk of the acreage needed for the extensions will be farmland and much of that will be second rate, from the point of view of agriculture. Let us assume its market value for its existing use averages £5,000 a hectare but that landowners are paid a generous £10,000 a hectare to compensate them and their tenants for being forced to sell (then having to buy replacement land, or retire, or establish a new livelihood). The state's annual land purchase bill for town extensions would then be some £30 million a year in England, small change in the Department of the Environment, Transport and the Region's £13 billion budget for 2001/2.

Once the boundary of an extension site had been designated, the local council, or new town trust, would purchase the land gradually in parcels, each large enough for several hundred homes. The costs of these purchases could be more than recouped by selling these parcels on to developers with permission to build on them. So the cost to the taxpayers of the town extensions policy would fall rapidly to zero as a stream of high-value land sales provided the revenue needed for further low-value land purchases

This scheme of things could only be workable, and affordable, if councils could designate sufficient land for entire town extensions without having to purchase all the land at the outset. They would also require the certainty that they can buy the necessary land at the appropriate time and at the appropriate price, which means farmland prices plus a modest compensation supplement.

Councils will need to be sure that the price of farmland they need to buy for extensions is not inflated by speculation. That should not be a problem. It is easy to find out what undeveloped, undevelopable farmland is worth because it is freely and frequently traded in an open market. It should also be fairly easy to devise a reasonable compensation formula that gives rural landowners something over and above this market price for the trouble of having their land compulsorily purchased. To make this scheme work, new town trusts would need the power to tell landowners that once their fields had been allocated for a statutory urban extension, the land must be sold to the council on request at the price set by the compensation formula. Until that time comes they can carry on farming, growing trees or doing anything else lawful that does not require planning permission.

A new Act of Parliament would probably be needed to bring this into effect, legitimising the public sector's right to designate the land, to control its price and to delay the purchase until a date of its choosing sometime in the next 25 years.* The scheme of town extensions would kill off most of the speculation in greenfield land around towns and cities. Why buy or sell land at anything above farmland prices, or take options on it, if you know it can only ever be developed after the council has bought it from you at the floor price? Those trusts which never got round to purchasing all of the land within 25 years of designating their extensions would be required to pay the remaining landowners a small sum of compensation; this would discourage the trusts from planning over-sized extensions. And to prevent really long-term speculation, government would have to make it clear that if further greenfield development was needed for urban expansion later in the twenty-first century, the same scheme, based on the same principles, would be relaunched.

Let us summarise the advantages of confining greenfield development to designated, masterplanned urban extensions. The private sector continues to do the things that it usually does rather better than the state, namely building and marketing homes and other types of building. Planning gain, currently an inadequate, unpredictable way of harnessing betterment, is vastly improved because most of the uncertainty and secrecy surrounding it are removed. The price the council paid for the greenfield land is known. The value of the land when fully developed can also be estimated; that will depend on the local property prices and the number and type of houses and other buildings to be constructed (which is set out in the extension's masterplan and design code). Once both figures are out in the open it becomes far

*New town corporations set up by the government have the power to do this under existing new towns and land compensation legislation. The Pointe Gourde principle states that when land is compulsorily purchased for the purposes of new town development, its value must be assessed as if there was no such scheme – as if, in other words, it were to remain farmland. However, Ted Totman, a partner with solicitors D.J. Freeman who has advised the Commission for the New Towns and specialises in these matters, told me the scheme advocated here would probably require primary legislation.

There is a potential problem for councils/new town trusts acquiring greenfield land already allocated for housing in the council's current development plan. A

easier to know how much betterment money will accrue to the community from the extension. It can be spent on infrastructure and community facilities within the extensions, on improvements and regeneration in the remainder of the town and – in the most affluent areas – on the regeneration of urban areas elsewhere in Britain. In short, this system captures the rise in land values which urban growth creates for the benefit of all.

A shift from piecemeal greenfield development to masterplanned urban extensions could transform the way housebuilders think and operate. When it came to choosing developers to build their extensions, wise councils would look beyond maximising their short-term gain. They would be more interested in the quality of what was built and its long-term value to the town or city. The record and reputation of the firms and consortia bidding to buy their greenfield land would be weighed up. Could they be trusted to deliver on mixed uses and a mixture of house types and of free market and social housing? Do their developments help people to use alternatives to the car? Are they building neighbourhoods which look and feel built to last?

Councils would tend to prefer developers which had committed, or were prepared to commit, to investing within their existing towns and cities. Several developers are already adept at spotting derelict urban sites and old buildings that can be made into unconventional, ingenious and exciting housing while netting fatter profits than greenfield development ever could. The remainder of the industry would need to acquire these skills and abandon the discredited art of grabbing fields, whacking up suburban boxes and moving on.

To succeed in a changed world, developers would have to learn how to create and market the urban virtues of diversity, character and liveliness in and around their developments while retaining the essential suburban virtues which most buyers demand – space, privacy, greenery, peace and security. They would want to change the reputation of city neighbourhoods in order to boost the prices of the homes they plan to build there. So they would find themselves drawn into improving such things as the public realm and public safety, local schools and shopping centres.[7]

I have argued that there should be a moratorium on all greenfield development, save for that covered by designated town and city extensions. But there needs to be an exception to this policy. Many villages have become dominated by commuters,

good deal of the area which would be designated for town extensions would be likely to fall into this category. That earlier allocation would have given the land some development value, so the landowner would have a case for being paid more than the 'farmland value plus compensation' sum. The same sort of problem would exist for any previously developed but now vacant land within the designated area; it could not be valued simply as undevelopable farmland. Planning gain obligations could, however, go a long way towards resolving these difficulties. Some early means would have to be found of quantifying and containing the extra costs.

retirees and weekenders. House prices are pushed so high by incomers pursuing the rural idyll that people who grew up and work in the area and want to remain in it are forced to move to towns and cities to find an affordable home. Villages that need to grow in order to survive as functioning communities ought to be allowed to.

There is, then, a good case for strictly controlled releases of greenfield land for new housing *for local people* next to villages. Between a third and a half of the homes on these small extension sites should be reserved for local people who cannot afford market prices; these houses would either be rented out by housing association landlords or covered by a low-cost home ownership scheme. The remainder of the new homes on the extension sites should be sold in a restricted market to people who can demonstrate a long-standing link to the area. They would only qualify as a prospective purchaser if they worked close by, or had worked nearby and were now retiring. They would buy the new houses under a covenant that ensured they could only sell them to someone with a similar local connection. These purchase conditions would reduce the market price of these village extension homes; that is their intention.

One further condition should apply; these new village houses should be built at fairly high densities of at least 40 per hectare. This would maximise the number of houses and fit them snugly into the existing fabric of village buildings. Real villages are places of cottages and terraces with front doors opening straight onto the street. They may be tiny but their texture is urban. If you want detached houses and large gardens, there are millions of those in suburbia.

Housebuilding firms and housing associations should jointly promote these village extensions. Each would cover no more than 2 or 3 hectares with a few dozen homes. To gain planning permission the promoters would have to demonstrate a local need for the new houses, to guarantee prices that local people could afford and to promise to attach covenants that would ensure they could not be sold on to commuters and weekenders. There is no need for the public sector to buy the land, for the tough conditions attached to the grant of planning permission would not allow for much rise in land value.

Happily, the government already allows and encourages such village extensions.[8] Local councils can grant planning permission for new village homes on small greenfield sites which would usually be sacrosanct, provided these homes are affordable to locals and there is a proven need for them. As greenfield development becomes more tightly constrained, rural homes will probably become more sought after and expensive. The countryside is well on the way to becoming a form of pricey suburbia. Village extensions built for local people could prevent its complete metamorphosis.

My idea is to give local councils and their new town trusts a virtual monopoly on greenfield development. We could then plan, monitor and manage the supply of new homes and get the right balance between development inside and outside of the towns and cities. This is not a policy of land nationalisation. Instead, it amounts to a municipalisation of a tiny fragment of the land. The greenfields designated for town extensions would be owned briefly by the trusts before being sold on to developers with strings attached.

But Britain's political class has little faith in the ability of the public sector in general, and local councils in particular, to be creative and competent. This lack of faith is not entirely unjustified. We cannot be sure that all of the councils granted town extensions would do a good job of overseeing their development. The performance of councils and trusts would need to be assessed by central government as the extensions were built. Any council that messed things up would be compelled to surrender the task to some Westminster and Whitehall-appointed body. The trusts would need to work closely with the private sector in planning and building extensions, making use of its superior skills in handling risks, in marketing and finding cost savings.

Many planners are already thinking along these lines. So are some developers, although few would embrace the entire package. But they are not proud of most post-war greenfield development. They have come to support masterplanning and public sector land assembly because they have seen places in Britain where both have worked fairly well, in new towns in the countryside and brownfield land deep within the conurbations. The government also endorses the concept of urban extensions sited on public transport corridors, and accepts that some will have to be built on Green Belt land.[9]

What are the alternatives to this scheme? Previous chapters have explained why carrying on as we have been is not an option. Another possibility for improving the balance between development inside and outside of cities is to tax greenfield development. Such proposals have attracted a lot of interest. The purpose of this kind of tax is to discourage construction on greenfields, thereby diverting development to urban sites while raising funds to subsidise urban regeneration at the same time. But a tax would make little impact unless it was set at fairly high levels. For it to be effective the government would have to place the money raised in each council's area into a regional or national urban regeneration fund. If the revenue only reached the local council, this might merely encourage it to grant more and more planning permissions for greenfield development in order to keep the money flowing in.

The tax could lower the amount of money councils raised from planning gain agreements. That would mean fewer subsidised homes for low-income families being built on greenfield land. It could also reduce the total number of new homes being built. For all of these reasons, a greenfield tax is failing to win friends and influence people.[10] The government did not come up with any plans to introduce one when it introduced its Urban White Paper in 2000, but it did promise to review the planning gain system described in the previous chapter and consider the option of impact fees to reflect the environmental damage done by new developments.[11]

The very worst place to put the millions of new homes Britain needs over the next few decades is in new settlements in the open countryside, in the tradition of Ebenezer Howard. Yet dozens of new settlements are what the Town and Country Planning Association, the body set up to promote Howard's new towns, proposed in a book published to celebrate its centenary year.[12] The authors were Professor Sir Peter Hall, Britain's leading town planning academic and member of the Urban Task Force and Colin Ward, a respected planner, thinker and veteran anarchist. The government has also decreed that a new town should be built near Cambridge to accommodate some of that dynamic city's growth, along with extensions from Cambridge's edge into its Green Belt.[13]

Hall and Ward argue that to avoid town cramming, most of the new housing needed over the next quarter century must be built on greenfields. The new homes should be concentrated in small towns of about 20,000 people, clustered along fast, frequent rail links like beads on a string. The authors believe towns of this size built at slightly higher densities than most post-war British suburbia could have a good mixture of uses and put their residents within walking distance of shops, primary and secondary schools and a railway station.

They sketch out a grand plan for the Greater Greater London region, the wider South East of England, where the demand for new homes is highest and the dangers of town cramming are largest. Most of the new development they propose is at least 50 miles from the capital, to discourage commuting to the metropolis and encourage self-containment. It is focused into three rail corridors, one along the East Kent coast, one linking Cambridge, Peterborough and Howard's Welwyn Garden City and the third stretching in a horseshoe from Rugby to Corby via Northampton and Wellingborough. Each grows into a linear city region with a population of up to 250,000. Existing towns and villages are expanded and new ones built, all interspersed by belts of greenery and interlaced by existing or reopened railway lines and new light rail links. Train and track improvements across the South East create a high-speed regional metro which enables people to cross the entire region – a distance of about 100 miles – almost as quickly as they now cross London by Underground.

Hall and Ward are clever, thoughtful, extremely well informed people. How could they get it so wrong? I suppose that, like Ebenezer Howard, they were so overwhelmed by the messiness and hopeless imperfectability of cities and the scale of urban problems that they wanted to start afresh (although, to be fair, they have some sensible things to say about urban regeneration). Many planners are ever drawn to the clean slate of greenfields.

The majority of working-age residents of any new small to medium-sized new town built today are bound to be commuters, unless there are enormous leaps in telecommuting or large companies are persuaded to move in, along with their entire workforces. Even if such a new community was sited on a fast and frequently served railway line, many people would still get to their jobs by car. If the new town discouraged them from doing so, they would probably not want to live in it. And if they wanted to visit facilities which their own, smallish new community could not provide – say, a large shopping or leisure centre – they might well choose to drive to the nearest old town, rather than taking the train. It appears that, in general, settlements need populations of above 50,000 before large reductions in car travel per capita can start to kick in, thanks to a wider range of local facilities and better public transport options.[14]

But the biggest flaw in the notion of freestanding new towns is that they can contribute nothing whatsoever to Britain's existing towns and cities.

Why are extensions to the urban areas we already have a much better idea? First, they put their residents close to all the workplaces of the existing town. That makes it easier for people to find satisfying jobs and climb career ladders without having to move or commute long distances. This becomes much more important in a society in which the norm is for both partners to work. Second, they put the new residents close to a wider range of shops, schools and other facilities from the outset.

Third, public transport is likely to be more viable, better supported and easier to intensify in an extension than in a freestanding new town. Hall and Ward envision walkable new communities in which every resident is within 15 to 20 minutes walk of a railway station with fast, frequent connections. But unfortunately many people will not be willing to walk for even a quarter of an hour, especially on a cold, rainy day. And trains which have to stop at small town and suburban stations every mile or two – the beads on the string – are not particularly fast.

Fourth, and most importantly, an extension can help its parent. It will be planned and its building overseen by an organisation which has the interests of the whole community in mind. An extension can raise money to be invested in the existing city, improving its image, environment, communal facilities and transport systems. Freestanding new towns can do none of these things. Let us hear no more about them.

Chapter 17

RENAISSANCE OR STILLBIRTH?

Do not despair — many are happy much of the time; more eat than starve, more are healthy than sick, more curable than dying; not so many dying as dead; and one of the thieves was saved. Hell's bells and all's well — half the world is at peace with itself, and so is the other half; vast areas are unpolluted; millions of children grow up without suffering deprivation, and millions, while deprived, grow up without suffering cruelties, and millions, while deprived and cruelly treated, none the less grow up. No laughter is sad and many tears are joyful.

Jumpers, a play by Tom Stoppard[1]

I STARTED WORK on this book less than a year after the 1997 General Election, and I finished it at about the same time as the next one was called. I worried, because throughout all of that period more and more was being said and done about the urban renaissance, about the war against poverty. What could there be left to add?

But I need never have fretted. The urban crisis is not a matter to be dealt with in one, two or three terms of government, as ministers have prudently acknowledged when they talk of needing 10 to 20 years to turn the worst neighbourhoods around. The scale of neglect and abandonment I saw for myself in Britain's cities should have persuaded me that there could be no quick fix. Indeed, there was every indication the problems were spreading. Urban degeneration has deep roots and a long history; there is grandeur in its intractability.

Policy for the Inner Cities, the landmark White Paper on urban regeneration published in 1977, marked a new awareness on the part of government, recognising that the inner areas of Britain's largest towns and cities suffered from a complex of 'economic decline, physical decay and adverse social conditions'.[2] In 2000 the government published another White Paper on urban regeneration.[3] Looking at the two documents, side by side, you realise that great chunks of the analysis from 1977 could simply have been repeated in 2000.* Indeed, many of the solutions proposed in the 1970s – joint machinery, local partnerships, a unified approach – are being tried again. Only the words 'inner city' seem particularly dated, because the urban malaise has spread wider in the meantime.

Early in 2001 the Prime Minister and four of his cabinet ministers visited a council estate in East London to launch the final version of the government's grand strategy for neighbourhood renewal, a strategy which had taken 3 years to prepare and was aimed at lifting the worst places in Britain out of exclusion and into the mainstream.[4] Three weeks earlier a striking piece of research had been published in the *British Medical Journal*. It demonstrated that the electoral wards of inner London with the greatest concentrations of poor people had barely changed in 100 years. Across these wards the modern geographical pattern of death from diseases known to be related to deprivation in early life – strokes, stomach and lung cancers – was almost identical to the pattern of poverty at the close of the nineteenth century. Indeed, the link between current mortality and deprivation 100 years ago was even closer than it was for today's distribution of deprivation. The researchers from

*But their appearance certainly differs. The 1977 White Paper is a slim, small, dull-looking document which cost 80p. The 2000 one is packed with colour photographs, maps and diagrams, looks like a GCSE geography text book, is about five times as long and costs £28.

Leeds, Bristol and Cardiff Universities had gone back to the street-by-street social class maps that Charles Booth, the great Victorian poverty investigator, had drawn up in his 17-volume study, *Life and Labour of the People of London*, and used them to make comparisons with today's data. They concluded: 'The maps and models . . . show that 100 years of policy initiatives have had almost no impact on the patterns of inequality in inner London and on the relationship between people's socio-economic position and their relative chances of dying.'[5]

Urban regeneration is a large and growing industry. It spends billions of pounds of private and public money each year. It has its own journals and websites. Many thousands of careers are built on it and some fortunes are made. Between 1979 and 1997 the three Conservative governments created ten major new urban regeneration programmes of various kinds, and in the four years after that the Labour administration set up as many more.[6] Throughout the twentieth century, however, such initiatives and efforts have generally failed to uplift neighbourhoods, transform their residents' life chances and turn around entire cities. There is a busy sub-industry which spends all of its time trying to work out why this is so, and whether the various regeneration activities could be better integrated. But the real blame for regeneration's failure lies with poverty's stubborn attachment to urban places, from individual housing estates and streets through to entire towns and cities. When people's life chances have been transformed, this has usually involved moving out.

If one's definition of poverty is relative – and receiving an income less than half of the national median is one widely accepted definition of relative poverty – then the wider the spread of low to high incomes becomes, the greater are the number and the proportion of people living in poverty. And poverty *is* relative. Poor people in a rich country like Britain do not starve; many of them eat too much for their own good. They live longer than their counterparts in the developing world. They have more creature comforts and diversions. But poverty still burdens them with anxiety, fear and misery. They know that they are missing out on life, that they are held in low regard, that the odds are stacked against their children having better lives than they did. Poverty imposes huge strains on the health service, on the criminal justice system, on society as a whole. It costs each earning household thousands of pounds a year in direct and indirect taxation which end up as welfare payments to the less fortunate. The potential benefits of raising families out of relative poverty are colossal, all the more so if this can be done by getting more poor people into paid employment.

Income inequality has indeed been spreading, leaving more and more in relative poverty. One reason for this is that economies are increasingly dominated by transactions of knowledge and information rather than of physical labour and material goods. Demand for intellectual skills such as analysis and expertise (and the ability to acquire fresh expertise) rises, and so does the demand for 'people' skills like leadership and networking. Rising demand begets rising salaries. People arriving on the jobs market differ greatly in these skills, much more so than in their ability to do low-skilled manual and clerical work.

LONDON POVERTY (1896 AND 1991) AND MORTALITY (1990s)

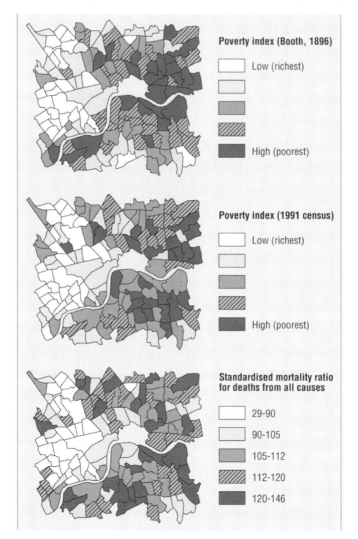

In the upper two maps, the darker the shading the higher is the proportion of people at the lower end of the socio-economic scale. In the map at the bottom, the darker the shading the higher the ratio of deaths from all causes. The blank areas in the maps correspond to the River Thames and the City of London. The maps extend from Fulham in the south-west corner to Stratford in the north east, from West Hampstead in the north-west corner to Lewisham in the south east.

The growth of the knowledge economy is stretching out the gap between rich and poor from the top end, heaping higher and higher rewards on the able. Inheritance, especially inheritance of houses, will exacerbate this inequality over the next few decades as the great increase and widening of personal wealth associated with mass home ownership flow down the generations.

But urban degeneration is stretching this gap *from the bottom end* of the distribution. Those who live in poor neighbourhoods are multiply cursed. It is more difficult for them to get work, credit and insurance than it is for the rest of us. They are more likely to be victims of crime and self-victims of alcohol and drugs. Their children are unlikely to get the exceptional, inspirational education required to overcome their home circumstances and live quite different lives from their parents. The poverty attached to place flows down the generations.

Amid this widening of incomes, the bulk of the population will probably continue to prosper.* People will try to use their prosperity to improve the environment they live in. For a broad upper band that will mean moving not only to a higher quality home but also to a better address, one with a higher status, better local schools and more pleasant surroundings. Good addresses, as we saw in Chapter 8, are positional goods whose supply can never satisfy demand, however fast the economy grows. So we can expect the price premium people are willing to pay for good addresses to become steadily higher. 'Social exclusion' may have become part of the language but 'exclusive' remains the property industry's favourite marketing tag. The patchiness of house price rises in the late 1990s seems to bear this differentiation out. Estate agents and lenders found 'hot spots' – small towns and particular areas of towns and cities – experiencing astonishing increases in property values. Meanwhile, prices across some entire regions were almost stagnant and in some locations they were falling.

As the spread of incomes widens, so does the huge spread of house prices. Already the cost of a square metre of living room can vary almost 100 fold across a distance of a couple of miles. Think of the price gap between a small terraced home in Salford on a street where half the homes are boarded up (£3,000) and a large apartment in a converted warehouse in central Manchester (£250,000 plus).

Rising prosperity also lengthens people's lives, shrinks household sizes and gives individuals more options to abandon old family relationships and attempt new ones. These changes fuel demand for new homes. The preferred place for these additional houses, as far as many builders and buyers are concerned, will continue to be greenfields. The natural course of things is for the chasms between social housing estate and owner-occupied street, inner city and suburb, impoverished conurbation and prosperous hinterland, north and south, to widen. In the absence of effective counter-measures, the number of wretched places at the very bottom of the hierarchy where homes are abandoned and entire neighbourhoods trashed will probably grow.

*There have been a couple of centuries of growth in living standards in industrialised nations. The best guess has to be that this will continue for the time being.

Abandoned housing: Beswick, East Manchester

Government, both local and central, would have no truck with such harshly pessimistic views. They maintain that in partnership with each other and with local communities they can prevent this from happening. Whitehall did, however, accept this geographical or 'ecological' analysis of poverty, the notion that it is not just a matter of individual circumstances but of the neighbourhood's physical and social environment shaping those circumstances. That is why the 1997 government created dozens of different anti-poverty zones of various types in deprived areas – 26 Health Action Zones, 15 Employment Zones, 73 Education Action Zones to name and number but three. It has, at the same time, boosted the previous government's area-based Single Regeneration Budget scheme and introduced other measures set out in its Urban White Paper and the National Strategy for Neighbourhood Renewal.* Towards the end of its first term, it was pushing up spending on regeneration.

This government argued that its predecessors had wasted hundreds of millions of pounds on schemes which delivered piecemeal physical improvements in bad neighbourhoods without tackling the fundamental people problems – unemployment, low employability, too much crime and fear of crime. It offered further reasons for the failure of previous regeneration efforts: a low standard of crucial public services such as education and health care, failure to secure the support and participation of local communities, lack of leadership and partnership between different parts of central and local government.

A number of policies and commitments were aimed at tackling these failures. There was heavy emphasis on improving key services through 'Local Strategic Partnerships' between public sector service providers, local communities, businesses and voluntary groups. Improving local people's self-esteem, self-reliance, employability, income and quality of life mattered most. More would be helped to start businesses or go to university. The top-down approach was to be replaced by community involvement, encouraging local people to come up with solutions. Millions of pounds would be spent on schemes aimed at soliciting local views and participation, on running surveys, holding meetings, training local residents on how to make their views count in committee meetings. More millions would go into local Community Chests, to encourage local people to devise their own schemes funded from the chests. Another important idea – granted further millions so as to be turned into action – was neighbourhood management; a single person or team, working at the sharp end, whom local people can turn to if they face a problem. Neighbourhood managers take responsibility in a small area for a range of services, including physical maintenance – the serious business of keeping up

*The government has faced persistent criticism from academics and think tanks for focussing so much of its attack on poverty on the most deprived neighbourhoods. They argue that all the millions of poor people living outside these places are missing out. See, for instance, C. Howarth, P. Kenway and G. Palmer,

appearances – either by dealing with more remote service providers and authorities or by doing things themselves.

To enable progress to be measured, there were a range of Public Service Agreement Targets for narrowing the gap in key social measures – GCSE passes, employment levels, burglary rates, ill health – between deprived areas and the UK average. But almost all of these targets apply across entire districts of tens or hundreds of thousands of people rather than neighbourhoods of thousands. And, to my mind, two critical objectives were lacking. 'By 2010, no major British city will have a declining population. And by 2015, no UK electoral ward should have more than three-quarters of its households earning less than half the median regional household income.'[7]

Now you could argue that this assault on urban deprivation and decay is mainly a matter of gilding the ghettoes. The government wants to be seen to be doing something about the worst neighbourhoods but only a small proportion of the population live in them and fewer and fewer of them bother to vote. The rest of the country is doing pretty well, thank you, and so long as the economic growth continues, the health service improves and the bulk of state schools raise their standards, New Labour should be assured of at least two terms. The party may have been born in the great cities a hundred years earlier but it ought to avoid being sentimental about them, for if Labour regards its heartlands as the inner cities and the tatty social housing estates, it cannot be interested in sustained power. Most voters are owner-occupiers living in suburbia or the new, more buoyant Britain of smaller towns and shires. These are the heartlands now. It was the electors of such places who put Labour in power in 1997 and who can do so again and again, provided their needs and aspirations are its top priority.

I think that is far too cynical. Party and government were serious about reducing poverty and transforming deprived neighbourhoods. The biggest departments of state were involved, ministerial and prime ministerial reputations put on the line and billions of pounds a year programmed to be spent in area-based, poverty-bashing programmes. All of this activity is purposeful, not cosmetic.

But as things stand, it is probably not going to work. At least, not nearly as well as was hoped for. The Prime Minister's vision 'of a nation where no-one is seriously disadvantaged by where they live' is, I fear, a mirage.[8] Why? To start with, people whose prospects and income are improved by such programmes are highly likely to leave for a better neighbourhood. If most residents were able to make such a move there would be nothing wrong in that, for the outcome would be lots of improved lives. But if only a minority are able to get out, while the majority stay on in a place that continues to be shunned by people with choices, then very little good will have been done. Remember that many deprived areas, like much of London's East End,

Responsibility for All: A National Strategy for Social Inclusion, London, New Policy Institute, Fabian Society, 2000, p. 4.

have stayed poverty stricken for more than a century because they remain a destination for poor incomers.

Second, the government ought not to pin great hopes on abandoning the top-down approach and asking local communities in deprived areas to take the lead in solving their own problems. There's little reason to be confident that these communities will be effective problem solvers. They are not well connected to the worlds of orthodox finance and political power. They are often weak and divided, sometimes dysfunctional, and when representatives and leaders do emerge, their mandate may be weak.[9] More prosperous neighbourhoods may have even less of a sense of community than deprived ones, but when faced with a challenge they are better placed to organise a response. Communities certainly should be encouraged to analyse and solve their problems and there surely is a moral duty on regeneration decision-makers in the public and private sector to consult all local people, to respond to what they hear, and to attempt to empower them as part of the process.[10] If it goes beyond lip service, the government's new bottom-up approach to regeneration is just and decent – but there is no guarantee that it will make much difference.

The third reason for pessimism is that the government does not give priority to bringing in new residents with earned incomes into declining areas. It seems to believe that their regeneration can be almost entirely endogenous. I think this is a big mistake. Throughout this book, I've argued that attracting people with choices to struggling neighbourhoods should have the highest priority. You can understand why government avoids this approach, for it seeks to challenge and change the city-shunning choices of the majority of voters. It could be portrayed as 'social engineering', although social engineering is an emotive, obfuscating label for the kind of activity that underlies all government; guiding the behaviour of more or less free human beings for the benefit of all.

To sum up, the government's approach to urban regeneration is commendable so far as it goes but it lacks a crucial organising principle. As such, it may merely amount to another layer of plasters and bandages atop all the others applied during previous attempts at regeneration. The neighbourhood remains a mummy. Beneath all the windings is a shrivelled corpse.

So more needs to be done. In the chapters preceding this one I set out some ideas for attracting people with choices – residents, parents, employers, employees and investors – back to cities and into districts and neighbourhoods currently in bad shape. Successful regeneration will depend on improvements in several aspects of city life, as I argued in Chapter 7. How can it all be made to fit together?

If developers, would-be residents, house purchasers and businesses are to be attracted, they need to be confident that improvements will be significant and sustained. Because these improvements cover varied but linked aspects of city life partnerships are required, bringing together the local council, other key public service providers, local business people, developers, landlords and community representatives. A plan is also needed, to set out an agreed vision of what those improvements will achieve and what the place will look like, along with a timed programme for their implementation and an explanation of how they will be maintained in the long term. Just as town extensions need flexible masterplans, so does regeneration within towns and cities.

Neighbourhoods of 4,000 to 7,000 residents will often be the optimum size at which to plan regeneration. Masterplans must also, however, be outward-looking, offering convincing links with the city centre and the remainder of the urban area and surrounding countryside. Between 4,000 and 7,000 residents provides a large enough catchment for several primary schools and one secondary school; good schools are critical for regeneration. It is also a large enough residential population to support key public and private facilities like doctors and dentists surgeries, schools, newsagents and mini-supermarkets within a 10-minute walk of every home, provided regenerated neighbourhoods are planned at the higher residential densities I advocated in Chapter 13. Public transport and car independency can flourish too. This size of neighbourhood also corresponds with electoral wards, the building blocks of local and parliamentary democracy.

A great deal of any masterplan is bound to be concerned with physical things such as buildings, transport and green spaces. The best of the area's physical assets, natural and man-made, will need to be protected, enhanced and perhaps redeployed, while new landmarks – they could be buildings, sculptures, or parks – need to go up in the early days of regeneration to symbolise the commitment to improvement. Where pieces of derelict and abused land held by several different landowners lie in clusters, they will need to be assembled into larger parcels for development. Unusable structures will have to be demolished, contamination cleared.

All of this is familiar activity for professional urban regenerators; it already happens. And many, perhaps most, of the practical tools required to regenerate already exist, thanks to Britain's long experience with trying to revive its cities. There are laws and regulations which enable local councils and government regeneration agencies to compulsorily purchase land, to insist waste and rubbish are cleared from it, to improve privately owned housing across entire areas. One promising way forward may be for councils, agencies and private sector developers to join in forming regeneration companies covering part or all of a city. The thinking is that such companies will be more focussed, swift, enterprising and intelligent than other types of partnership in achieving the kind of regeneration that demands a great deal of redevelopment. Such company-like structures are judged to have worked well in the 1990s in the rebuilding of Hulme, Manchester, and of that city's IRA bomb-devastated centre. Regeneration companies were a key recommendation of the Urban Task Force's report and at the time of completing this book the government had overseen the creation of six – covering

Liverpool and Sheffield's centres, a large swathe of East Manchester and all of Corby in Northamptonshire, Sunderland and Leicester.

But some important new tools are required. I've put forward my own and other people's proposals for what those might be in this book, and the final report of the Task Force delivered no less than 105 recommendations to government. By early 2001 ministers had implemented, or begun to implement, several of the Task Force's most important recommendations whilst explicitly rejecting a few of its major proposals (of which more below).[11] That left several powerful recommendations on which the government had promised consultations, or reviews, or even legislation. It was examining the planning gain system, discussed and condemned in Chapter 15, and considering whether developers might pay impact fees to compensate for the environmental damage caused by their projects. The government accepted that the compulsory purchase system, which allows land to be acquired by local or central government in the public interest when owners don't want to sell, needed improving. It promised to bring forward legislation 'when parliamentary time allows' to simplify and consolidate existing laws, to speed up procedures and make compensation arrangements simpler and fairer.[12] It was also considering giving local councils new money raising and subsidisation powers to boost urban regeneration; we'll return to those later in the chapter. There were, then, important things for government to be getting on with early in the second Labour term, with the Prime Minister's old friend Lord Falconer – of Dome renown – as minister in charge of regeneration.

Even if several powerful new tools for urban regeneration are created by government, there is still no guarantee of an urban renaissance. The nation starts from a very low base of urban quality – Lord Rogers claimed that Britain's cities were among the worst in Europe[13] – and it appears to have a shortage of the requisite skills. The Task Force identified this a key deficiency, especially for urban design, and made several recommendations for improvements.[14] We can be reasonably confident that if regeneration begins to succeed more widely and attracts increasing resources then the number of people in the private and public sectors with key skills – in architecture, planning, land decontamination, land assembly and brownfield development – will grow in response to demand and dissemination of 'how to do it' knowledge.

But there are two more challenging shortages undermining a great revival of cities: in local leadership and organisation and in securing adequate long-term public sector spending. Let's look at the former first.

Towns and cities will not revive without strong local political leadership. It is difficult to see how their fortunes could be turned around without more respect and trust for the town hall – both from the citizens and business people below and from central government above. Unfortunately local government has been described as the two most boring words in the English language. The number of people voting in local elections has been falling, and in England's metropolitan boroughs only around a quarter of the electorate turn out. In deprived inner city wards it is not unusual for only one in eight to bother to cast a vote.[15] Most people don't really know what their local council actually does, what it's for, how it's run – although the Conservative government's haphazard local government reorganisations of the 1980s and 1990s did, at least, simplify matters by leaving all sizeable towns and cities in Britain with one local council responsible for all of the major functions rather than two.*

The New Labour government embarked on a programme to 'modernise' local government but by 2001 the indicators of performance used to measure the quality of council services showed that in many areas there had been little or no improvement.[16] The competence of some urban councils to deliver critical services remained seriously in question and the noxious fog of apathy and low esteem which came to surround local government during the Tory years had not been dispelled. Hackney in East London was in the headlines – forced to make cuts in services because of a multi-million pound overspend, failing to collect much of the council tax and rent it was due, afflicted by political infighting in the dominant Labour group, by strikes, uncollected refuse, unswept streets.

341

I met several dedicated, talented people working in local government while researching this book. And I found keen awareness in several town halls of the need to keep people with choices living in the big towns and cities. How could it be otherwise, given the large numbers of middle and high ranking council officers who choose to live outside the urban boroughs that employ them? There seems no reason to believe that central government, or agencies of central government, or anyone else, is better placed to take responsibility for essential local public services than local authorities or to lead regeneration. Indeed, town halls are inherently better qualified for the task. Why? Because of the local accountability that comes from being chosen in local ballots, the proximity of the leadership to the people served, the opportunity for synergies in one local organisation delivering a wide range of services and, above all, the existence of local identity or pride of place.

So what can be done to help – or compel – town halls to raise their game? The reforms introduced in the Local Government Act 2000 were meant to breathe new political life into town halls. The act compelled English councils to change their old decision-making structures, generally based on committees of elected councillors. They had either to adopt cabinet government, in which a small number of senior councillors each took responsibility for particular services or groups of services and

---

*Until the mid-1980s, there were two tiers of council across England – large shire counties and smaller districts in the more rural areas and metropolitan counties and smaller metropolitan boroughs in the big conurbations (with London boroughs and the Greater London Council together covering the capital). The different types of council had largely separate functions. The metropolitan counties and the GLC were abolished in the conurbations in the 1980s. Then, in the 1990s, many large and medium-sized towns and cities outside the big conurbations were given unitary councils, responsible for the full range of local government functions, and independence from county councils. In 2000 the government created an elected Greater London Assembly, bringing two-tier government back to London. The great virtue of unitary councils, which cover all of Scotland and Wales, is that they give local government a clearer identity in the public mind.

worked together under one leader, or have a directly elected mayor. If they wanted the latter they first had to obtain the people's consent through a referendum. The idea of mayors sounds promising. It gives talented politicians the prospect of a reasonable salary, heavy responsibilities and a high profile while working in local government full time. Mayorships could encourage a higher calibre of local politicians to enter town halls. Directly elected mayors should also inject more personality and colour into local politics. That could only be a good thing since ignorance and boredom dominate the public's views of local government. The astonishing run-up to the election of Greater London's first executive mayor with its big political personalities, scandal (Jeffrey Archer had to withdraw as the Conservative's candidate) and split (Ken Livingstone, standing as an independent, trounced Labour's controversially selected candidate Frank Dobson) ought to have got things off to an excellent start.*

But only 34 per cent of the capital's electorate bothered to vote in 2000 in the mayoral election, the same proportion as had voted in the referendum on whether to have a London mayor and assembly back in 1998. A cloud was cast over the entire elected mayors project by the low London turnout, the hurdle of having to hold a referendum first and by the failure of government – always nervous of losing control from the centre – to collectively and wholeheartedly endorse the idea. Nevertheless, by the beginning of 2001 a handful of councils including mighty Birmingham had firm plans for mayoral referenda later that year.

The new Local Government Act did some other good things. It contained machinery for enforcing higher ethical standards for elected councillors and council employees. And – an unusually decentralising and slightly brave move, this – it gave councils a wide-ranging new power to promote the economic, social and environmental well-being of local people. The hope was that this would encourage councils to be more imaginative and innovative in trying to improve local people's quality of life and their environment. Under this power they could do pretty well anything provided it did not conflict with other legislation or amount to a new form of local taxation. Councils could form partnerships with companies, trusts and charities, take over the functions of other bodies, use their resources for whatever they saw fit.

What more could be done to raise the reputation of local councils and raise their competence to improve urban living? First, there needs to be a move towards more proportional representation in local elections, to prevent councils from being permanent one party states without any effective political opposition. There are dozens of urban councils where one dominant party – usually Labour – has a very much larger share of council seats than its share of the votes. People are less likely to vote when there is little prospect of power changing hands.[17]

*It was actually an earlier piece of legislation that created the Greater London Assembly and its elected mayor. The Greater London Authority Act 1999 pioneered elected city bosses and the new local transport taxes – road tolls and workplace parking charges.

Second, there should be fewer local elections – the more often people are asked to vote, the less enthusiastic they become – and they should be concerted. The boroughs in the largest English conurbations excepting London hold elections for a third of their seats every year. Instead, there should be nationwide elections for every council seat held on one day every four years. General elections should not be allowed to take place on that date, or within six weeks either side of it. This would be local democracy day, when voters would be encouraged to contemplate local issues and their own council's – not central government's – performance.

Third, councils ought to raise more of the money they spend through local taxes *whose rate they determine* with a corresponding reduction in the money granted by central government. Shifting the balance in this way would make them more relevant and accountable to the local people who elected them and whom they serve. It would also give councils more responsibility and subject them to more scrutiny. The temperature of local political debate would be raised, with deep divisions between local politicians who wanted to raise local taxes to do particular things and those who wanted to cut them. Local taxes will be no more popular than national ones, but it's likely that shifting the balance from the latter to the former a little will put pressure on town halls and political parties to improve local governments. The overall system of council finance would, however, have to remain broadly redistributive, with government grant continuing to supply the bulk of town hall income and extra money going to the more deprived areas.

There are several ways of increasing the tax-raising power of local government, none of which would be easy or popular. One is to increase the local, property-based council tax on households (which itself should be reformed to make it less regressive) while reducing income tax or Value Added Tax correspondingly. Another which has sometimes been argued for is to introduce a small local sales tax set by each council. The simplest, best idea is to allow councils to take control of the business rate (a local tax which used to be based on estimated rent levels for commercial and other non-domestic buildings) back from central government and determine its level in town halls.

Labour's 1997 election manifesto argued: 'There are sound democratic reasons why, in principle, the business rate should be set locally, not nationally.'[18] But once in power it lost conviction and its 2001 manifesto made no such declaration. Instead, the Labour government made a cautious proposal that individual councils might be allowed to very gradually raise the business rate by up to 5 per cent above the nationally set level. This could only be done with the demonstrable support of local businesses, in order to fund additional spending on projects agreed with them. Legislation would have to be passed by Parliament to give councils this power. As for the overall balance between locally and nationally-determined levels of taxation, the government said it had no plans to change things.

One can understand the government's great reluctance to devolve any real taxation power to local authorities. It would involve sacrificing some overall control of fiscal policy, although given the ratio of national to local taxation that loss would be minuscule. A few town halls might, furthermore, push their local taxes recklessly and damagingly high, as happened in some cities and boroughs in the early 1980s. But voters should have the freedom to make the mistake of voting in local politicians with bad tax policies (or any kind of bad policy); how else can they, or politicians, learn to get things right?

The fourth requirement for a town hall revival is for councils to develop ways of engaging with people to improve the quality of life at street level, addressing neighbourhood problems such as noise, dereliction, squalor, disorder and vandalism. This means consulting residents, using their ideas, enlisting their help, releasing stifled talent and enterprise – the bottom-up approach the government advocates for deprived neighbourhoods. It could involve devolving some town hall power down to area committees of local councillors, something the Local Government Act 2000 allows. But as we noted above, places with big problems tend to have weak, sometimes conflicting communities. It may fall to a council to help build an effective community – one which could work with the town hall, but might come to challenge it in various ways. Neighbourhood wardens, neighbourhood management and community chests may provide some answers. In Birmingham the city council had the idea of giving £50,000 to an advisory panel of citizens and councillors in each ward to spend on improvements they had decided on in their own patch.

I've argued that a sustained revival of cities will require a multitude of quality of life improvements across fairly large areas, the construction or refurbishment of substantial quantities of good quality, owner-occupier housing for sale and – above all – the retention or attraction of people with jobs and choices about where they live. There is ample potential for resentment and conflict in this, and some danger of repeating the mistakes of the past when large-scale slum clearances and redevelopment shattered communities. Local people who have endured deprivation and a grim environment for years may not take kindly to the notion that salvation depends on bringing in lots of better-off newcomers.

The case of Newcastle's West End was discussed in Chapter 15. In Lambeth, south London, council tenants were asked to vote on a proposed £440 million scheme that would have demolished their run-down estate of 900 homes and replaced it with a higher-density development of 2,500 pricey private flats plus 600 new or refurbished homes for them. The razing and replacement of the Ethelred Estate were devised by two major market housebuilders and the council. It would have brought many people with choices into a deprived part of London. However, much of the new social housing would be smaller than the existing council homes and some of the tenants would have faced relocation to other estates. In 2000 60 per cent of them voted to reject the scheme and it was dropped.

The lesson here is that regeneration schemes which attempt to bring more affluent households into deprived areas will often require high leadership and political skills in persuading, explaining, creating and sustaining partnerships and consensus, providing vision, making peace. We need those urban revival skills in revived town halls.

The final shortage which could hold back an urban renaissance concerns money. The physical and social restructuring of large tracts of big cities will depend on huge flows of private sector money from individuals and firms, money spent from day to day and invested for the future. All central and local government can do is to skilfully and effectively prime the pump. Taxpayers' money is needed to purchase and reclaim derelict land, to lure in private sector developers, particularly housebuilders. It is needed to fight crime and reduce the fear of crime, to improve public transport and standards in urban schools, to create beautiful, eye-catching new public buildings, parks, woodlands and lakes.

The government plans to spend increasing sums on regeneration. Given the scale of the task, they may not be large enough. But there are ways in which it could raise more money for reviving towns and cities by taxing – and thereby discouraging – things which in themselves harm cities.

The act of owning derelict land and buildings in urban areas needs to be seen as anti-social, as damaging, and penalised accordingly. Landlords often hang on to this dead property for years or decades, hoping for a higher price, waiting for the local land market to pick up, sometimes keeping it on their books at an inflated value. So buildings and urban land left derelict for some time should be taxed in some way, with the local council able to take a charge on the land when the tax is not paid. The tax would push the price of derelict land down, making it more affordable to public sector bodies or private developers with plans for putting it to use. The revenues from taxation of dereliction and abandonment could be devoted to urban regeneration. The owners of empty homes could be charged the full council tax (instead of paying only half or – if they allow it to become technically uninhabitable – none at all).

Local councils, regional development agencies and other public sector regenerators must be exempted from such taxation, since they would be engaged in assembling derelict sites and buildings in order to have them redeveloped. But housing associations owning empty homes in towns and cities should be taxed if, after a year has elapsed, they cannot put their vacant properties into use or redevelop them. That would encourage them to either improve their management of voids or dispose of them to a statutory agency. Councils should, for their part, remove any planning designations that prevent derelict urban land and buildings from being put into use – with the proviso than any particularly anti-social use of land or a building which was likely to prevent surrounding sites from being developed, or any proposed development which conflicted with the council's regeneration masterplan, should be granted only temporary planning permission.

It is conceivable that a vacant land tax would lead to growing quantities of derelict sites being dumped in the laps of local councils in the most depressed, weakest towns and cities, with little prospect of them ever attracting development. In that case grind the old buildings and streets to rubble, cover them in compost and topsoil, and use them as nature reserves, for growing timber or grazing sheep.

In 1999 the Urban Task Force recommended a vacant land tax, and full council tax liability for the owners of empty homes. In 2000 the government unfortunately rejected both. It did, however, commit itself to a few fiscal initiatives aimed at incentivising regeneration (as opposed to penalising dereliction). For instance, all property transactions in the most deprived urban areas were to be exempted from stamp duty. The government also announced that it was considering, and seeking views on, two other regeneration-boosting tax schemes which the Task Force had encountered in the USA and recommended for Britain.[19] One was a local tax reinvestment programme. A local council or regeneration partnership planning to regenerate an urban area would be allowed to retain all of any growth in revenues from business rates and council tax which flowed from new business premises and homes being built there. The extra funds would be ploughed back into further regeneration projects. The second suggested scheme would allow a local council to levy a small, extra business rate within a defined town centre. But first it would have to reach agreement with local businesses on the improvements the extra revenue would pay for, and they would have to back the idea in a referendum. In April 2001 the Prime Minister promised a re-elected Labour administration would introduce the legislation needed to set up these Business Improvement Districts.*

I want to propose a rather more radical money-raising idea. In April 2000 the government removed the final fragment of the income tax relief available on interest payments for home mortgages. But one very large tax break for home-owners survives; they pay no capital gains tax when they sell their properties at a profit.

If a modest tax – at, say, 10 per cent – was imposed on people's above-inflation gains from private house sales, this would bring in several billion pounds of additional revenue a year which could, quite appropriately, be devoted to urban regeneration. Such a tax would have some other useful features. It would dampen down the general rise in house prices which can damage the wider economy. Most people use money from the sale of their houses to finance their next purchases, so if they had to pay tax on any capital gains this would reduce the prices they could afford for their new homes. Such a tax would also tend to redistribute wealth, extracting money mainly from the better-off half of the population – home-owners in regions and districts where property prices have been rising – and using it to finance regeneration in poorer urban neighbourhoods across Britain.

---

*Government moves so very slowly. In June 1997 John Prescott, the Deputy Prime Minister, spent an hour or so walking the streets of Mid-town Manhattan, seeing for himself the improvements that Business Improvement Districts had brought there. He received an impressive presentation on their virtues – I know, because I was one of a small group of journalists who walked with him. Four years later his boss Tony Blair

**Priming the pump – the Lowry Museum,
Salford Quays, Greater Manchester**

Money that home-owners had invested in maintaining and improving their homes over the years would, in fairness, have to be exempted from the tax. (If my loft extension cost me £10,000, then it is only fair that the first £10,000 of any increase in my house's value is free from capital gains tax when I come to sell it.) This would have the interesting side effect of shrinking the black economy. In order to claim any exemption, home-owners would be required to keep records of, and *bona fide* receipts from, all of the various building firms and trades people who worked on their houses over the years. A large proportion of this kind of work is done for cash so as to conceal non-payment of various taxes. Placing a capital gains

promised legislation for BIDs in Britain – but how long before that is enacted? Fortunately, dozens of UK councils have already joined with local businesses to set up voluntary town centre management

partnerships which achieve many of the things BIDs are for.

tax on house sales and ensuring vendors claimed only legitimate exemptions would put pressure on myriad small builders and traders to emerge from the shadows and stop cheating on tax.

Such a capital gains tax would, of course, be deeply unpopular and furiously resisted. The government would probably not wish to touch it with a barge pole because of the opposition it would face from the Opposition and the press. Perhaps it would get a better hearing if the house price boom of the late 1990s was followed by another slump.

Some important policies required for an urban renaissance across Britain are likely to be unpopular. As citizens, we should buy into big towns and cities. As consumers, we will continue to shun them. Businesses denied greenfield sites threaten to move to regions and nations where countryside is more readily available. Car commuters who clog urban roads as they drive into towns and cities campaign against congestion-curbing, pro-public transport policies that make their motoring more awkward or expensive. Housebuilders claim that local and central government side with the NIMBYists in denying ordinary people needed new greenfield homes. State schools serving privileged catchments oppose policies that divert funds into education in disadvantaged places, or lure their more affluent parents and pupils away.*

So while I sometimes worried that the urban renaissance might arrive before this chapter was written, it often seemed to be a political impossibility. Yes, there were ways of improving matters, plenty of good ideas, lots of energetic, skilful, clever people talking up Britain's cities and urban life – but this was not enough. I suppose someone who quit the inner city for deepest suburbia on London's rim is bound to have had doubts about the cause. Those doubts did not diminish as I discovered that a succession of the urban regeneration enthusiasts I was interviewing appeared to have made similar choices. There was the dynamic young property developer transforming run-down Victorian buildings on the fringes of Manchester and Liverpool city centres into superb apartments, shops and offices. He was good news and had a great line in urban renaissance patter but he lived in leafy Worsley, 7 miles out from central Manchester 'because my wife thinks it's better for the children, and I wouldn't argue with her'. There was the enterprising redeveloper of social housing who had grown up in poverty on a council estate and now cherished her countryside farmhouse. And the persuasive pamphleteer on urban crime prevention seconded to the Social Exclusion Unit who lived in an Oxfordshire village.

Perhaps, I sometimes felt, the death of huge fragments of cities had to happen. It was just another of those big, unpleasant, unavoidable things – like the extinction

*Pauline Latham, Chair of the Foundation and Voluntary Aided Schools Association, has complained that the former grant maintained schools (those that 'opted out' of local authority control under the previous government) were losing funding because the government was giving more priority to schools and pupils in

of species, poverty and the nights drawing in for another winter. Most people had seen their lives improve during half a century of counter-urbanisation. Those millions of lives wasted in rotting, left-behind urban places were a great shame but many of their inhabitants would cope and a fair few blessed with luck, wits and energy would escape.

Our urban failure may have been reckless and tragic. But played out alongside the enormous increase in overall living standards, the tragedy amounted to a mere sideshow. The opinions I collected from my neighbours in Bromley, and everything I discovered as I read, travelled, questioned and listened, told me that most people felt they were bettering their lives by moving away from urban environments they loathed. Who was I to say the great escape is a great mistake?

But it is. We could have done better; we need to do better and maybe we will. In 2001 it seemed likely that we would let great chunks of towns and cities rot until only a few residents were left and property had little value; then we would perhaps clear large areas and start again. If only we could be more sensitive, more civilised and develop the means of curing sickening neighbourhoods before they become terminal cases. Getting our cities right may be the secret of making Britain a fairer, kinder nation that really is more at ease with itself.

We can take hope because we have seen this kind of transformation before. A complex of problems emerges and enlarges. A few people and organisations advocate radical solutions which are attacked by those who fear their interests are threatened. The government tinkers with half-measures, the problems get worse, the debate intensifies, and eventually the arguments are won and the solutions that sounded so impossibly interventionist and awkward 10, 20, 30 years earlier become obvious and unavoidable. We can also take hope from the fluidity of culture and fashion. One of my themes has been the longevity of our anti-urban prejudices. But perhaps the regeneration of cities will acquire a critical cultural mass, and then these prejudices might wither more rapidly than we could ever suppose.

You decide. We moved. We left our archetypal inter-war suburban house for another semi, a fine Edwardian one, on a main road – a Red Route, no less – with double decker buses thundering by and a tower block on the opposite side. A busy railway station and large shopping centre lie within spitting distance. We moved a little closer to the core of the great metropolis and right next to the throbbing heart of Bromley. Not quite the inner city, I grant you, but a step in the right direction.

I want to end not in a halfway house in Bromley but with the place I began: Glasgow's Possilpark, where several hundred of the city's tens of thousands of council flat demolitions took place. I went back there with the photographer David Rose, who took the pictures that illustrate this book. It looked as strange and as

deprived areas. 'Just because a school is in a so-called leafy suburb, that does not mean the parents are wealthy ... Many will have stretched themselves to the limit to buy houses in the catchment areas of these schools.' 'Sink schools "should be shut to help others"', *The Guardian*, 29 September 1999.

desolate as ever. 'A book?' said a passer-by, curious as to what we were up to. 'Aye, well, I'll nae fookin' read it.'

The powers-that-be in Glasgow now plan to build about 700 middle-income family homes for sale on this derelict land, and a similar number on the site of an old infectious diseases hospital less than a mile away. All this is an area dominated by municipal housing, some of which has been rebuilt once, twice or thrice, and deprivation. The Glasgow Alliance, a city boosting body whose membership includes the city council, the health board, the Scottish Executive, the local training and enterprise council and an organisation representing business, is leading the plan and negotiating to bring private sector housebuilders onto the site. If they come, it will be because they can get the land (which belongs to the state and the council) fairly cheap.

Andrew Fyfe, the Alliance's chief executive, says the point of this break with the past is to retain people with good jobs who are currently leaving the city. It is not unusual for young couples and singles working in Glasgow to buy their first home within the city, and then maybe their second. But by the time they plan children and want a house with a garden, they are usually looking outside the city limits. This is partly because of Glasgow's multitude of poverty-related afflictions; the city has long vied with Liverpool for the title of Britain's biggest urban failure (although most would judge that Liverpool won). It is partly because there are relatively few houses with gardens in good neighbourhoods for sale in Glasgow, and few new ones being built.

Fyfe is adamant that the 1,400 or so units at Possilpark and Ruchill Hospital will be primarily middle-class family houses, priced at more than £100,000, rather than cheap starter homes or subsidised housing for rent. He sees several new, or completely refurbished, local primary schools being an essential part of the development package. He hopes the few, poverty-stricken shops nearby can be uplifted by all of the new homes; maybe there will be a new supermarket from one of the mainstream chains (rather than one of those discount stores which characterise poor areas). And then there is the need for a fast, direct public transport service to the city centre just one and a half miles away.

The incomers are going to have to feel safe in the new homes and streets; otherwise they will never come. But, says Fyfe, it must not be a walled community. It is all formidably difficult, a huge experiment. The proposal has encountered major opposition from the declining population of this depressed neighbourhood. The Alliance has had to embark on a large consultation exercise and rethink its proposals in at attempt to gain the support of local people and organisations. It has been compelled to put much more effort into making the redevelopment benefit the lives of existing residents.

A bold plan for an abandoned site in the conurbation that was proud to call itself the Second City of Empire is not a bad place to end this story without end. Until now, cities have been quite indivisible from civilisation. The development of the arts and thought, of trading and banking, of scientific and technological progress, pretty much everything, in fact, over and above subsistence farming, fishing, hunting and gathering, has depended on cities.[20] The essence of the urban – large concentrations of buildings and streets – has survived, unchanged, for 5,000 years. Around the globe, urban areas continue to grow faster than ever. Around 2000 a half-way mark was crossed; most of the earth's 6 billion plus people now live in towns and cities rather than villages and countryside.

Glasgow has a special place in all of this. In the nineteenth century iron, coal, deep water, cheap labour, capital and enterprise all fused and one of the word's first industrial supercities, or 'shock' cities as they have so rightly been called, grew up beside the Clyde. Through the second half of the twentieth century Glasgow was one of Britain's, and perhaps one of the world's, great victim cities. The city fathers built and rebuilt, pushing outwards, trying to put right the disorder and overcrowding imposed by Glasgow's early industrial blossoming. All the while the city was falling behind the rest of Britain, bleeding tens of thousands of manufacturing jobs, failing to attract anywhere near enough new employment in the services sector and the more modern, growing industries. The new kind of jobs it needed were going elsewhere in Scotland, in England, in the new, fast growing Britain of small and medium-sized settlements outside the great Victorian conurbations. As employment in Glasgow shrank the city's population fell by hundreds of thousands. And while many thousands of new jobs – but not enough – were created within the city, far too many of them were being taken by people who chose or planned to live well outside it.

Can the likes of Glasgow hope for anything better than managed retreat? Or does the future belong to smaller towns, faceless housing estates, the 'town provinces' and 'urban regions' H.G. Wells foresaw? These are places – or perhaps they are one huge place – which depart from 5,000 years of urban form. Not blobs on the map but networks, zones that can only be understood and used through car journeys and electronic communication and for which the word settlement seems hopelessly old-fashioned.

We still have choices.

West End, Glasgow, overleaf

ENDNOTES

CHAPTER 1 THE ABANDONED CITY

1 M. Berman, *All That is Solid Melts into Air: The Experience of Modernity*, New York, Simon and Schuster, 1982.
2 Office for National Statistics, *Regional Trends 32*, London, The Stationery Office, 1997.
3 Figure derived from estimates of net annual urban emigration in T. Champion, D. Atkins, M. Coombes and S. Fotherington, *Urban Exodus*, London, Council for the Protection of Rural England, 1998.
4 DETR, *Indices of Deprivation 2000*, London, DETR, 2000. These indices, devised by a team at Oxford University, cover measures of deprivation for income, employment, health and disability, education, skills and training, housing and geographical access to services. The most deprived ward of all 8,414 in England (each with a population of about 6,000) is Benchill in Wythenshawe. The indices at ward and local authority district level are available on the DETR's website, www.detr.gov.uk.
5 Author's analysis using *Indices of Deprivation 2000*.
6 A. Ashworth, *Once in a House on Fire*, London, Picador, 1998, is a memoir of growing up in poverty in inner city Manchester through the 1970s and 1980s under the care – or rather lack of care – of two violent step-fathers and an abused, depressed mother. It is a haunting escape story which climaxes with the writer leaving home to take up an undergraduate place at Oxford.
7 M. Rahman, G. Palmer, P. Kenway and C. Howarth, *Monitoring Poverty and Social Exclusion 2000*, York, Joseph Rowntree Foundation, 2000.
8 CPRE and Civic Trust, *Going to Town*, London, CPRE, 1997.
9 Urban Task Force, *Towards an Urban Renaissance: Report of the Urban Task Force*, London, E. & F.N. Spon, 1999.

CHAPTER 2 DARKSHIRE AND COKETOWN – THE APPROACH TO 1900

1 Ebenezer Howard, *To-morrow: A Peaceful Path to Real Reform*, London, Swan Sonnenschein, 1898.
2 John Ruskin, *The Two Paths*, London, Smith, Elder & Co., 1859.
3 Benjamin Disraeli, *Coningsby, or The New Generation*, Harmondsworth, Penguin, 1983, p. 177 (first published 1844).
4 In 1801 34 per cent of the population of England and Wales was urban, in 1901 it was 78 per cent. C. Law, 'The growth of urban population in England and Wales, 1801 to 1911', *Transactions of the Institute of British Geographers*, 1967, Vol. 41, pp. 125–143.
5 E. Wrigley and R. Schofield, *The Population History of England: A Reconstruction*, London, Arnold, 1981.
6 It first appeared in the Eyre Estate's development of St John's Wood, London around 1815. See F. Thompson, *The Rise of Suburbia*, Leicester, Leicester University Press, 1982.
7 F. Engels, *The Condition of the Working Class in England from Personal Observation and Authentic Sources*, Oxford, Oxford University Press, 1993, p. 58 (first published 1845).
8 Figures from Census returns quoted in H. Dyos, *Victorian Suburb*, Leicester, Leicester University Press, 1961, p. 55.
9 Dyos estimated that there were 416 different building firms, some of them one-man bands, building the 5,670 homes put up in Camberwell between 1878 and 1880. Only 36 of these firms built more than 30 houses each. Ibid.

10 R. Rodger, 'The invisible hand: market forces, housing and the urban form in Victorian cities', in D. Fraser and A. Sutcliffe (eds), *The Pursuit of Urban History*, London, Edward Arnold, 1983.

11 Dyos, *Victorian Suburb*, p. 145.

12 It has been estimated that 120,000 people were displaced from their urban homes by railway development between 1840 and 1900. J. Kellet, *The Impact of Railways on Victorian Cities*, London, Routledge, 1969.

13 C. Peel, *The New Home*, London, Archibald Constable & Co, 1898.

14 Prince Albert showed an interest in model dwellings for the working classes, supporting the construction of a demonstration block of four small flats as part of the Great Exhibition of 1851. The Prince of Wales sat on the Royal Commission on the Housing of the Working Classes, which reported in 1885.

15 Quoted in D. Stenhouse, *Understanding Towns*, Hove, Wayland Publishers, 1977.

16 J. Morton, *Cheaper than Peabody: Local Authority Housing from 1890–1919*, York, Joseph Rowntree Foundation, 1991, p. 2.

17 Ibid.

18 H.G. Wells, *Anticipations of the Reaction of Mechanical and Scientific Progress upon Human Life and Thought*, London, Chapman and Hall, 1902.

19 P. Kropotkin, *Fields, Factories and Workshops*, London, Hutchinson, 1899.

20 E. Howard, op. cit.

CHAPTER 3   ENTER THE STATE – 1900 TO 1951

1 M. Foot, *Aneurin Bevan: A Biography*, Vol. 2: *1945 to 1960*, London, Four Square, 1973, p. 78.

2 J.B. Priestley, *English Journey*, Harmondsworth, Penguin, 1977, p. 377 (first published 1934).

3 P. Hall *et al.*, *The Containment of Urban England*, London, Allen & Unwin, 1973, Vol. 1, p. 83.

4 Railway promoters built a branch line which terminated at my own suburb of Hayes, near Bromley, in 1882, but mass housing development did not begin until more than 40 years later when the line was electrified. See A. Jackson, *Semi-Detached London: Suburban Development, Life and Transport, 1900–39*, London, Allen & Unwin, 1973, p. 234.

5 T. Sharp, *Town Planning*, London, Pelican, 1940, p. 143.

6 P. Abercrombie, *Greater London Plan 1944*, London, HMSO, 1945. And J. Forshaw and P. Abercrombie, *County of London Plan*, London, Macmillan, 1943.

7 New Towns Committee, *Final Report* (Cmnd 6876), London, HMSO, 1946, p. 10.

8 A. Holmans, *Housing Policy in Britain*, London, Croom Helm, 1987, pp. 74–75.

9 D. Donnison and C. Ungerson, *Housing Policy*, London, Harmondsworth Penguin, 1982, p. 142.

10 Priestly, *English Journey*, p. 10.

11 For a fine history and appreciation of the suburban semi, see P. Oliver, I. Davis and I. Bentley, *Dunroamin: The Suburban Semi and Its Enemies*, London, Barrie and Jenkins, 1981.

12 Ed. C. Williams-Ellis, *Britain and the Beast*, London, J.M. Dent and Sons, 1938.

13 Ibid., p. 15.

14 This is the view advanced by Alan Holmans, an economist who advised the Department of the Environment. Op. cit.

CHAPTER 4   OVERSPILL AND HIGH RISE – 1951 TO 1976

1 H.G. Wells, *Anticipations of the Reaction of Mechanical and Scientific Progress upon Human Life and Thought*, London, Chapman and Hall, 1902, p. 60.

2 The 1981 Census found that between 1971 and 1981 every metropolitan county (the local council areas which covered the seven biggest English conurbations) lost population, but all of the non-metropolitan counties gained population.

3 T. Champion, D. Atkins, M. Coombes and S. Fotherington, *Urban Exodus*, London, Council for the Protection of Rural England, 1998, p. 10.

4 Department of the Environment, *Policy for the Inner Cities*, London, HMSO, 1977.

5 R. Atkinson and G. Moon, *Urban Policy in Britain: The City, the State and the Market*, London, Macmillan, 1994, pp. 64–86.

6   A government circular of 1955 asked city councils outside of London to consider establishing Green Belts. Most decided they wanted them.

7   ECOTEC, *Reducing Transport Emissions Through Planning*, London, HMSO, 1993, p. 49.

8   D. Sudjic, *The 100 Mile City*, London, André Deutsch, 1992.

9   P. Hall, *Cities of Tomorrow* (updated edition), Oxford, Blackwell, 1996, p. 304.

10 M. Halcrow, *Keith Joseph: A Single Mind*, London, Macmillan, 1989, p. 31.

11 National Sustainable Tower Blocks Initiative, *Streets in the Sky*, 2000. See www.towerblocks.org.uk.

12 Between 1966 and 1970 an average of 72,000 council houses a year were being built in England and Wales and 74,000 flats and maisonettes. A. Holmans, *Housing Policy in Britain*, London, Croom Helm, 1987, p. 114.

13 N. Dennis, *People and Planning: The Sociology of Housing in Sunderland*, London, Faber, 1970.

14 M. Foot, *Aneurin Bevan: A Biography*, Vol. 2: *1945 to 1960*, London, Four Square, 1973, p. 82.

15 Quoted in an article in *The Guardian*, 16 June 1971.

16 See Champion *et al.*, 1998, *Urban Exodus*, p. 16. And R. Bate, R. Best and A. Holmans (eds), *On the Move: The Housing Consequences of Migration*, York, Joseph Rowntree Foundation/York Publishing Services, 2000.

CHAPTER 5   THINGS CAN ONLY GET BETTER – 1976 TO 2000

1   M. Thatcher, *The Downing Street Years*, London, HarperCollins, 1993, p. 145.

2   The Social Exclusion Unit, *Bringing Britain Together: A National Strategy for Neighbourhood Renewal*, London, The Stationery Office, 1998.

3   C.B. Hillier Parker/Department of the Environment, Transport and Regions (DETR), *The Impact of Large Foodstores on Market Towns and District Centres*, London, The Stationery Office, 1998.

4   Thatcher, *The Downing Street Years*, p. 638.

5   T. Champion, D. Atkins, M. Coombes and S. Fotherington, *Urban Exodus*, London, Council for the Protection of Rural England, 1998.

6   I. Turok and N. Edge, *The Jobs Gap in Britain's Cities: Employment Loss and Labour Market Consequences*, York, The Policy Press/Joseph Rowntree Foundation, 1999.

7   DETR, *Transport Statistics for Metropolitan Areas 1998*, London, The Stationery Office, 1998.

8   West Midlands Joint Data Team, 'Journey to work trends for the major urban areas of the West Midlands, 1995', cited in Birmingham City Council, *Birmingham Transport Strategy*, Birmingham, 1998.

9   DTZ Debenham Thorpe, *Retail Landscape 2000*, June 1999.

10 Department of the Environment (DoE), *Projection of Households in England to 2016*, London, HMSO, 1995.

11 DoE, *Urbanisation in England: Projections 1991–2016*, London, HMSO, 1995.

12 Thatcher, *The Downing Street Years*, p. 147.

13 Percentages for 1979 and 2000 supplied by the DETR.

14 J. Goodwin and C. Grant (eds), *Built to Last?: Reflections on British Housing Policy* (Second edition), London, ROOF Magazine, 1997, p. 179.

15 D. Page, *Building for Communities: A Study of New Housing Association Estates*, York, Joseph Rowntree Foundation, 1993.

16 Ibid., p. 46.

17 A. Evans, B. Pannell, J. Stewart and P. Williams, *Trends, Attitudes and Issues in British Housing*, London, Council of Mortgage Lenders, 1998, p. 9.

18 DETR News Release HB-12, *Housebuilding: October 2000*, 7 December 2000.

19 DETR, *Quality and Choice: A Decent Home for All*, Housing Green Paper, London, DETR, 2000.

20 Ibid.

21 For extensive discussion of the issue, see S. Lowe, S. Spencer and P. Keenan (eds), *Housing Abandonment in Britain: Studies in the Causes and Effects of Low Demand Housing*, York, Centre for Housing Policy Conference Papers, University of York, 1998. And DETR, *National Strategy for Neighbourhood Renewal, Report of Policy Action Team 7: Unpopular Housing*, London, DETR, 1999.

22 Ibid., paragraphs 1.21, 1.26 and 4.51–4.53.

23 Urban Task Force, *Towards an Urban Renaissance*, London, E. & F.N. Spon, 1999, p. 194. See also DETR, 1999, op. cit., paragraphs 1.8, 3.9 and 3.14.

24   R. Bate, R. Best and A. Holmans (eds), *On the Move: The Housing Consequences of Migration*, York, Joseph Rowntree Foundation/York Publishing Services, 2000.

25   P. Keenan, 'Residential mobility and low demand: a case history from Newcastle' in S. Lowe *et al.*, op. cit.

26   From estimates of English housing affected by low demand in H. Pawson and G. Bramley, *Low Demand and Unpopular Neighbourhoods*, London, DETR, 2000.

27   DETR, 1999, op. cit., paragraphs 4.22, 4.75–4.83 and Table 3. More than half of local authorities found selective demolition to be a successful way of dealing with low demand; they regarded it more highly than conversion and rehabilitation. The policy action team report cites research carried out for the DETR by consultants Pieda which found that demolition is increasing and that for every ten dwellings demolished, only six are replaced.

28   Nationwide, *Nationwide House Price Index: Quarterly Review*, Autumn 2000. Website: www.nationwide.co.uk.

29   *Homes for a World City*, report of the Mayor's Housing Commission, pre-publication text available from the Greater London Authority's website www.london.gov.uk, November 2000.

30   Ibid.

31   Social Exclusion Unit, 1998, op. cit.

32   Social Exclusion Unit, *A New Commitment to Neighbourhood Renewal: National Strategy Action Plan*, London, Cabinet Office, 2001.

33   B. Robson, M. Parkinson, M. Boddy and D. MacLennan, *The State of English Cities*, London, DETR, 2000.

34   DETR, *Our Towns and Cities: The Future*, London, The Stationery Office, 2000.

35   SERPLAN, *Sustainable Development Strategy for the South East*, London, SERPLAN, 1998.

36   DETR, *Regional Planning Guidance for the South East of England, Public Examination May–June 1999, Report of the Panel*, London, DETR, 1999.

37   Ibid., 4.12, p. 20.

38   DETR, *Planning Policy Guidance Note 3: Housing*, London, The Stationery Office, 2000.

39   Ibid., paragraph 57.

CHAPTER 6   TEN OPPORTUNITIES

1   House of Commons Environment, Transport and Regional Affairs Committee, Eleventh Report, Volume I, *Proposed Urban White Paper*, London, The Stationery Office, 2000.

2   Address to the Housing Corporation Conference, Brighton, February 1999, quoted in *Housing Today*, 18 February 1999.

3   Author's estimate from ONS population estimates for local authority districts in England, Scotland and Wales.

4   DETR, *Indices of Deprivation 2000: Regeneration Research Summary Number 31*, London, DETR, 2000. This ranking of districts is based on the average deprivation 'score' for all of their wards, based on income, employment, education, skills and training, quality of housing and geographical access to services.

5   DETR, *The State of English Cities*, London, DETR, 2000. And General Register Office for Scotland, *Mid-1999 Population Estimates, Scotland*, www.gro-scotland.gov.uk.

6   Social Exclusion Unit, *National Strategy for Neighbourhood Renewal: A Framework for Consultation*, London, Cabinet Office, 2000.

7   Council of Mortgage Lenders, *Home Ownership, House Purchases and Mortgages: International Comparisons*, London, CML, 2000.

8   Council of Mortgage Lenders Press Release, *Buy-to-Let Still Rising*, London, Council of Mortgage Lenders, 21 February 2001.

9   DETR, *Our Towns and Cities: The Future*, London, The Stationery Office, 2000.

10   DETR, *The State of English Cities*, London, DETR, 2000.

11   A MORI poll commissioned by the information technology group Mitel in 1999 found that some 5 per cent of the UK working population spent part of their working week at home. Strategic Workstyles, an Oxford consultancy, has forecast that 25 per cent of jobs could involve teleworking while the Telework, Telecottage and Telecentre Association estimates that the proportion could be as much as half of all non manual jobs.

12   Stanford Institute for the Quantitative Study of Society, *Study of the Social Consequences of the Internet: Preliminary Report*, Stanford, University of California at Stanford, 2000.

13　D. Coyle, *The Weightless World*, Oxford, Capstone, 1997, pp. 192–211. This author argues that the growth in the information economy will favour urban living. *The Economist* journalist F. Cairncross, in *The Death of Distance: How the Communications Revolution Will Change Our Lives*, London, Orion, 1997, concludes that while many more people will work from home, cities will become safer and retain their role as centres of entertainment and culture as well as becoming places where people want to live.

CHAPTER 7　PUSHES AND PULLS

1　Countryside Commission, *Public attitudes to the Countryside*, Cheltenham, Countryside Commission, 1997.

2　J. Murdoch, *Counterurbanisation and the Countryside: Some Causes and Consequences of Urban to Rural Migration*, Papers in Environmental Planning Research 15, Cardiff, Department of City and Regional Planning, Cardiff University, 1998.

3　M. Wiener, *English Culture and the Decline of the Industrial Spirit, 1850–1980*, Cambridge: Cambridge University Press, 1981.

4　A. Freeman, A. Holmans and C. Whitehead, *Is the UK Different? International Comparisons of Tenure Patterns*, London, Council of Mortgage Lenders, 1996, pp. iv and v.

5　D. Dorling, *A New Social Atlas of Britain*, London, Wiley, 1995. Fifty people per hectare is equivalent to about twenty-one homes per hectare, which seems like a fairly low, suburban density. But the measurement here is of the gross density across an entire ward (the smallest administrative and electoral area, of some 6,000 people) including land occupied for shops, schools, commerce and industry, parks and so forth. In terms of the densities within residential areas, it equates to roughly forty homes per hectare.

6　J. Todorovic and S. Wellington, *Living in Urban Areas: Attitudes and Aspirations*, London, DETR, 2000.

7　Council of Mortgage Lenders, *Second Homes: A Market Report*, London, Council of Mortgage Lenders, 2000.

8　M. Gwilliam, C. Bourne, C. Swain and A. Prat, *Sustainable Renewal of Suburban Areas*, York, York Publishing Services/Joseph Rowntree Foundation, 1998.

9　DETR, *Indices of Deprivation 2000: Regeneration Research Summary Number 31*, London, DETR, 2000, pp. 26–41.

10　R. Rogerson, *Quality of Life in Britain*, 1997. Available from the Quality of Life Research Group, Department of Geography, University of Strathclyde, 50 Richmond Street, Glasgow G1 1XN.

11　R. Burrows and D. Rhodes, *Unpopular Places? Area Disadvantage and the Geography of Misery in England*, York, Joseph Rowntree Foundation/The Policy Press, 1998.

12　Office of National Statistics, *1998-based Short-term Subnational Population Projections for Local Authority Areas in England*, Dataset PT100PJ2, London, ONS, 2000, and ONS, *Population Mid-1999: Local Authority Components of Population Change*, Dataset PP99T9, London, ONS, 2000.

13　URBED, *But Would You Live There? Shaping Attitudes to Urban Living*, report for the Urban Task Force, London, DETR, 1999.

14　Summary of findings available from the author at 103 Masons Hill, Bromley, Kent BR2 9HT.

15　*The Counterurbanisation Cascade: An Analysis of the 1991 Census Special Migration Statistics for Great Britain*, Department of Geography Seminar Paper 66, Newcastle, University of Newcastle upon Tyne, 1996.

16　A. Hooper, K. Dunmore and M. Hughes, *Home Alone*, London, Housing Research Foundation/National Housebuilding Council, 1998.

17　T. Champion, D. Atkins, M. Coombes and S. Fotherington, *Urban Exodus*, London, Council for the Protection of Rural England, 1998, p. 16.

CHAPTER 8　THE MILTON KEYNES EFFECT

1　E. Banfield, *The Unheavenly City: The Nature and Future of Our Urban Crisis*, Boston, Little, Brown and Company, 1970.

2　Steen Eiler Rasmussen, *The Unique City*, Cambridge, Mass., MIT Press, 1982 (first published 1934).

3　F. Hirsch, *The Social Limits to Growth*, Cambridge, Mass., Harvard University Press, 1976.

4　Banfield, op. cit.

5　Ibid., p. 210.

6　Ibid., p. 76.

7   Ibid., p. 235.

8   See A. Evans, 'Economic influences on social mix', *Urban Studies*, 1976, Vol. 13, pp. 247–260. He concluded that the case for intervening against the strong natural tendency of addresses to sort by income is weak. I disagree.

9   Social Exclusion Unit, *A New Commitment to Neighbourhood Renewal: National Strategy Action Plan*, London, Cabinet Office, 2001, p. 19.

10 D. Page and R. Boughton, *Mixed Tenure Housing Estates*, London, Notting Hill Housing Association, 1997.

11 Perri 6, *Escaping Poverty: From Safety Nets to Networks of Opportunity*, London, Demos, 1997.

12 R. Atkinson and K. Kintrea, *Reconnecting Excluded Communities: The Neighbourhood Impacts of Owner Occupation*, Edinburgh, Scottish Homes Research Report 61, 1998.

13 B. Jupp, *Living Together: Community Life on Mixed Tenure Estate*, London, Demos, 1999.

14 *Small and Medium Enterprise Statistics for the United Kingdom, 1997*, London, SME Statistics Unit, Department of Trade and Industry, July 1998. In 1997, 6.37 million jobs out of a total of 21.1 million jobs were in establishments employing fewer than 10 people.

15 Perri 6, *Escaping Poverty*, p. 39.

16 Statement to the House of Commons by John Prescott, Deputy Prime Minister, 7 March 2000. DETR Press Notice 164, London, DETR, 2000.

17 *Milton Keynes Economic Partnership 1998 Employment Survey*, Milton Keynes and North Bucks Chamber of Commerce, 1999.

18 English Partnerships reported that in 1997/98 its grants facilitated the construction of 7,540 homes. In 1998/99 the number was 4,175.

19 House of Commons Select Committee on Environment, Transport and the Regions, *The Implications of the European Commission Ruling on Gap Funding Schemes for Urban Regeneration in England*, Sixteenth Report, London, The Stationery Office, 2000.

20 London Docklands Development Corporation, *Regeneration Statement*, London, LDDC, 1998 and LDDC, *Housing in the Renewed London Docklands*, London, LDDC, 1998.

21 D. Barnes, A. North and M. Walker, *LETS on Low Income*, London, New Economics Foundation, 1996.

CHAPTER 9 HOW TO MINGLE

1   John Gwynn, *London and Westminster Improved (to which is prefixed a discourse on Publick Magnificence)*, London, 1766.

2   Community Centres and Associations Survey Group of the National Council of Social Service, London, Allen and Unwin, 1943.

3   Gerald Eve/Department of the Environment, *The Relationship Between House Prices and Land Supply*, London, HMSO, 1992. This research found that even in the South East, where house prices are highest, only around a third of prices can be attributed to land supply constraints. It would require significant additional releases of land for housing, on a national scale, and over a considerable time period to have any measurable impact on average house prices.

4   The high point of council housing in Scotland was in 1966, when 54 per cent of households were local authority tenants. Personal communication, Tenure Unit of the Scottish Executive's Housing Department, April 2000.

5   Article by Chris Holmes in *Housing Today*, 28 September 1997.

6   DETR, *Quality and Choice: A Decent Home for All*, London, DETR, 2000.

7   DETR, *Rethinking Construction: The Report of the Construction Task Force*, London, DETR, 1998.

8   Ikea has joined with the construction group Skanska to build several hundred largely prefabricated, highly standardised apartments in Sweden at edge of town locations. The developments consist of several L-shaped two-storey buildings set in their own grounds, each containing six one- and two-bedroom apartments. Construction takes as little as six weeks, ceilings are a generous 2.8 metres high, and the apartments are for rent or for sale, costing only £10,000 to £16,000 (at early 2001 exchange rates). See www.boklok.com.

9   Shelter, *An Urban and Rural Renaissance: Planning for the Communities of the Future, Conclusions of Shelter's National Inquiry into Housing Need in Urban and Rural Areas*. Vol. 1, London, Shelter, 1998, p. 5.

10   Shelter, *Building for the Future: The Homes We Need and How to Pay for Them*, London, Shelter, 2000.
11   Shelter, *Roof Briefing*, Issue No. 40, London, Shelter, June 2000.
12   DETR, Circular 6/98, *Planning and Affordable Housing*, London, DETR, 1998.
13   Shelter, 1998, *Urban and Rural Renaissance*, p. 8.
14   K. Young and J. Kramer, *Strategy and Conflict in Metropolitan Housing: Suburbia versus the GLC, 1965–1975*, London, Heinemann, 1981.
15   DETR, *Local Housing Needs Assessment: A Good Practice Guide*, London, DETR, 2000.
16   DETR, *Planning Policy Guidance Note 3 – Housing*, London, DETR, 2000, paragraphs 12 to 20.
17   I. Cole, S. Kane and D. Robinson, *Changing Demand, Changing Neighbourhoods: The Response of Social Landlords*, London, Housing Corporation/Centre for Regional Economic and Social Research, Sheffield Hallam University, 2000.
18   A. Power and K. Mumford, *The Slow Death of Great Cities?: Urban Abandonment or Urban Renaissance*, York, Joseph Rowntree Foundation/York Publishing Services, 1999.
19   Social Exclusion Unit, *National Strategy for Neighbourhood Renewal: A Framework for Consultation*, London, Cabinet Office, 2000.
20   M. Young and G. Lemos, *The Communities We Have Lost and Can Regain*, London, Lemos and Crane, 1997.
21   M. Gwilliam, C. Bourne, C. Swain and A. Prat, *Sustainable Renewal of Suburban Areas*, York, Joseph Rowntree Foundation/York Publishing Services, 1998.

359

CHAPTER 10   EDUCATION, EDUCATION, REGENERATION

1   Department for Education and Employment, *Excellence in Cities*, London, DfEE, 1999.
2   P. Mortimore and G. Whitty, *Can School Improvement Overcome the Effects of Disadvantage?*, London, Institute of Education, University of London, 1997, p. 4.
3   Department for Education and Employment White Paper (DfEE), *Excellence in Schools*, London, The Stationery Office, 1997, p. 12.
4   Ofsted, *Secondary Education: A Review of Secondary Schools in England, 1993–1997*, London, The Stationery Office, 1998.
5   D. Robertson and J. Symons, *Do Peer Groups Matter? Peer Group Versus Schooling Effects on Academic Attainment*, Discussion Paper No. 311, London, Centre for Economic Performance, London School of Economics, 1996. If there is rigorous setting into ability bands within a school, then the less able children will be denied benign peer group effects.
6   D. Leech and E. Campos, *Is Comprehensive Education Really Free? A Study of Secondary School Admission Policies on House Prices*, Warwick Economic Research Paper Number 581, Warwick, Warwick University, 2000.
7   A. Gibson and S. Asthana, Schools, markets and equity: access to secondary education in England and Wales, paper presented to the American Educational Research Association Annual Meeting, Montreal, 21 April 1999.
8   S. Gorard and J. Fitz, 'Investigating the determinants of segregation between schools', and A. Gibson and S. Asthana, 'What's in a number? Commentary on Gorard and Fitz's "Investigating the determinants of segregation between schools"', in *Research Papers in Education*, Vol. 15, Issue 2, 2000.
9   Ofsted, *Secondary Education*, p. 19.
10   Author's estimate from DfEE, *School and College Performance Tables 2000*, DfEE website, www.dfee.gov.uk.
11   All percentages from DfEE, *School and College Performance Tables 2000*, op. cit.
12   DfEE, *Excellence in Cities*, London, DfEE, 1999, p. 9.
13   R. Rogerson, *Quality of Life in Britain*, available from Quality of Life Research Group, Department of Geography, University of Strathclyde, 50 Richmond Street, Glasgow G1 1XN, 1997.
14   Mulholland Research Associates Limited, *Report of Research Findings on the Housing Market*, London, House Builders' Federation/Halifax, 1997.
15   DfEE, *Statistics of Education: Schools in England 2000*, London, The Stationery Office, 2000.
16   Interviewed by Donald Macintyre, *The Independent*, London, 3 December 1998.
17   DfEE, *Schools: Building on Success*, London, DfEE, 2001.
18   DfEE Press Releases 2000/0503, 2000/0568 and 2000/0559, all published in 2000.

2   Bartlett School of Planning and Llewelyn-Davies, *The Use of Density in Urban Planning*, London, DETR, 1998.

3   For an excellent discussion of the issues, see P. Hall, *Sustainable Cities or Town Cramming?*, London, Town and Country Planning Association, 1999.

4   Urban Task Force, *Towards an Urban Renaissance*, London, E. & F.N. Spon, 1999, p. 187.

5   More than 4 million homes are projected to be built in the UK over the next 25 years, of which 40 per cent – or at least 1.6 million – will be built on greenfield sites if the government meets its target for 60 per cent of new homes to be constructed on recycled urban land. The difference between the land consumed by 1.6 million homes at 30 and at 60 homes per hectare is 533 square kilometres.

6   Analyses based on the 'walkability' of residential neighbourhoods and the need for key services, including public transport, to be within convenient walking distances of homes are increasingly applied by urban planners and architects for new development and regeneration. Since the economic viability of these services depends on the number of customers using them, population and housing densities in the surrounding neighbourhood are important factors. These have long been issues in town planning; the American architect Peter Calthorpe's writings on developments within easy walking distance of public transport stations are often cited as sparking renewed interest during the increasingly car-dependent 1980s and 1990s. See P. Calthorpe, 'The Pedestrian Pocket', in D. Kelbaugh (ed.), *The Pedestrian Pocket Book*, New York, Princeton Architectural Press, 1989.

7   Urban Task Force, op. cit., 1999, p. 61.

8   Llewelyn-Davies, *Four World Cities: A Comparative Study of London, Paris, New York and Tokyo. Summary Report*, London, Llewelyn-Davies, 1996, p. 12.

9   Countryside Commission, *Linking Town and Country*, Cheltenham, The Countryside Commission, 1999.

10 The Wildlife Trusts/Urban Wildlife Partnership, *Proposed Urban White Paper: Memorandum*, Newark, The Wildlife Trusts, 2000.

11 M. Gwilliam, C. Bourne, C. Swain and A. Prat, *Sustainable Renewal of Suburban Areas*, York, Joseph Rowntree Foundation, 1998.

12 In Scottish cities the housing built at around this period for this market is more likely to consist of tenements or flats. On Tyneside it will often be terraces of two-storey flats.

13 Housing Corporation, 'Corporation launches Housing Regeneration Company pilots', *Housing Regeneration Update 6* newsletter, Update Extra, December 2000.

14 London Borough of Camden, *Unitary Development Plan Deposit Draft*, L.B. Camden, 1993.

CHAPTER 14   EROSION OF CITIES OR ATTRITION OF CARS

1   J. Jacobs, *The Death and Life of Great American Cities*, New York, Random House, 1961.

2   H.G. Wells, *A Modern Utopia*, London, Chapman and Hall, 1905, p. 47.

3   Quoted in I. Hardhill and A. Duddleston, *Employment Opportunities and Constraints*, paper available from the Department of Social Sciences, Nottingham Trent University, 1996.

4   Le Corbusier, *The City of Tomorrow*, translated by Frederick Etchells, Cambridge, Mass., MIT Press, 1971, p. 3 (first published 1924).

5   DETR, *Transport Statistics Great Britain, 2000 Edition*, London, The Stationery Office, 2000, p. 98.

6   Committee on the Medical Effects of Air Pollution, *Quantification of the Effects of Air Pollution on Health in the UK*, London, Department of Health, 1998. And DETR, *Transport Statistics Great Britain 2000*, London, The Stationery Office, 2000, p. 98.

7   The quantity of billion vehicle kilometres per year travelled on Britain's roads (the sum of the distance travelled by every vehicle) rose from about 100 in 1960 to 467 in 1999. During the 1990s the trend rate of growth appeared to fall to well below 3 per cent a year. DETR, *Transport Statistics Great Britain 2000*, London, The Stationery Office, 2000, p. 92. And Royal Commission on Environmental Pollution, *Transport and the Environment*, 18th Report, London, HMSO, 1994, p. 19.

8   Academics and the rest of us have long come to suspect this. It was the government's Standing Advisory Committee on Trunk Road Assessment (SACTRA) which examined the arguments for this rigorously, accepted them broadly and concluded that government decision-making on whether the benefits of new road schemes to society outweighed their costs was therefore flawed. SACTRA, *Trunk Roads and the Generation of Traffic*, London, HMSO, 1994.

9   DETR, *Town Centres: Defining Boundaries for Statistical Monitoring*. London, The Stationery Office, 1998.

10 In *Edge City: Life on the New Frontier*, New York, Doubleday, 1991, Joel Garreau proposes the '600 foot law', this being the maximum distance an American will willingly walk before getting into a car. He advised developers: 'In either a downtown or an edge city, if you do everything you can to make casual use of the auto-mobile inconvenient at the same time that you make walking pleasant and attractive you maybe, just maybe, can up the distance an American will willingly walk to 1,500 feet [500 yards]. And this at the substantial risk of everybody saying forget it and choosing not to patronize your highly contrived environment at all.'

11 The sources used here were: *National Travel Survey: 1996–98 Update* (1999), *Focus on Personal Travel* (1998), *A Bulletin of Public Transport Statistics: Great Britain 2000* (2000), *Transport Statistics Great Britain 2000* (2000), all from The Stationery Office, London.

12 Excluding overseas air travel.

13 For a discussion of this issue, see S. Owens and M. Breheny, 'Exchange: the compact city and transport energy consumption', *Transactions of the Institute of British Geographers*, 1995, Vol. 20, pp. 381–386.

14 Department of Transport, *Roads for Prosperity*, London, HMSO, 1989.

15 DETR, *Transport 2010: The Ten Year Plan*, London, DETR, 2000.

16 DETR, *Transport Statistics Great Britain 2000*, London, TSO, 2000, Table 4.10, p. 93.

17 CPRE, *Rural Traffic Fear Survey*, London, CPRE, 1999. And *Traffic Trauma or Tranquillity*, London, CPRE, 1999.

18 Bus deregulation is made interesting in C. Wolmar, *Stage Coach: A Classic Rags-to-Riches Tale from the Frontiers of Capitalism*, London, Orion, 1998.

19 DETR, *Transport Statistics Great Britain 2000*, London, The Stationery Office, 2000. And DETR, *Focus on Public Transport, Great Britain, 1999*, London, The Stationery Office, 1999.

20 The Transport Act 2000, covering England and Wales, received the Royal Assent on 30 November 2000.

21 DETR, *Transport 2010*, p. 62.

22 Information provided to the author by First Group plc.

23 Estimates of passenger growth and shift from cars supplied by Leeds City Council and Ipswich Borough Council.

24 From Greater Manchester Passenger Transport Executive's estimate of 2.6 million a year, based on stud-ies by University of Salford and Oscar Faber TPA.

25 Author's estimate, based on total capital costs and line lengths of the systems which opened in England in the 1990s.

26 S. Cairns, C. Hass-Klau and P. Goodwin, *Traffic Impact of Highway Capacity Reductions: Assessment of the Evidence*, London, Landor Publishing, 1998.

27 DETR, *Planning Policy Guidance Note (PPG) 13 – Transport*, London, DETR, 2001, paragraphs 65, 67, 68, 77 and 80.

28 DETR Press Release, 4 August 1999.

29 The Transport Act 2000. The intention to enable congestion charging was established in a White Paper, DETR, *A New Deal for Transport: Better for Everyone*, London, The Stationery Office, 2000.

30 G. Monbiot, *Captive State: The Corporate Takeover of Britain*, London, Macmillan, 2000.

31 DETR, *Planning Policy Guidance Note (PPG) 13 – Transport*, London, DETR, 1994.

32 DETR, *PPG 13*, 2001. The guidance turns out to be equivocal on the matter of the quantities of car park-ing developers should provide. It sets down maximum numbers of car parking spaces for various categories of major new development, then gives local planners plenty of scope to ignore them. On the one hand, the government recognises that the more parking spaces are provided, the more drivers will be encouraged to use their cars and forsake alternatives – so it calls for restraint. On the other, it recognises that 'adequate' car parking is required in and around town centres in order to encourage new investment there.

33 DETR, *Transport 2010*, p. 65.

34 Urban Task Force, *Towards an Urban Renaissance*, London, E. & F.N. Spon, 1999, p. 101.

35 DETR, *Our Towns and Cities: The Future*, London, TSO, 2000, p. 141.

36 N. Schoon, 'I'll get you on the bus says Prescott,' article in *The Independent*, 6 June 1997.

37 DETR, *Traffic Speeds in Central and Outer London: 1996–97*, Statistics Bulletin (98)17, London, DETR, 1998.

38   The sources used here were: Government Statistical Service, *National Travel Survey 1996–98 Update* (SB(99)21), London, GSS, 1999; DETR, *Focus on Personal Travel 1998*, London, The Stationery Office, 1998; London Research Centre and Conseil Regional Ile de France, *London Paris: A Comparison of Transport Systems*, London, HMSO, 1992.

CHAPTER 15   TOWN AND COUNTRY

1   Writing under the name 'Common Sense' in the *Monthly Magazine*, Feb. 1811. Phillips was a radical politician and publisher.

2   *Tarbuck's Handbook of House Property*, 1875, quoted in H. Dyos, *Victorian Suburb*, Leicester: Leicester University Press, 1961.

3   This is an estimate for the value added by land development in the UK in 1998, kindly supplied to me by Milan Khatri, senior economist at the Royal Institute of Chartered Surveyors.

4   The Labour government re-elected in 2001 was minded to reform the planning system to make it friendlier to free enterprise. But given the limited progress of 'modernisation' in this sphere during the first term, the bulk of the system seemed likely to survive.

5   *The Economic Determinants of Household Formation: A Literature Review* by Glen Bramley, Moira Munro and Sharon Lancaster, London, DETR, July 1997. For a short, useful summary, see *Household Formation: A Suitable Case for Policy* by Glen Bramley and Sharon Lancaster, p. 21 in *Housing Finance*, the quarterly economics journal of the Council of Mortgage Lenders, London, No. 38, May 1998. See also DETR, *Projections of Households in England to 2021*, London, DETR, 1999, pp. 27–30.

6   DETR, *Projections of Households in England to 2021*, London, DETR, 1999. See p. 36.

7   Department of the Environment, *Household Growth: Where Shall We Live?*, London, The Stationery Office, 1996, pp. 47–48.

8   For a brief, lucid discussion of these issues, see *Housing*, the Tenth Report of the Environment, Transport and Regional Affairs Committee of the House of Commons, London, The Stationery Office, 1998.

9   DETR, 1999, *Projections of Households in England to 2021*, p. 13.

10 Some 43 per cent of the new homes built in the UK in 2000 were detached and 14 per cent semi-detached, while 15 per cent were terraces, 23 per cent flats and maisonettes and the remainder bungalows. National House Building Council, *New House Building Statistics, Q4 2000*, Amersham, NHBC, 2000.

11 Urban Task Force, *Towards an Urban Renaissance: Final Report of the Urban Task Force*, E. & F.N. Spon, 1999, p. 186.

12 Building costs: according to the Royal Institute of Chartered Surveyors' Building Costs Information Service's *Quarterly Review of Building Charges, January 2001*, London, BCIS/RICS, 2001, the average costs of building a house were £514 a square metre (excluding land purchase). The typical three-bedroom house covers 80 to 85 square metres. Infrastructure costs – roads, sewers and so forth – add another 30 to 50 per cent.
    Price of new housing: the median selling price of new homes in the UK was £107,000 in the final quarter of 2000, according to the National House-Building Council's *New House-Building Statistics*. Only in one UK region, Merseyside, was the median price of a new home below £80,000. Three-bedroom homes are the largest category.

13 Figure for 1994 to 1998 from DETR News Release 495, *Land Use Change in England*, 27 July 2000.

14 *Farmland Update October 2000*, London, Strutt & Parker, 2000.

15 David Rudlin, URBED, 1998, *Tomorrow: A Peaceful Path to Urban Reform. The Feasibility of Accommodating 75 Per Cent of New Homes in Urban Areas*, Friends of the Earth. If a quarter of the 3.8 million additional homes which the government projects England requires between 1996 and 2021 are built on greenfield sites then, at a rather high gross (overall) density of 30 units per hectare 300 square kilometres of land would be required.

16 M. Breheny and P. Hall (eds), *The People: Where Will They Go? National Report of the TCPA Inquiry into Housing Need and Provision in England*, Town and Country Planning Association, London, 1996. This suggested only 30 to 40 per cent of new homes could be built within existing urban areas.

17 DETR, Government Statistical Service Update (May 2000) of Information Bulletin 500, *National Land Use Database – Final Estimates of Previously Developed Land in England*, London, DETR, 2000. Website: www.nlud.org.uk.

18 Ibid.

19 Llewelyn-Davies and the University of Westminster, *Conversion and Redevelopment: Process and Potential,* London, DETR, 2000.

20 D. Rudlin, *Tomorrow: A Peaceful Path to Urban Reform,* London, Friends of the Earth, 1998, p. 39.

21 Llewelyn-Davies, *Sustainable Residential Quality: New Approaches to Urban Living,* London, London Planning Advisory Committee, 1997. Llewelyn-Davies estimated that 1,400 hectares of 'backland' – the rear portions of large suburban gardens – could theoretically be used for housing. Assuming fairly high densities of 36 homes per hectare, that gives a figure of 50,000 homes.

22 Figure for England for 1999/2000 supplied to Empty Homes Agency by DETR, 2000.

23 See Urban Task Force, 1999, Chapter 7, for a full discussion of the government's target and its attainability. The 57 per cent for the years 1995 to 1998 was announced in News Release 495, DETR, 27 July 2000.

24 UK Round Table on Sustainable Development, *Housing and Urban Capacity,* London, DETR, 1997.

25 Some 52 per cent of all the South East land outside the urban areas is covered by national or international designations related to its intrinsic quality and 24 per cent by Green Belts, according to ROSE *Housing Capability Study: Inventory of Designations Report,* 1998. Available from the London and South East Regional Planning Conference Secretariat, 14 Buckingham Gate, London SW1E 6LB.

26 DETR, *National Land Use Database,* May 2000. In London and the South East 1.6 per cent of the total area of developed land was vacant, derelict or covered by vacant buildings. In the North East, the North West and Yorkshire and Humberside, the percentages were 5.3, 4.3 and 6.7, respectively.

27 Urban Task Force, 1999, p. 181.

CHAPTER 16   NEW NEW TOWNS

1 House of Commons Environment, Transport and Regional Affairs Committee, Tenth Report, *Housing,* Vol. 1, p. lxxiii, para. 279, London, The Stationery Office, 1998.

2 Harvey Sherlock, *Cities Are Good for Us,* London, Paladin, 1991.

3 P. Hall and C. Ward, *Sociable Cities: The Legacy of Ebenezer Howard,* Chichester, John Wiley, 1998.

4 See Urban Task Force, 1999, pp. 213 and 214 and UK Round Table on Sustainable Development, *Review of Urban Capacity Studies* by consultants Llewelyn-Davies, London, 1997.

5 Based on 40 per cent of 3.8 million additional homes being built on Green Belt extensions with an overall (gross) housing density of 19 units per hectare.

6 'There may in very exceptional circumstances be a justification for building homes adjacent to the existing urban area on the Green Belt . . . where the alternative may be the development in villages beyond which would encourage commuting,' said the MPs of the Commons Environment, Transport and Regional Affairs Committee in their report on housing (ET&RA Committee, 1998, para. 264, p. lxx). But such circumstances are not very exceptional.

7 John Gummer, MP, former Secretary of State for the Environment, told the House of Commons Environment, Transport and the Regions Committee, that a successful developer had told him: 'Don't give us the option [of greenfield development]. We'll go for the easy solution, every time. Make it clear that we either use recycled land or we shall not get any development at all. Then we shall find a way.'

8 Department of the Environment, Transport and the Regions, *Planning Policy Guidance Note 3: Housing,* London, DETR, 2000. See Appendix B, Providing for rural exception housing, p. 25.

9 DETR, 2000, *PPG 3,* paragraph 68, page 21.

10 The Urban Task Force could not support such a tax; see Urban Task Force, 1999, p. 221. Nor was the House of Commons Environment, Transport and Regional Affairs Committee persuaded. House of Commons, 1998, p. lxviii, para. 251.

11 DETR, *Our Towns and Cities: The Future,* London, TSO, 2000, p. 42.

12 Hall and Ward, op. cit.

13 Government Office for the East of England/DETR, *Regional Planning Guidance for East Anglia to 2016,* RPG 6, London, The Stationery Office, 2000, pp. 31–36. This regional planning guidance gives extensions to Cambridge – which would intrude into the city's tightly drawn Green Belt – priority over the new settlement, but it fails to explain why any of the city's growth should be deployed into a new settlement rather than extensions.

14   ECOTEC/Department of the Environment, *Reducing Transport Emissions Through Planning*, London, HMSO, p. vii.

CHAPTER 17   RENAISSANCE OR STILLBIRTH?

1   T. Stoppard, *Jumpers*, London, Faber and Faber, 1972.
2   Department of the Environment, *Policy for the Inner Cities*, London, HMSO, 1977, p. 2.
3   DETR, *Our Towns and Cities: The Future*, London, The Stationary Office, 2000.
4   Social Exclusion Unit, *A New Commitment to Neighbourhood Renewal: National Strategy Action Plan*, London, Cabinet Office, 2001.
5   D. Dorling, R. Mitchell, M. Shaw and S. Orford, 'The Ghost of Christmas Past: health effects of poverty in London in 1896 and 1991', *British Medical Journal*, 2000, Vol. 321, pp. 1547–1551.
6   The Urban Task Force's report (*Towards an Urban Renaissance*, London, E. & F.N. Spon, 1999, pp. 136–137) provides a handy summary of all of the urban regeneration initiatives between 1981 and 1998. Since then the Labour government has added the New Deal for Communities, the Neighbourhood Renewal Fund, Regional Development Agencies, the Millennium Communities initiative and four local Urban Regeneration Companies as well as several departmentally-based zones (Health Action Zones, Education Action Zones and so forth).
7   The first was suggested by the Environment, Transport and Regional Affairs Committee of the House of Commons, *17th Report, Departmental Annual Report 2000 and Expenditure Plans 2000–20001*, London, The Stationery Office, 20000. The second objective is my suggestion.
8   Social Exclusion Unit, *A New Commitment to Neighbourhood Renewal*, London, Cabinet Office, 2001, p. 5.
9   A paper which deals with these issues has, as its title, an angry remark made by a resident to a community leader who was seen to have become detached. See A. McCulloch, 'You've fucked up the estate and now you're carrying a briefcase', in P. Hoggett (ed.), *Contested Communities*, Bristol, Polity Press, 1997.
10   None of which is easy. For an example of how not to do it, it's worth reading George Monbiot's account of what he condemns as an entirely fake urban regeneration consultation exercise in Southampton. G. Monbiot, *Captive State: The Corporate Takeover of Britain*, London, Macmillan, 2000.
11   DETR, *Our Towns and Cities: The Future*, London, TSO, 2000, pp. 139–154.
12   Ibid., p. 59.
13   N. Schoon, 'Run-down, sprawling and decayed. Are our cities the worst in Europe?', *The Independent*, 14 January 1999.
14   Urban Task Force, *Towards an Urban Renaissance*.
15   DETR, *Turnout at Local Elections*, London, DETR, 2000. In Knowsley, Merseyside – one of the most deprived English boroughs – the turnout at the council election in 2000 was only 18 per cent.
16   Audit Commission, *1999/2000 Performance Indicators for English Local Authorities*, London, Audit Commission, 2000.
17   DETR, *Turnout at Local Elections*.
18   The Labour Party, *New Labour Because Britain Deserves Better*, London, The Labour Party, 1997, p. 34.
19   DETR, *Modernising Local Government Finance: A Green Paper*, London, DETR, 2000, pp. 26–31.
20   Jane Jacobs' second, less successful book argued that even the origins of agriculture lie in the very first cities and towns, established as trading and craft settlements in the Middle East. Her hypothesis gathered little support. J. Jacobs, *The Economy of Cities*, London, Jonathan Cape, 1970 (first published 1969).

INDEX

369